Imagining the Future

One particularly adaptive feature of human cognition is the ability to mentally preview specific events before they take place in reality. Familiar examples of this ability—often referred to as episodic future thinking—include what happens when an employee imagines when, where, and how they might go about asking their boss for a raise, or when a teenager anguishes over what might happen if they ask their secret crush on a date. In this book, the editors bring together current perspectives from researchers from around the globe who are working to develop a deeper understanding of the manner in which the simulations of future events are constructed, the role of emotion and personal meaning in the context of episodic simulation, and how the ability to imagine specific future events relates to other forms of future thinking such as the ability to remember to carry out intended actions in the future.

This book was originally published as a special issue of *The Quarterly Journal of Experimental Psychology*.

Karl K. Szpunar is Assistant Professor of Psychology and Principal Investigator of the UIC Memory Lab at the University of Illinois at Chicago, USA.

Gabriel A. Radvansky is Professor of Psychology and Principal Investigator of the Memory Lab at the University of Notre Dame, South Bend, IN, USA.

Imagining the Future
Insights from Cognitive Psychology

Edited by
Karl K. Szpunar and Gabriel A. Radvansky

Routledge
Taylor & Francis Group

LONDON AND NEW YORK

First published 2017 by Routledge

2 Park Square, Milton Park, Abingdon, Oxfordshire OX14 4RN
52 Vanderbilt Avenue, New York, NY 10017

Routledge is an imprint of the Taylor & Francis Group, an informa business

First issued in paperback 2018

British Library Cataloguing in Publication Data
A catalogue record for this book is available from the British Library

ISBN 13: 978-0-415-78940-0 (hbk)
ISBN 13: 978-0-367-14266-7 (pbk)

Typeset in Adobe Caslon
by RefineCatch Limited, Bungay, Suffolk

Publisher's Note
The publisher accepts responsibility for any inconsistencies that may have
arisen during the conversion of this book from journal articles to book chapters,
namely the possible inclusion of journal terminology.

Disclaimer
Every effort has been made to contact copyright holders for their permission to
reprint material in this book. The publishers would be grateful to hear from any
copyright holder who is not here acknowledged and will undertake to rectify
any errors or omissions in future editions of this book.

Contents

CONTENTS

Citation Information

The chapters in this book were originally published in *The Quarterly Journal of Experimental Psychology*, volume 69, issue 2 (February 2016). When citing this material, please use the original page numbering for each article, as follows:

Chapter 1
Cognitive approaches to the study of episodic future thinking
Karl K. Szpunar and Gabriel A. Radvansky
The Quarterly Journal of Experimental Psychology, volume 69, issue 2 (February 2016), pp. 209–216

Chapter 2
Frequency, characteristics, and perceived functions of emotional future thinking in daily life
Catherine Barsics, Martial Van der Linden, and Arnaud D'Argembeau
The Quarterly Journal of Experimental Psychology, volume 69, issue 2 (February 2016), pp. 217–233

Chapter 3
The degree of disparateness of event details modulates future simulation construction, plausibility, and recall
Valerie van Mulukom, Daniel L. Schacter, Michael C. Corballis, and Donna Rose Addis
The Quarterly Journal of Experimental Psychology, volume 69, issue 2 (February 2016), pp. 234–242

Chapter 4
Visual perspective in remembering and episodic future thought
Kathleen B. McDermott, Cynthia L. Wooldridge, Heather J. Rice, Jeffrey J. Berg, and Karl K. Szpunar
The Quarterly Journal of Experimental Psychology, volume 69, issue 2 (February 2016), pp. 243–253

Chapter 5
Prevalence and determinants of direct and generative modes of production of episodic future thoughts in the word cueing paradigm
Olivier Jeunehomme and Arnaud D'Argembeau
The Quarterly Journal of Experimental Psychology, volume 69, issue 2 (February 2016), pp. 254–272

Chapter 6
Do future thoughts reflect personal goals? Current concerns and mental time travel into the past and future
Scott N. Cole and Dorthe Berntsen
The Quarterly Journal of Experimental Psychology, volume 69, issue 2 (February 2016), pp. 273–284

CITATION INFORMATION

Chapter 7

Remembering the past and imagining the future: Selective effects of an episodic specificity induction on detail generation
Kevin P. Madore and Daniel L. Schacter
The Quarterly Journal of Experimental Psychology, volume 69, issue 2 (February 2016), pp. 285–298

Chapter 8

You'll change more than I will: Adults' predictions about their own and others' future preferences
Louis Renoult, Leia Kopp, Patrick S. R. Davidson, Vanessa Taler, and Cristina M. Atance
The Quarterly Journal of Experimental Psychology, volume 69, issue 2 (February 2016), pp. 299–309

Chapter 9

The relationship between prospective memory and episodic future thinking in younger and older adulthood
Gill Terrett, Nathan S. Rose, Julie D. Henry, Phoebe E. Bailey, Mareike Altgassen, Louise H. Phillips, Matthias Kliegel, and Peter G. Rendell
The Quarterly Journal of Experimental Psychology, volume 69, issue 2 (February 2016), pp. 310–323

Chapter 10

Scripts and information units in future planning: Interactions between a past and a future planning task
Aline Cordonnier, Amanda J. Barnier, and John Sutton
The Quarterly Journal of Experimental Psychology, volume 69, issue 2 (February 2016), pp. 324–338

Chapter 11

Thinking about the future can cause forgetting of the past
Annie S. Ditta and Benjamin C. Storm
The Quarterly Journal of Experimental Psychology, volume 69, issue 2 (February 2016), pp. 339–350

Chapter 12

Retrieval-induced forgetting is associated with increased positivity when imagining the future
Saskia Giebl, Benjamin C. Storm, Dorothy R. Buchli, Elizabeth Ligon Bjork, and Robert A. Bjork
The Quarterly Journal of Experimental Psychology, volume 69, issue 2 (February 2016), pp. 351–360

Chapter 13

Understanding deliberate practice in preschool-aged children
Jac T. M. Davis, Elizabeth Cullen, and Thomas Suddendorf
The Quarterly Journal of Experimental Psychology, volume 69, issue 2 (February 2016), pp. 361–380

Chapter 14

Autonoetic consciousness: Reconsidering the role of episodic memory in future-oriented self-projection
Stanley B. Klein
The Quarterly Journal of Experimental Psychology, volume 69, issue 2 (February 2016), pp. 381–401

For any permission-related enquiries please visit:
http://www.tandfonline.com/page/help/permissions

Notes on Contributors

Donna Rose Addis is a Professor at the School of Psychology, The University of Auckland, New Zealand.

Mareike Altgassen is Assistant Professor at the Donders Institute for Brain, Cognition, and Behaviour, Radboud University, Nijmegen, Netherlands.

Cristina M. Atance is an Associate Professor at the School of Psychology, University of Ottawa, Canada.

Phoebe E. Bailey is a Senior Lecturer based at the School of Social Sciences and Psychology, University of Western Sydney, Sydney, Australia.

Amanda J. Barnier is a Professor at Department of Cognitive Science, Macquarie University, Australia.

Catherine Barsics is a Postdoctoral Researcher at the University of Geneva, Geneva, Switzerland.

Jeffrey J. Berg is a Graduate student at the Department of Psychology, New York University, New York, USA.

Dorthe Berntsen is Professor at the Department of Psychology and Behavioral Sciences, Center on Autobiographical Memory Research, Aarhus University, Denmark.

Elizabeth Ligon Bjork is a Professor at the Department of Psychology, University of California, Los Angeles, USA.

Robert A. Bjork is a Distinguished Research Professor at the Department of Psychology, University of California, Los Angeles, USA.

Dorothy R. Buchli is Assistant Professor at the Department of Psychology, Mercer University, USA.

Scott N. Cole is a Lecturer in Psychology at York St. John University, UK.

Michael C. Corballis is Emeritus Professor at the School of Psychology, The University of Auckland, New Zealand.

Aline Cordonnier is a PhD student at the Department of Cognitive Science, Macquarie University, Sydney, NSW, Australia.

Elizabeth Cullen is based at the School of Psychology, University of Queensland, Brisbane, Australia.

Arnaud D'Argembeau is a Professor at Department of Psychology, University of Liege, Belgium.

Patrick S. R. Davidson is an Associate Professor at the School of Psychology, University of Ottawa, Canada.

NOTES ON CONTRIBUTORS

Jac T. M. Davis is a PhD student at the School of Psychology, University of Queensland, Australia.

Annie S. Ditta is a PhD student based at the Department of Psychology, University of California at Santa Cruz, USA.

Saskia Giebl is a PhD student based at the Department of Psychology, University of California, Los Angeles, USA.

Julie D. Henry is a Professor at the School of Psychology, University of Queensland, Brisbane, Australia.

Olivier Jeunehomme is based at the Department of Psychology, University of Liège, Belgium.

Stanley B. Klein is a Professor based at the Department of Psychological and Brain Sciences, University of California at Santa Barbara, USA.

Matthias Kliegel is Full Professor and Chair of Cognitive Aging at the Department of Psychology, University of Geneva, Switzerland.

Leia Kopp is a PhD student at the School of Psychology, University of Ottawa, Canada.

Kevin P. Madore is a PhD student at the Department of Psychology, Harvard University, USA.

Kathleen B. McDermott is Professor of Psychological and Brain Sciences at the Department of Psychology, Washington University in St Louis, USA.

Louise H. Phillips is Chair in Psychology at the School of Psychology, University of Aberdeen, UK.

Gabriel A. Radvansky is Professor of Psychology and Principal Investigator of the Memory Lab at the University of Notre Dame, South Bend, IN, USA.

Peter G. Rendell is a fully accredited Supervisor at the School of Psychology, Australian Catholic University, Melbourne, Australia.

Louis Renoult is a Lecturer in Psychology at the School of Psychology, University of East Anglia, UK.

Heather J. Rice is a Lecturer based at the Department of Psychology, Washington University in St Louis, USA.

Nathan S. Rose is an Assistant Professor in the Department of Psychology at Notre Dame University, USA.

Daniel L. Schacter is William R. Kenan, Jr Professor at the Department of Psychology, Harvard University, USA.

Benjamin C. Storm is Associate Professor at the Department of Psychology, University of California at Santa Cruz, USA.

Thomas Suddendorf is a Professor at the School of Psychology, University of Queensland, Australia.

John Sutton is Deputy Director of the Department of Cognitive Science, Macquarie University, Sydney, NSW, Australia.

Karl K. Szpunar is Assistant Professor of Psychology and Principal Investigator of the UIC Memory Lab at the University of Illinois at Chicago, USA.

Vanessa Taler is an Associate Professor based at the School of Psychology, University of Ottawa, Canada.

Gill Terrett is a Supervisor at the School of Psychology, Australian Catholic University, Melbourne, Australia.

Martial Van der Linden is a Professor at the Department of Psychology, University of Geneva, Switzerland.

Valerie van Mulukom was a Postdoctoral Researcher at the University of Oxford, UK. She is now a Research Associate at Coventry University, UK.

Cynthia L. Wooldridge is an Associate Professor at the Department of Psychology, Washburn University, USA.

Cognitive approaches to the study of episodic future thinking

Karl K. Szpunar[1] and Gabriel A. Radvansky[2]

[1]Department of Psychology, University of Illinois at Chicago, Chicago, IL, USA
[2]Department of Psychology, University of Notre Dame, Notre Dame, IN, USA

The concept of episodic future thinking—the ability to simulate events that may take place in the personal future—has given rise to an exponentially growing field of research that spans a variety of sub-disciplines within psychology and neuroscience. In this introduction to the special issue, we provide a brief historical overview of factors that have shaped research on the topic and highlight the need for additional behavioural work to uncover cognitive mechanisms that support episodic future thinking and differentiate it from other related modes of future-oriented cognition. We conclude by discussing the manner in which the various contributions to the special issue fill the gaps in our knowledge and make some of our own suggestions for future work.

Memory does not primarily exist to think about the past. It primarily exists to help us know what to do in the present and to plan for the future. On a daily basis, people spend considerable time turning their attention away from the immediate environment and focusing instead on events that have yet to transpire, such as upcoming meetings, trips, or chores. Indeed, a recent study estimated that healthy human adults think about the future an average of about 60 times per day with many of those instances of future thinking focusing on specific events (D'Argembeau, Renaud, & Van der Linden, 2010). Social and clinical psychologists have long been interested in this ubiquitous feature of human cognition and in particular in the manner in which the ability to evaluate the future often fails. For instance, social psychologists have identified various biases in mental simulation that limit our ability to predict how future events will make us feel (e.g., Gilbert & Wilson, 2007). At the same time, clinical psychologists have focused on how thinking about the future may change in the context of mood and anxiety disorders (e.g., Miloyan, Bulley, & Suddendorf, in press). Despite all this, it was not until relatively recently that cognitive psychologists and neuroscientists began to pay attention to the future and how the human brain/mind supports our ability to think about it. Here we provide a brief exposition of what episodic future thinking is, how it has been studied, and what we know about it to date. This provides a basis for understanding and integrating the various topics covered by articles in this special issue of the *Quarterly Journal of Experimental Psychology*, which has been created to introduce and attract research to this emerging topic in the field.

Background and motivation

Interest in thinking about the future from a cognitive perspective has its roots in the seminal observations of Tulving (1985), who wrote about an amnesic individual with no episodic memory who was unable to remember events from his personal past or imagine events that might take place in his personal future. About a decade later, Suddendorf and Corballis (1997) suggested that the capacity to engage in mental time travel into the personal past and future may be a uniquely human ability, an idea that continues to be hotly debated in the literature (Corballis, 2013; Suddendorf, 2013). A few years later, Atance and O'Neill (2001) formally dubbed the ability to simulate personal future events as *episodic future thinking*. More recently, additional findings from neuroscience and cognitive psychology have continued to support the claim that episodic memory and episodic future thinking may represent two sides of a single overarching capacity. For instance, Klein, Loftus, and Kihlstrom (2002) described a case of amnesia in which an individual lost the ability to remember the personal past and imagine the personal future, but retained the ability to think about the past and future in non-personal ways. Okuda et al. (2003) used positron emission tomography (PET) imaging to show that thinking about the past and future evoke similar patterns of neural activity. Finally, D'Argembeau and Van der Linden (2004) demonstrated that events imagined as occurring in the near past or future (e.g., past or next day or month) are mentally represented in more detail than events imagined as occurring in the distant past or future (e.g., past or next year), while Spreng and Levine (2006) found that people generally tend to spend more time thinking about past and future events that are temporally near than about those that are temporally distant.

Despite the growing evidence connecting the personal past and future, interest in the cognitive and neural mechanisms that give rise to the ability to simulate future events did not galvanize until early 2007. At that time, three articles were published that provided unique insights into the close connection between the personal past and future at a neural level. Whereas Tulving (1985) and Klein et al. (2002) both reported deficits of episodic memory and episodic future thinking in amnesia, their patients were respectively characterized by brain damage that was distributed across the entire brain or indeterminate. Because of this, it was difficult to know for certain which part(s) of the brain were responsible for the coinciding deficits of episodic memory and future thinking. Hassabis, Kumaran, Vann, and Maguire (2007) provided clarity on this issue by demonstrating similar patterns of deficits in memory and future thinking in patients whose brain damage was largely limited to the hippocampus. Although the role of hippocampus in future thinking continues to be refined (for a detailed review, see Addis & Schacter, 2012), the results of Hassabis et al. (2007) shed light on which neural structure is vital to the relation between episodic memory and future thinking.

At about the same time as Hassabis et al. (2007), two separate neuroimaging studies were published that showed that a close neural relation between memory and future thinking could be attributed to thoughts about specific past and future events. While prior work by Okuda et al. (2003) had found similar patterns of activity associated with thinking about the past and future, that study used a blocked design during which participants were told to generate many thoughts about the past or future. Moreover, there was little assurance that the past and future events that were generated were specific as opposed to general or semantic. Addis, Wong, and Schacter (2007) and Szpunar, Watson, and McDermott (2007) expanded on the Okuda et al. (2003) work and made use of event-related experimental designs that ensured that participants thought about specific past and future events. Importantly, these latter studies reported results supporting the idea that specific or episodic memories and future thoughts engage a common core network of brain regions (for a recent review, see Benoit & Schacter, 2015).

Taken together, the neuropsychological and neuroimaging results relating the past and future led to the suggestion that one adaptive function of human memory, and of episodic memory in particular, may be to provide the building blocks for constructing mental representations of the future (Schacter & Addis, 2007), along with other memory-based modes of cognition (Buckner & Carroll, 2007). This hypothesis has served as the driving force for much of the research that followed. For instance, studies of individuals with underdeveloped and impoverished episodic memory, such as young children and older adults, have reported deficits in episodic future thinking (e.g., Addis, Wong, & Schacter, 2008; Atance, 2008). Similarly, studies of individuals with varying degrees of episodic memory impairment, such as Alzheimer's disease (Addis, Sacchetti, Ally, Budson, & Schacter, 2009), mild cognitive impairment (Gamboz et al., 2010), schizophrenia (D'Argembeau, Raffard, & Van der Linden, 2008), and posttraumatic stress disorder (Brown et al., 2013), have likewise demonstrated deficits of episodic future thinking. In general, limitations in the ability to extract details of past experiences from episodic memory are associated with an inability to generate detailed simulations of future events, supporting the hypothesis that episodic memory serves as the basis for episodic future thinking.

In terms of functional brain imaging, dozens of studies have now replicated the finding that a common core set of brain regions support both episodic memory and episodic future thinking (Benoit & Schacter, 2015). Presently, cognitive neuroscientists are working diligently to identify how various regions within this network support various aspects of simulated events (e.g., people, places, objects, scenarios; Hassabis et al., 2014; Szpunar, St. Jacques, Robbins, Wig, & Schacter, 2014), the extent to which various regions may serve multiple functions in the context of simulating future events (e.g., encoding, detail recombination, retrieval; e.g., Addis & Schacter, 2012), the manner in which this common core network may interact with other networks in the brain to achieve goal-directed cognition (Spreng, Stevens, Chamberlain, Gilmore, & Schacter, 2010), and how structural and functional abnormalities associated with this network may be associated with limitations in the ability to simulate the future (Hach, Tippett, & Addis, 2014).

An important point to take away from the last decade of research on episodic future thinking is the strong contribution from the cognitive neurosciences. This is not too surprising given that the close neural overlap associated with thinking about specific past and future events seems to have driven interest in the area. Regardless, many such studies have also made important advances in terms of the cognitive paradigms that have been developed to study episodic future thinking. For instance, initial studies of episodic future thinking largely used cueing techniques that employed pre-existing stimuli, such as common nouns, to evoke memories and simulations of the future, much like studies of autobiographical memory (Crovitz & Schiffman, 1974). More recently, researchers have become interested in ensuring that participants are generating truly novel future events. The experimental recombination procedure, which involves randomly rearranging participant-generated lists of familiar people, places, and objects into unique simulation cues, was developed specifically for this purpose (Addis, Pan, Vu, Laiser, & Schacter, 2009; for variations on this technique, see Szpunar, Addis, & Schacter, 2012).

In addition to developing novel research paradigms, researchers have also amassed a number of techniques for assessing the quality of simulated events. For instance, Hassabis et al. (2007) developed a measure for assessing the spatial coherence associated with imagined events, an approach that has since been applied to the study of episodic future thinking in other populations (e.g., D'Argembeau et al., 2008). Other lines of research have borrowed methods from the autobiographical memory literature, such as the autobiographical interview (Levine, Svoboda, Hay, Winocur, & Moscovitch, 2002), which assesses the extent to which participant descriptions of past and future events are characterized by specific or extraneous details (e.g., Addis et al., 2008).

Nonetheless, primarily cognitive studies of episodic future thinking are somewhat lagging behind the quantity of research emerging from the neurosciences. This difference may often be recognized in the number of presentations on the topic at national conferences devoted to the neurosciences and cognitive psychology. The purpose of this special issue is to spur further interest in the development of cognitive studies of episodic future thinking. In this context we highlight recent approaches to the study of episodic future thinking from a largely cognitive standpoint, such as studies of the frequency with which people think about the future (D'Argembeau et al., 2010), techniques for enhancing the specificity with which people are able to simulate the future (e.g., Madore, Gaesser, & Schacter, 2014), and considerations of how episodic future thinking may interact with other modes of future-oriented cognition (e.g., Brewer & Marsh, 2009). Next, we provide additional details of specific contributions.

Contents of this issue

As we alluded to earlier, prior work has demonstrated that people spend considerable portions of their days thinking about the future (D'Argembeau et al., 2010). Among the many interesting findings to emerge from this survey is that most thinking about the future has a strong emotional component. The special issue begins with an extension of this survey approach for identifying characteristics of future thinking in everyday life. In particular, Barsics, Van der Linden, and D'Argembeau (2016) present results from a diary/laboratory study that further serves to outline the frequency and characteristics of emotional future event simulations in daily life.

Next, van Mulukolm, Schacter, Corballis, and Addis (2016) focus on the role of constructive processes in episodic future thinking via the experimental recombination procedure (Addis, Pan et al., 2009). Specifically, the authors demonstrate a novel use of the paradigm whereby personal information from different social spheres may be variably re-organized to manipulate the ease with which participants are able to generate novel simulations of future events. This unique twist on the experimental recombination procedure may hold promise for further developing our understanding of factors involved in making judgments about the future as the approach provides an easy way to manipulate the perceived plausibility of simulated events.

Keeping with the theme of constructive processes involved in episodic future thinking, McDermott, Wooldridge, Rice, Berg, and Szpunar (2016) set out to assess whether a similar or different constructive process may underlie the construction of memories and future events. In particular, the authors assess the visual perspective(s) that people adopt as they remember the past and imagine the future in the third person: perspectives that could have never occurred or could never occur and so must be constructed (for further reading, see Rice & Rubin, 2011). Interestingly, the distribution of third-person perspectives in the context of remembering and episodic future thinking turns out to be highly similar, suggesting that a common constructive mechanism underlies mental time travel into the past and future.

It is important to keep in mind that not all thoughts about the future require effortful processing, and that direct or spontaneous processes also play a role in this context. Jeunehomme and D'Argembeau (2016) demonstrate that the commonly used word-cueing paradigm can elicit thoughts about memories and future events that come to mind in a direct manner. Notably, Jeunehomme and D'Argembeau highlight that future thoughts that come to mind with little effort are likely to have been previously thought about before (Ingvar, 1985; Szpunar, Addis, McLelland, & Schacter, 2013) and are emotionally laden. Cole and Berntsen (2016) focus on the related concept of involuntary instances of memory and future thinking, whereby future events come to mind not only with little to no effort but also in times when the individual is not necessarily attempting to think about the future. Interestingly, Cole and Berntsen demonstrate that involuntary thoughts about the future may nonetheless be goal directed, a finding that resonates well with recent findings in the cognitive neuroscience literature that instances of mind wandering, often characterized by thoughts about the future (Stawarczyk, Cassol, & D'Argembeau, 2013), evoke activity in both core regions involved in representing simulated events (Benoit & Schacter,

2015) and frontoparietal control regions involved in goal-directed cognition (Fox, Spreng, Ellamil, Andrews-Hanna, & Christoff, 2015; see also Spreng et al., 2010).

As mentioned earlier, recent advances in the behavioural study of episodic future thinking include attempts to enhance the extent to which people are able to simulate the future in a specific manner. Madore et al. (2014) showed that a brief cognitive interview about a recent experience can be used to induce people to generate more detailed simulations of the future and that such an induction can selectively increase specific details associated with the event as opposed to extraneous details that may be irrelevant to the event (cf. Levine et al., 2002). Madore and Schacter (2016) replicate and extend their earlier findings using a novel set of stimuli and control conditions, thereby demonstrating the generality of their results. The authors discuss the potential implications of their induction procedure for improving performance on daily future-oriented tasks, such as planning.

Indeed, the extent to which detailed simulations of the future may enhance planning and other future-oriented tasks represents a topic of growing interest in the field (Szpunar, Spreng, & Schacter, 2014). This movement to broaden our understanding of episodic future thinking and its relation to future-oriented cognition is well represented by several contributions to the special issue. For instance, Renoult, Kopp, Davidson, Taler, and Atance (2016) focus on the biases that typically pervade predictions about the future, which many have argued are based on incomplete simulations of the future (e.g., Gilbert & Wilson, 2007). Terrett et al. (2016) assess the role of episodic future thinking in improving prospective memory performance. This latter study replicates considerable prior work showing that episodic simulation can in fact enhance the extent to which people remember to perform specific actions in the future (e.g., Brewer & Marsh, 2009). Importantly, Terrett et al. provide the first evidence that the relations between these modes of future thinking can differ as a function of age.

Whereas most researchers have focused on the manner in which memory facilitates the ability to think about the future, a number of contributions to this special issue address the manner in which memory may actually limit future thinking or the manner in which future thinking may limit memory for related events. Cordonnier, Barnier, and Sutton (2016) show that memory, albeit in the form of scripted knowledge, may constrain simulations generated in the context of a future planning task (for other recent discussions of the role of scripted or semantic knowledge to future thinking, see Irish & Piguet, 2013; Klein, 2013; Szpunar, 2010). Ditta and Storm (2016) test the assumption that memory and future thinking are closely related to one another by assessing the extent to which the generation of future events may actually reduce the accessibility of related autobiographical experiences in memory. Giebl, Storm, Buchli, Bjork, and Bjork (2016) extend this viewpoint by assessing correlations between an index of retrieval induced forgetting and the propensity for individuals to generate positive as opposed to negative future events. While the results of this latter study raise interesting insights into the positivity biases that commonly characterize memory (Walker & Skowronski, 2009) and future thinking (Szpunar et al., 2012), the reported data also call for more work to establish causal links between measures of retrieval-induced forgetting and positivity in event cognition.

Although the special issue does not focus much attention on developmental perspectives in episodic future thinking (Atance, 2008), Davis, Suddendorf, and Cullen (2016) introduce a unique approach to the study of episodic future thinking in young children. In particular, the authors assess the role of practice in improving performance in future tasks. As with most demonstrations of episodic future thinking in young children, practice must be demonstrated in an objective manner as the ability to discuss future-oriented behaviour is still developing at an early age. Overt practice of task performance may turn out to represent an important advance for gaining insights into the prospective abilities of young children.

The special issue concludes with a contribution from Klein (2016) that asks readers to consider the central role of autonoetic consciousness—the capacity to be aware of subjective time—in episodic future thinking. Specifically, Klein argues, as Tulving (1985) had done so previously, that the ability to sense subjective time is central in enabling the capacity to mentally travel into the personal past or future. Although research on the concept of autonoetic consciousness is lacking in the literature (but see, Nyberg, Kim, Habib, Levine, & Tulving, 2010; Piolino et al., 2003), it is our hope that this thoughtful piece from Klein will help to inspire researchers to develop novel techniques that may be used to advance the study of this ubiquitous mental phenomenon.

Moving forward

The study of episodic future thinking represents a unique opportunity for psychologists and neuroscientists from various disciplines to come together in the study of a psychological concept that has far-reaching implications for adaptive behaviour. A deeply rooted understanding of the cognitive mechanisms that support episodic future thinking will serve to advance our understanding of how the brain supports this important capacity, limitations associated with using this capacity to predict the future, its developmental trajectory, and the manner in which this capacity may break down in various neuropsychological, mood, and anxiety disorders. The contents of this special issue are intended to highlight novel developments in the study of episodic future thinking from a cognitive perspective and also to hopefully serve as an impetus for additional work.

It is also hoped that a greater awareness of this topic will be helpful to researchers who study other areas of cognition that may be aided by the insights from work on episodic future thinking. For example, this area of research could be helpful to studies of event cognition (cf. Radvansky & Zacks, 2014). As one example of this, research on narrative creation, the production of fictional alternative worlds, is likely to involve processes that are similar to the imagining of future events. Also, when using event models to help solve problems, one would need to imagine future states to help determine whether they could be helpful or harmful to the process of problem solving. Finally, the ability or inability people have in imagining possible future events surely plays a role in how people come to make decisions and to reason about the world (but see, Rosenbaum et al., in press). Even this sampling shows the potential value of work on episodic future thinking to a wide variety of aspects of cognition in general.

REFERENCES

Addis, D. R., Pan, L., Vu, M. A., Laiser, N., & Schacter, D. L. (2009). Constructive episodic simulation of the future and the past: Distinct subsystems of a core brain network mediate imagining and remembering. *Neuropsychologia*, 47, 2222–2238.

Addis, D. R., Sacchetti, D. C., Ally, B. A., Budson, A. E., and Schacter, D. L. (2009). Episodic simulation of future events is impaired in mild Alzheimer's disease. *Neuropsychologia*, 47, 2660–2671.

Addis, D. R., & Schacter, D. L. (2012). The hippocampus and imagining the future: Where do we stand? *Frontiers in Human Neuroscience*, 5, 173.

Addis, D. R., Wong, A. T., & Schacter, D. L. (2007). Remembering the past and imagining the future: Common and distinct neural substrates during event construction and elaboration. *Neuropsychologia*, 45, 1363–1377.

Addis, D. R., Wong, A. T., & Schacter, D. L. (2008). Age-related changes in the episodic simulation of future events. *Psychological Science*, 19, 33–41.

Atance, C. M. (2008). Future thinking in young children. *Current Directions in Psychological Science*, 17, 295–298.

Atance, C. M., & O'Neill, D. K. (2001). Episodic future thinking. *Trends in Cognitive Sciences*, 5, 533–539.

Barsics, C., Van der Linden, M., & D'Argembeau, A. (2016). Frequency, characteristics, and perceived functions of emotional future thinking in daily life. *Quarterly Journal of Experimental Psychology*.

Benoit, R. G., & Schacter, D. L. (2015). Specifying the core network supporting episodic simulation and episodic memory by activation likelihood estimation. *Neuropsychologia*, 75, 450–457.

Brewer, G. A., & Marsh, R. L. (2009). On the role of episodic future simulation in encoding of prospective memories. *Cognitive Neuroscience*, 1, 81–88.

Brown, A. D., Root, J. C., Romano, T. A., Chang, L. J., Bryant, R. A., & Hirst, W. (2013). Overgeneralized autobiographical memory and future thinking in combat veterans with posttraumatic stress disorder. *Journal of Behavior Therapy and Experimental Psychiatry, 44*, 129–134.

Buckner, R. L., & Carroll, D. C. (2007). Self-projection and the brain. *Trends in Cognitive Sciences, 11*, 49–57.

Cole, S. N., & Berntsen, D. (2016). Do future thoughts reflect personal goals? Current concerns and mental time travel into the past and future. *Quarterly Journal of Experimental Psychology.*

Corballis, M. C. (2013). Mental time travel: A case for evolutionary continuity. *Trends in Cognitive Sciences, 17*, 5–6.

Cordonnier, A., Barnier, A., & Sutton, J. (2016). Scripts and information units in future planning: Interactions between a past and a future planning task. *Quarterly Journal of Experimental Psychology.*

Crovitz, H. F., & Schiffman, H. (1974). Frequency of episodic memories as a function of their age. *Bulletin of the Psychonomic Society, 4*, 517–518.

D'Argembeau, A., Raffard, S., & Van der Linden, M. (2008). Remembering the past and imagining the future in schizophrenia. *Journal of Abnormal Psychology, 117*, 247–251.

D'Argembeau, A., Renaud, O., & Van der Linden, M. (2010). Frequency, characteristics, and functions of future-oriented thoughts in daily life. *Applied Cognitive Psychology, 35*, 96–103.

D'Argembeau, A., & Van der Linden, M. (2004). Phenomenal characteristics associated with projecting oneself back into the past and forward into the future: Influence of valence and temporal distance. *Consciousness and Cognition, 13*, 844–858.

Davis, J., Suddendorf, T., & Cullen, E. (2016). Understanding deliberate practice in preschool aged children. *Quarterly Journal of Experimental Psychology.*

Ditta, A. S., & Storm, B. C. (2016). Thinking about the future can cause forgetting of the past. *Quarterly Journal of Experimental Psychology.*

Fox, K. C. R., Spreng, R. N., Ellamil, M., Andrews-Hanna, J. R., & Christoff, K. (2015). The wandering brain: Meta-analysis of functional neuroimaging studies of mind-wandering and related spontaneous thought processes. *NeuroImage, 111*, 611–621.

Gamboz, N., de Vito, S., Brandimonte, M. A., Pappalardo, S., Galeone, F., Iavarone, A., & Della Sala, S. (2010). Episodic future thinking in amnesic mild cognitive impairment. *Neuropsychologia, 48*, 2091–2097.

Giebl, S., Storm, B. C., Buchli, D., Bjork, E. L., & Bjork, R. A. (2016). Retrieval-induced forgetting is associated with increased positivity when imagining the future. *Quarterly Journal of Experimental Psychology.*

Gilbert, D. T., & Wilson, T. D. (2007). Prospection: Experiencing the future. *Science, 317*, 1351–1354.

Hach, S., Tippett, L. J., & Addis, D. R. (2014). Neural changes associated with the generation of specific past and future events in depression. *Neuropsychologia, 65*, 41–55.

Hassabis, D., Kumaran, D., Vann, S. D., & Maguire, E. A. (2007). Patients with hippocampal amnesia cannot imagine new experiences. *Proceedings of the National Academy of Sciences, 104*, 1726–1731.

Hassabis, D., Spreng, R. N., Rusu, A. A., Robbins, C. A., Mar, R. A., & Schacter, D. L. (2014). Imagine all the people: How the brain creates and uses personality models to predict behavior. *Cerebral Cortex, 24*, 1979–1987.

Ingvar, D. H. (1985). "Memory of the future": An essay on the temporal organization of conscious awareness. *Human Neurobiology, 4*, 127–136.

Irish, M., & Piguet, O. (2013). The pivotal role of semantic memory in remembering the past and imagining the future. *Frontiers in Behavioral Neuroscience, 7*, 27.

Jeunehomme, O., & D'Argembeau, A. (2016). Prevalence and determinants of direct and generative modes of production of episodic future thoughts in the word cueing paradigm. *Quarterly Journal of Experimental Psychology.*

Klein, S. B. (2013). The complex act of projecting oneself into the future. *Wiley Interdisciplinary Reviews: Cognitive Science, 4*, 63–79.

Klein, S. B. (2016). Autonoetic consciousness: Reconsidering the role of episodic memory in future-oriented self-projection. *Quarterly Journal of Experimental Psychology.*

Klein, S. B., Loftus, J., & Kihlstrom, J. F. (2002). Memory and temporal experience: The effects of episodic memory loss on an amnesic patient's ability to remember the past and imagine the future. *Social Cognition, 20*, 353–379.

Levine, B., Svoboda, E., Hay, J. F., Winocur, G., & Moscovitch, M. (2002). Aging and autobiographical memory: Dissociating episodic from semantic retrieval. *Psychology and Aging, 17*, 677–689.

Madore, K. P., Gaesser, B., & Schacter, D. L. (2014). Constructive episodic simulation: Dissociable effects of a specificity induction on remembering, imagining, and describing in young and older adults. *Journal of Experimental Psychology: Learning, Memory, and Cognition, 40*, 609–622.

Madore, K. P., & Schacter, D. L. (2016). Remembering the past and imagining the future: Selective effects of an episodic specificity induction on detail generation. *Quarterly Journal of Experimental Psychology.*

McDermott, K. B., Wooldridge, C., Rice, H. J., Berg, J. J., & Szpunar, K. K. (2016). Visual perspective in remembering and episodic future thought. *Quarterly Journal of Experimental Psychology.*

Miloyan, B., Bulley, A., & Suddendorf, T. (in press). Episodic foresight and anxiety: Proximate and ultimate perspectives. *British Journal of Clinical Psychology.*

Nyberg, L., Kim, A. S., Habib, R., Levine, B., & Tulving, E. (2010). Consciousness of subjective time in the brain. *Proceedings of the National Academy of Sciences, 107,* 22356–22359.

Okuda, J., Fujii, T., Ohtake, H., Tsukiura, T., Tanji, K., Suzuki, K., Kawashima, R., Fukuda, H., Itoh, M., & Yamadori, A. (2003). Thinking of the future and past: The roles of the frontal pole and the medial temporal lobes. *NeuroImage, 19,* 1369–1380.

Piolino, P., Desgranges, B., Belliard, S., Matuszewski, V., Lalevee, C., De La Sayette, V., & Eustache, F. (2003). Autobiographical memory and autonoetic consciousness: Triple dissociation in neurodegenerative diseases. *Brain, 126,* 2203–2219.

Radvansky, G. A. & Zacks, J. M. (2014). *Event cognition.* New York: Oxford University Press.

Renoult, L., Kopp, L., Davidson, P. S. R., Taler, V., & Atance, C. M. (2016). You'll change more than I will: Adults' predictions about their own and others' future preferences. *Quarterly Journal of Experimental Psychology.*

Rice, H. J., & Rubin, D. C. (2011). Remembering from any angle: The flexibility of visual perspective during retrieval. *Consciousness and Cognition, 20,* 568–577.

Rosenbaum, R.S., Kwan, D., Floden, D., Levine, B., Stuss, D. T., & Craver, C. F. (in press). No evidence of risk-taking or impulsive behaviour in a person with episodic amnesia: Implications for the role of the hippocampus in future-regarding decision-making. *Quarterly Journal of Experimental Psychology.*

Schacter, D. L., & Addis, D. R. (2007). The cognitive neuroscience of constructive memory: Remembering the past and imagining the future. *Philosophical Transactions of the Royal Society B: Biological Sciences, 362,* 773–786.

Spreng, R. N., & Levine, B. (2006). The temporal distribution of past and future autobiographical events across the lifespan. *Memory & Cognition, 34,* 1644–1651.

Spreng, R. N., Stevens, W. D., Chamberlain, J. P., Gilmore, A. W., & Schacter, D. L. (2010). Default network activity, coupled with the frontoparietal control network, supports goal-directed cognition. *NeuroImage, 53,* 303–317.

Stawarczyk, D., Cassol, H., & D'Argembeau, A. (2013). Phenomenology of future-oriented mind-wandering episodes. *Frontiers in Psychology, 4,* 425.

Suddendorf, T. (2013). Mental time travel: Continuities and discontinuities. *Trends in Cognitive Sciences, 17,* 151–152.

Suddendorf, T., & Corballis, M. C. (1997). Mental time travel and the evolution of the human mind. *Genetic, Social, and General Psychology Monographs, 123,* 133–167.

Szpunar, K. K. (2010). Episodic future thought: An emerging concept. *Perspectives on Psychological Science, 5,* 142–162.

Szpunar, K. K., Addis, D. R., McLelland, V. C., & Schacter, D. L. (2013). Memories of the future: New insights into the adaptive value of episodic memory. *Frontiers in Behavioral Neuroscience, 7,* 47.

Szpunar, K. K., Addis, D. R., & Schacter, D. L. (2012). Memory for emotional simulations: Remembering a rosy future. *Psychological Science, 23,* 24–29.

Szpunar, K. K., Spreng, R. N., & Schacter, D. L. (2014). A taxonomy of prospection: Introducing an organizational framework for future-oriented cognition. *Proceedings of the National Academy of Sciences, 111,* 18414–18421.

Szpunar, K. K., St. Jacques, P. L., Robbins, C. A., Wig, G. S., & Schacter, D. L. (2014). Repetition-related reductions in neural activity reveal component processes of mental simulation. *Social Cognitive and Affective Neuroscience, 9,* 712–722.

Szpunar, K. K., Watson, J. M., & McDermott, K. B. (2007). Neural substrates of envisioning the future. *Proceedings of the National Academy of Sciences, 104,* 642–647.

Terrett, G., Rose, N. S., Henry, J. D., Bailey, P. E., Altgassen, M., Phillips, L. H., Kliegel, M., & Rendell, P. G. (2016). The relationship between prospective memory and episodic future thinking in younger and older adulthood. *Quarterly Journal of Experimental Psychology.*

Tulving, E. (1985). Memory and consciousness. *Canadian Psychology, 26,* 1–12.

van Mulukolm, V., Schacter, D. L., Corballis, M. C., & Addis, D. R. (2016). The degree of disparateness of event details modulates future simulation construction, plausibility, and recall. *Quarterly Journal of Experimental Psychology.*

Walker, W. R., & Skowronski, J. J. (2009). The fading affect bias: But what the hell is it for? *Applied Cognitive Psychology, 23,* 1122–1136.

Frequency, characteristics, and perceived functions of emotional future thinking in daily life

Catherine Barsics[1,2], Martial Van der Linden[1,2,3], and Arnaud D'Argembeau[1,3]

[1]Swiss Center for Affective Sciences, University of Geneva, Geneva, Switzerland
[2]Cognitive Psychopathology and Neuropsychology Unit, University of Geneva, Geneva, Switzerland
[3]Department of Psychology: Cognition and Behavior, University of Liège, Liège, Belgium

While many thoughts and mental images that people form about their personal future refer to emotionally significant events, there is still little empirical data on the frequency and nature of emotional future-oriented thoughts (EmoFTs) that occur in natural settings. In the present study, participants recorded EmoFTs occurring in daily life and rated their characteristics, emotional properties, and perceived functions. The results showed that EmoFTs are frequent, occur in various contexts, and are perceived to fulfil important functions, mostly related to goal pursuit and emotion regulation. When distinguishing between anticipatory and anticipated emotions (i.e., emotions experienced in the present versus emotions expected to occur in the future), a positivity bias in the frequency of EmoFTs was found to be restricted to anticipated emotions. The representational format and perceived function of EmoFTs varied according to their affective valence, and the intensity of anticipatory and anticipated emotions were influenced by the personal importance and amount of visual imagery of EmoFTs. Mood states preceding EmoFTs influenced their emotional components, which, in turn, impacted ensuing mood states. Overall, these findings shed further light on the emotional properties of future-oriented thoughts that are experienced in daily life.

Over the last decade, important progress has been made in understanding the representations and processes that support our ability to mentally explore possible futures (Atance & O'Neill, 2001; Gilbert & Wilson, 2007; Schacter et al., 2012; Seligman, Railton, Baumeister, & Sripada, 2013). This capacity—often termed "future thinking" or prospection— is central to many aspects of human cognition and behaviour, from decision making and planning, to self-regulation and the sense of identity (Bechara & Damasio, 2005; Conway, 2005; Suddendorf & Corballis, 2007; Taylor, Pham, Rivkin, & Armor, 1998; for a recent perspective on the functions of future thinking, see Szpunar, Spreng, & Schacter, 2014). The proportion of daily time we spend projecting

The authors thank Marjorie Texier for her help with data collection.

This research was supported by the National Center of Competence in Research (NCCR) Affective Sciences financed by the Swiss National Science Foundation [grant number 51NF40-104897] and hosted by the University of Geneva. Arnaud D'Argembeau is a Research Associate of the National Fund for Scientific Research (FRS-FNRS), Belgium.

ourselves into the short-term and long-term future is far from being trivial: On average, during a typical day, young adults might experience 60 future-oriented thoughts—that is, about one future-oriented thought every 16 minutes (D'Argembeau, Renaud, & Van der Linden, 2011). Some of these thoughts and mental images refer to emotionally significant events, either positive situations that we strive to achieve or negative situations that we would rather avoid (D'Argembeau et al., 2011). When anticipating such future events, people might experience intense emotional reactions (Van Boven & Ashworth, 2007), which in turn might influence motivation, behavioural intentions, and ultimately behaviour (Baumgartner, Pieters, & Bagozzi, 2008). Therefore, the emotional component of future-oriented thoughts is a key aspect of prospection that probably plays critical roles in goal pursuit. Relatively little is known about the characteristics of emotional future-oriented thoughts that are experienced in daily life, however, and the purpose of the present study is to shed further light on this phenomenon.

One of the earliest findings on prospection was the identification of a valence effect on the characteristics of future-oriented thoughts. Compared with representations of negative future events, positive future events are more frequent and imagined faster (MacLeod & Byrne, 1996; Newby-Clark & Ross, 2003), are associated with more sensorial details, clearer representations of contextual information, and greater feelings of preexperiencing, and include more social contents (D'Argembeau & Van der Linden, 2004; De Vito, Neroni, Gamboz, Della Sala, & Brandimonte, 2015; Painter & Kring, 2015; Rasmussen & Berntsen, 2013). In addition, details associated with positive future-event simulations are better remembered than details associated with negative ones (Szpunar, Addis, & Schacter, 2012), and repeated simulations of emotional events make them seem more plausible (Szpunar & Schacter, 2013). Overall, these findings are consistent with the idea that most people are optimistic and tend to conceive their personal future in a favourable light (Sharot, Riccardi, Raio, & Phelps, 2007; Taylor & Brown, 1988; Weinstein, 1980; but see Harris & Hahn, 2011).

The aforementioned findings have been observed in laboratory studies in which positive and negative future-oriented thoughts were produced in response to experimentally provided cues. On the other hand, little is known about the nature of emotional prospection occurring in natural settings. Results from diary studies have indicated that involuntary future event representations are as common as involuntary autobiographical memories in daily life, and that future event representations are more positive and idyllic than memories for past events (Berntsen & Jacobsen, 2008; Finnbogadóttir & Berntsen, 2013). In addition, positive future-oriented thoughts are more frequent (D'Argembeau et al., 2011), more specific, and associated with more visual images than negative future-oriented thoughts (D'Argembeau et al., 2011; Finnbogadóttir & Berntsen, 2011). Beyond this valence effect, the emotional properties of the future-oriented thoughts that occur in daily life remain unknown. In particular, these previous studies did not assess whether the reported prospections were actually accompanied by emotions when they occurred. Here we focus on this emotional component of prospection, and we designate thoughts that are accompanied by an emotional response when they occur as "emotional future-oriented thoughts" (EmoFTs).[1] The first aim of the present study was therefore to examine thoroughly the conditions of occurrence, characteristics (notably the emotional properties), and perceived functions of EmoFTs arising in natural settings, using a diary method (Bolger, Davis, & Rafaeli, 2003).

To further investigate the affective dimension of prospection, we distinguished between two

---

[1]It should be noted that EmoFT is conceived here as a broad phenomenon that include any type of future-oriented thought that is accompanied by an emotion. Therefore, this notion could potentially include more specific types of future-oriented thoughts that have been previously described in the literature, such as worries (which predominantly involve verbal thoughts about negative events that we are afraid might happen in the future; Borkovec, Ray, & Stober, 1998).

kinds of future-oriented emotions: *anticipatory emotions*, which refer to the emotions experienced in the present in response to the prospect of future events, and *anticipated emotions*, which refer to the emotions that are expected to be experienced in the future, if and when imagined events occur (Baumgartner et al., 2008). Research has shown that anticipatory and anticipated emotions are both useful to predict behaviour. In a recent study, Carrera, Caballero and Muñoz (2012) indeed found that anticipatory emotions better predict the subjective probability that a behaviour will actually be performed (i.e., behavioural expectations; Davis & Warshaw, 1992), whereas anticipated emotions better predict the amount of effort that one is willing to exert to attain a goal (i.e., behavioural intentions; Ajzen, 1991). While anticipated emotions, or affective forecasts, have been under the scope of a wide range of studies (e.g., Hoerger, Chapman, Epstein, & Duberstein, 2012; Wilson & Gilbert, 2005; for a recent review, see Miloyan & Suddendorf, 2015), the anticipatory emotions that accompany the simulation of future events have received less attention. Hence, an important aim of this study was to better characterize both anticipatory and anticipated emotions associated with EmoFTs and to investigate whether they are related to other properties of future-oriented thoughts. In particular, considering previous work on the role of visual images in inducing emotional responses (Daselaar et al., 2008; Holmes & Mathews, 2010), we expected that visual imagery would be a significant predictor of anticipatory emotions. Furthermore, we sought to investigate whether the previously reported effects of valence on the characteristics of future-oriented thoughts (see above) is similar for anticipatory and anticipated emotions.

We also investigated the perceived functions of EmoFTs. Whereas the various functions of autobiographical memories have been much investigated and are now well characterized (e.g., Bluck, Alea, Habermas, & Rubin, 2005; Harris, Rasmussen, & Berntsen, 2014; Rasmussen & Berntsen, 2009), relatively little work has been devoted to the functions of future-oriented thoughts. Future-oriented thoughts are perceived

as serving various functions, such as decision making, action planning, and emotional regulation (D'Argembeau et al., 2011), and it has been found that thoughts involving positive events have more social and self-regulating functions than thoughts involving negative events (Rasmussen & Berntsen, 2013). Drawing on the distinction between anticipatory and anticipated emotions as defined above, our aim here was to investigate whether valence effects in the perceived functions of prospections are similar for these two kinds of emotions.

Another aim of the present study was to investigate the potential impact of the emotional properties of EmoFTs on mood. Insofar as intense emotional reactions can accompany prospections (Van Boven & Ashworth, 2007), one might hypothesize that these emotions would influence ensuing mood states. Recent results have shown that the affective content of mind-wandering episodes (i.e., thoughts that are unrelated to current perception and ongoing actions; Smallwood & Schooler, 2006) sampled in daily life not only is predicted by previous mood, but has also an impact on subsequent mood (Poerio, Totterdell, & Miles, 2013). More precisely, Poerio et al. (2013) found that the affective content of mind wandering is congruent with feelings preceding the episode of mind wandering and in turn predicts subsequent mood, even after previous mood has been taken into account. Considering that the content of a non-negligible part of mind wandering refers to the anticipation and planning of future events (Smallwood, Nind, & O'Connor, 2009; Song & Wang, 2012; Stawarczyk, Cassol, & D'Argembeau, 2013), these findings suggest that the affective content of EmoFTs might significantly impact ensuing mood states. In the present study, we investigated this possibility by asking participants to assess the mood states that they experienced before and after the occurrence of EmoFTs.

In summary, the main purpose of this study was to shed light on the emotional components of future-oriented thoughts that arise in natural settings. More specifically, we aimed at providing a detailed account of the frequency, conditions of occurrence, characteristics, and perceived functions of emotional prospections in daily life, notably as a function of emotional valence. Furthermore, we aimed at characterizing the anticipatory and anticipated emotions associated with EmoFTs and their relation with other properties of future-oriented thoughts. A final goal of this study was to investigate the relationships between mood states and the emotional components of EmoFTs. To address these questions, we used a diary method in which participants recorded and described a sample of emotional future-oriented thoughts that they experienced in their daily life, before rating them on the characteristics of interest.

EXPERIMENTAL STUDY

Method

Participants

Eighty-nine individuals, recruited via personal contacts and referrals, volunteered to take part in the study. The data of 13 participants were not included in the analysis because they did not follow the instructions properly: A substantial part of their thoughts were rated more than 16 hours after their occurrence (see Materials and Procedure section). The 76 remaining participants (44 females) were aged between 19 and 29 years ($M = 23.7$, $SD = 2.5$). They were undergraduate students at the Universities of Liège and Geneva or young workers from these areas (27 lived in Belgium and 47 in Switzerland). Overall, 44 were undergraduate students, 30 were young workers, and two were unemployed. They had completed between 10 and 23 years of education ($M = 16.2$, $SD = 2.8$).

Materials and procedure

The study consisted of three phases. The first phase aimed at determining the number of EmoFTs experienced during three typical days in participants' lives. The second phase involved more detailed assessments (thematic content, phenomenological characteristics, and perceived functions) of a total of 10 EmoFTs. In the third phase, participants completed a series of questionnaires assessing

personality traits and psychopathological symptoms; these questionnaires are not the object of the present article and thus are not discussed further.

Prior to the first phase, an information session took place in order to familiarize participants with the concept of EmoFT. It was explained to them that the research focused on EmoFTs that are experienced in everyday life. EmoFTs were described as thoughts that are oriented towards the future and that are accompanied by a positive or a negative emotion (e.g., joy, fear, pride, anger, fear, or sadness). It was specified that EmoFTs could be thoughts that refer to events that might happen at a specific time in the future (e.g., an event scheduled next Saturday) or more general thoughts about the future that do not refer to specific events (e.g., thinking about possible professional orientations in abstract terms); intentional thoughts or thoughts that come spontaneously to mind; thoughts referring to a near future (e.g., later the same day or within a week) or to a more distant future (e.g., in several months or years); or thoughts whose content might be more or less plausible. In summary, it was clearly explained that we were interested in any kind of future-oriented thought that participants might experience, provided that these thoughts were accompanied by a positive or a negative emotion when they occurred. This information was given both orally and in a written format, and was discussed with the participant until the experimenter was sure that it was clearly understood, allowing for the instructions to be given for the first phase of the study.

In the first phase of the study, participants were asked to record all EmoFTs that they experienced during three consecutive days, starting from when they woke up on the first day up to when they fell asleep on the third day. In order to do so, they were provided with a mark sheet, which they were requested to carry along, as well as a pen, during the whole phase. Participants were instructed to tick the mark sheet each time an EmoFT came to their mind. In cases where it was not possible to record an EmoFT immediately when it happened (e.g., while driving), they had to record the thought as soon as possible. Participants were allowed to start this first phase at their convenience, provided that they choose typical days of their life for recording EmoFTs (e.g., holidays were excluded as a recording period). Following this first phase, participants came back to the laboratory to return their completed mark sheet. In addition, they were asked to rate on 7-point Likert scales the extent to which they omitted to record certain thoughts because they did not have the time or desire to record them ($1 =$ not at all, $3 =$ a little, $5 =$ a lot, $7 =$ tremendously) and the extent to which they experienced more future-oriented thoughts than usual because they had to record them ($1 =$ not at all, $3 =$ a little, $5 =$ a lot, $7 =$ tremendously).

The instructions for the second phase of the study were then provided. This time, participants were asked to record in more detail a total of 10 EmoFTs—whose aforementioned definition they were reminded of. There was no time limit for carrying out this task; instead, participants were instructed to report the first 10 distinct EmoFTs that came to their mind, from the moment they awoke on a typical day of their lives. In order to do so, they were provided with a small booklet to record and rate each of their EmoFTs. For each EmoFT, they first had to briefly describe its content, in such a way that a reader would be able to understand what it was about. In case some thoughts would be judged by the participants as too intimate to disclose them, it was specified that they could merely indicate "private" instead of describing their content. Participants were also asked to describe the context of occurrence of each EmoFT: the time and the place where the EmoFT occurred, their current activity, and whether or not they were alone.

Participants then rated, on a series of 7-point Likert scales, the following characteristics of their thought: its representational format (i.e., the amount of visual images and inner speech; $1 =$ not at all, $7 =$ extremely), the way the thought came to their mind ($1 =$ completely spontaneous, $7 =$ completely voluntary/intentional), the extent to which the thought was triggered by the environment ($1 =$ not at all, $7 =$ extremely), its personal importance ($1 =$ not at all; $7 =$ extremely), its relation to identity ($1 =$ not at all, $7 =$ extremely), and its recurrence (i.e., the extent to which the thought had been previously experienced; $1 =$ never, $7 =$ very frequently). Participants also rated the extent to which the thought helped them to handle a present or future situation ($1 =$ not at all, $7 =$ completely), their inclination to suppress the thought ($1 =$ not at all, $7 =$ completely) or to share it with others ($1 =$ not at all probable; $7 =$ extremely probable), and the probability of occurrence of the thought's content ($1 =$ not at all probable; $7 =$ extremely probable). Participants also reported the future time period to which the thought referred, which could be either specified (e.g., in one day, next week, in two months) or undefined.

In addition to these characteristics, two aspects of the affective dimension of each thought were assessed. First, participants rated, on a scale ranging from -3 (very negative) to $+3$ (very positive), the valence and intensity of anticipatory emotions (i.e., emotions currently experienced when thinking about something that could happen in the future; see Baumgartner et al., 2008), and they indicated what specific emotion or emotions were involved (i.e., joy, fear, pride, anger, fear, sadness; an "other" category was also available in order to allow the participant to specify any other emotion). Second, participants completed the exact same ratings regarding anticipated emotions (i.e., the emotions that a person imagines experiencing in the future if and when the event occurs; see Baumgartner et al., 2008). Furthermore, mood states prior and after the EmoFT occurred were both assessed, on a scale ranging from -3 (very negative) to $+3$ (very positive).

Finally, participants were requested to report the perceived function(s) of their EmoFT by selecting among distinct categories, which were related to goal pursuit (i.e., making a decision, planning an action, setting oneself a goal) or emotional regulation (either with a present-oriented focus, i.e., reassuring oneself or feeling better, or with a future-oriented focus, i.e., preparing oneself to deal with an anticipated emotion). Two supplementary categories were also available to allow the participant to specify another perceived function, or to indicate that no function was identified.

The experimenter reviewed all the items of the questionnaire with each participant to ensure their full understanding. Participants were asked to describe and rate each thought immediately after the thought came to their mind, or as soon as possible. In any case, they had to report the time at which they performed the thought rating. When participants came back to the laboratory to return their completed mark sheet, they were asked to rate on 7-point Likert scales the extent to which they omitted to record certain thoughts because they did not have the time or desire to record them (1 = not at all, 3 = a little, 5 = a lot, 7 = tremendously), and the extent to which they experienced more future-oriented thoughts than usual because they had to record them (1 = not at all, 3 = a little, 5 = a lot, 7 = tremendously).

Coding of content

The content of each EmoFT was coded by first author according to its specificity. An EmoFT was coded as specific if it referred to a unique event, precisely located in time and lasting no more than a day (e.g., going to the theatre tomorrow evening; cf. Williams et al., 1996). Events taking place over extended time periods (e.g., next summer vacation in France), as well as abstract thoughts devoid of any reference to particular events (e.g., thinking about being a parent), were coded as nonspecific. Finally, undescribed private events were categorized as unclassifiable. A random selection of 10% of the thoughts was scored by an independent rater, which showed a good inter-rater reliability ($K = .90$; agreement = 95%).

Results

Frequency, specificity, and context of occurrence of EmoFTs

In the first phase of the study, participants reported having experienced, on average, 28 emotional future-oriented thoughts during the

three consecutive days ($SD = 18.7$; minimum = 2, maximum = 83). Participants felt that they omitted to record a few EmoFTs ($M = 3.01$, $SD = 1.22$, on the 7-point rating scale) because they did not have the time or did not feel like doing it; a similar tendency was also apparent for the second phase of the study ($M = 3.54$, $SD = 1.36$). Besides, participants also reported having experienced a few more thoughts than usual because they were requested to record them, both in the first phase ($M = 2.87$, $SD = 1.50$) and in the second phase ($M = 2.34$, $SD = 1.26$) of the study. During the second phase, each of the 76 participants recorded 10 thoughts. However, 31 of these thoughts were excluded from the analyses because they had been rated too long (i.e., more than 16 hours) after their occurrence, and five additional thoughts were also discarded because their description indicated that they were actually not future oriented. The data therefore consisted of a total of 724 EmoFTs.

Among these thoughts, 49% referred to specific events, whereas 45% were nonspecific; 6% could not be classified according to specificity. Twenty-two percent of EmoFTs did not refer to any specified time period. Among the EmoFTs involving a specified time period, 31% referred to an event occurring later the same day, 31% to an event occurring within one week, 14% to the period between one week and one month, 17% to the period between one month and one year, 5% to the period between one year and five years, and 2% to more than 5 years. Therefore, the vast majority of EmoFTs associated with a specified time period referred to the near future, with 76% of EmoFTs referring to events that might happen within the next month.

When examining the context of occurrence of EmoFTs, we found that 52% of thoughts occurred at home, 25% at work, 11% in transport, 8% in public places, and 3% at a relative's or friend's place (the remaining 1% could not be classified according to these categories). With regard to ongoing activities, we found that 29% of EmoFTs happened while participants were working, 21% while resting, 13% while eating,

Table 1. *Mean number of anticipatory and anticipated emotions reported for each type of emotion category*

Emotion category	Anticipatory	Anticipated	t	df	p
Fear	3.64 (1.90)	1.57 (1.36)	10.06	75	<.001
Joy	4.66 (1.85)	5.89 (1.92)	5.93	75	<.001
Pride	0.97 (1.18)	1.70 (1.70)	5.18	75	<.001
Anger	0.86 (1.15)	0.80 (1.11)	0.44	75	.66
Shame	0.36 (0.60)	0.53 (0.74)	2.13	75	.04
Sadness	1.21 (1.40)	1.22 (1.31)	0.10	75	.92
Others	1.12 (1.24)	0.92 (1.15)	1.43	75	.16

Note: Standard deviations in parentheses.

grooming, or dressing, 11% while watching movies, playing video games or surfing on the web, 8% while driving, 7% during conversations, 6% while doing housework, and 3% during leisure activities (the remaining 2% could not be classified according to these categories). Forty-nine percent of EmoFTs were experienced when participants were alone.

Emotional characteristics and perceived functions of EmoFTs

On average, anticipatory emotions (emotions currently experienced in association with the thoughts) were rated as slightly positive ($M = 0.35$, $SD = 2.04$), and, when looking at the specific type of emotion experienced, we found that 49% of EmoFTs were accompanied by joy or enthusiasm, 38% by fear, anxiety, or a feeling of stress, 13% by sadness, 10% by pride, 9% by anger, 4% by shame, and 12% by other kinds of emotions. With regard to anticipated emotions (the emotion that one would expect to experience if the future-oriented thought materialized), we also found that EmoFTs were rated as slightly positive ($M = 1.05$, $SD = 2.15$). Anticipated emotions were described as joy or enthusiasm (63% of EmoFTs), pride (18%), fear, anxiety, or a feeling of stress (17%), sadness (13%), anger (9%), or shame (6%); 10% of anticipated emotions did not fall into these categories.

To investigate whether some specific type of emotions were more frequently reported as anticipatory or anticipated emotions, we computed for each participant the number of reported anticipatory and anticipated emotions for each emotion category. A series of paired samples *t*-tests

(see Table 1) indicated that joy, pride, and shame were more frequently reported as anticipated emotions than as anticipatory emotions, whereas fear was more frequently reported as an anticipatory emotion than as an anticipated emotion; there was no significant difference in the reports of anger, sadness, and other emotions.

As regards the perceived functions of EmoFTs, 40% were judged as helpful to plan actions, 21% were related to intention formation (i.e., setting oneself a goal), 21% were considered to allow emotion regulation with a present-oriented focus (i.e., to reassure oneself or feel better), 17% were judged as helpful to allow emotion regulation with a future-oriented focus (i.e., preparing oneself to deal with an anticipated emotion), and 10% were regarded as helpful to make a decision; 20% of EmoFTs were considered devoid of any apparent function, and 5% were reported to involve other kinds of functions, for instance daydreaming (note that these percentages do not sum up to 100% because participants were allowed to select more than one function for each thought, if appropriate).

EmoFTs associated with positive versus negative anticipatory emotions

One of the aims of the present study was to compare the characteristics of EmoFTs that were associated with positive versus negative anticipatory emotions. To do so, each of the 724 EmoFTs was classified as positive or negative based on ratings of anticipatory emotions (positive and negative EmoFTs were associated with a rating > 0 or < 0, respectively), which left 19 thoughts unclassified

Table 2. *Characteristics of emotional future-oriented thoughts associated with positive versus negative anticipatory emotions*

Thought property	Positive	Negative	df	t	p
Number of thoughts	4.92 (1.82)	4.36 (1.93)	75	1.37	.17
Specific thoughts (%)	51 (31)	48 (31)	73	0.82	.42
Nonspecific thoughts (%)	44 (31)	47 (32)	73	0.95	.48
Unclassifiable thoughts (%)	6 (17)	5 (15)	73	0.22	.83
Characteristics (mean ratings)					
Inner speech	3.53 (1.28)	4.50 (1.39)	73	5.72	<.001
Visual images	4.76 (1.32)	3.63 (1.28)	73	7.68	<.001
Intentionality	3.75 (1.32)	3.69 (1.29)	73	0.39	.69
Environment	3.45 (1.37)	3.89 (1.36)	73	2.12	.04
Personal importance	5.00 (1.07)	5.11 (1.39)	73	0.90	.37
Handling	3.86 (1.30)	4.40 (1.36)	73	2.52	.01
Identity	3.32 (1.28)	3.36 (1.16)	73	0.28	.78
Sharing	4.44 (1.34)	3.95 (1.32)	73	2.94	<.01
Recurrence	3.82 (1.07)	3.86 (1.02)	73	0.36	.72
Suppression	1.48 (0.74)	3.43 (1.26)	73	12.99	<.001
Probability	5.88 (0.94)	5.18 (1.09)	73	5.16	<.001
Function (%)					
Decision making	7 (14)	13 (20)	73	2.21	.03
Planning	41 (29)	42 (29)	73	0.19	.85
Intention formation	19 (23)	20 (20)	73	0.18	.86
P.-O. emotional regulation	28 (27)	16 (22)	73	2.84	<.01
F.-O. emotional regulation	8 (16)	28 (29)	73	5.40	<.001
Other	5 (15)	4 (14)	73	0.50	.62
None	21 (24)	18 (23)	73	0.69	.49

Note: Means; standard deviations in parentheses. $df = 73$ since two participants did not experience any negative emotional future-oriented thoughts (EmoFTs); P.-O. = present oriented; F.-O. = future oriented.

(note that although these thoughts were rated 0, they were nonetheless reported as accompanied by some anticipatory emotions, as indicated by the participants' selection of specific emotions; in some cases, these EmoFTs implicated both positive and negative emotions that might have offset each other; in other cases, it is possible that the selected emotion was not deemed sufficiently intense to receive a positive or negative value on the rating scale). A series of paired samples *t*-tests was conducted in order to examine whether EmoFTs associated with positive anticipatory emotions differed from EmoFTs associated with negative anticipatory emotions regarding frequency, specificity, phenomenological characteristics, and perceived functions.

As shown in Table 2, overall, participants did not report more positive than negative EmoFTs, and specificity also did not differ between two types of thoughts. Regarding their representational format,

negative EmoFTs were more likely to take the form of inner speech, while positive EmoFTs were more frequently experienced as visual images. On average, the thoughts were slightly more likely to come to mind spontaneously rather than voluntarily, and the two kinds of thoughts did not differ in this respect. Negative EmoFTs were judged to be triggered by the environment to a larger extent than positive EmoFTs and were also rated as more helpful to handle a present or future situation. Overall, the thoughts were judged quite important but only moderately related to the participants' sense of self-identity; there was no difference between positive and negative EmoFTs on these dimensions. The reported EmoFTs had sometimes already been experienced on a previous occasion and were judged as moderately likely to be shared, with positive EmoFTs being rated as more likely to be shared than negative ones. The inclination to suppress EmoFTs was judged higher for negative than for positive thoughts. On the other hand, the perceived probability of EmoFTs was higher for positive than for negative thoughts. Finally, with regard to the perceived functions of the thoughts, negative EmoFTs were judged as more related to decision making and as more helpful to prepare oneself to deal with a potential future emotion (i. e., future-oriented emotional regulation), whereas positive EmoFTs were judged as more helpful to reassure oneself or to feel better (i.e., present-oriented emotional regulation).

EmoFTs associated with positive versus negative anticipated emotions

We also examined whether the characteristics of EmoFTs varied depending on the valence of anticipated emotions. Each of the 724 EmoFTs was classified as associated with a positive or negative anticipated emotion, according to the corresponding ratings (positive and negative EmoFTs were associated with a rating > 0 or < 0, respectively), which left 15 thoughts unclassified. As indicated in Table 3, participants reported more thoughts involving positive rather than negative anticipated emotions, while specificity did not differ between the two kinds of thoughts. Emotional future-oriented thoughts were more likely to take the form of visual images when associated with a positive anticipated emotion, whereas, conversely, they were more frequently experienced as inner speech when associated with a negative anticipated emotion. Positive EmoFTs were rated as more likely to be shared with others and as more probable than negative EmoFTs, whereas the latter were judged as more likely to be suppressed than the former. There were no difference between positive and negative EmoFTs in terms of their intentional aspect, relation to the environment, importance, relevance to identity, recurrence, and role in handling present or future situations. With regard to the perceived functions of the thoughts, negative EmoFTs were judged as more helpful to prepare oneself to deal with a potential future emotion (i.e., future-oriented emotional regulation), while positive EmoFTs were rated as more helpful to reassure oneself or to feel better (i.e., present-oriented emotional regulation).

Predicting the intensity of anticipatory and anticipated emotions

Our next goal was to investigate to what extent the intensity of anticipatory and anticipated emotions (as indexed by the absolute values of the ratings) can be predicted by properties of EmoFTs, such as their representational format, personal importance, and recurrence. Due to their hierarchical structure (i.e., the sampled EmoFTs were nested within participants), data were analysed using multilevel modelling (Goldstein, 2011), with EmoFTs as Level 1 units and participants as Level 2 units; these analyses were performed using MLwiN (Rasbash, Charlton, Browne, Healy, & Cameron, 2009). The intercorrelations between the variables at Level 1 (i.e., within participants) are shown in Table 4. As can be seen, the intensity of both anticipatory and anticipated emotions increased with visual imagery, personal importance, and relevance to identity; in addition, the intensity of anticipatory emotions decreased for EmoFTs involving more inner speech.

Next, we constructed a series of multilevel models to investigate the independent contribution of the different EmoFT properties that showed a significant bivariate association with anticipatory emotion intensity. Although some of the predictor variables included in these models were intercorrelated (see Table 4), simulation experiments have shown that the fixed-effect parameter estimates and standard errors are relatively unbiased in multilevel modelling for this magnitude of correlations among Level 1 predictors (Shieh & Fouladi, 2003). We first looked at the effect of the subjective significance of EmoFTs and found that both personal importance and relevance to identity were significant predictors of anticipatory emotions when they were entered simultaneously in a random intercept model (personal importance: standardized coefficient = 0.21, $SE = 0.04$, $Z = 5.25$, $p < .001$; relevance to identity: standardized coefficient = 0.09, $SE = 0.04$, $Z = 2.25$, $p = .02$). Adding variables assessing the representational format of EmoFTs (visual imagery and inner speech) in this model resulted in a significantly better fit (likelihood ratio, LR = 27.17, $df = 2$, $p < .001$); however, only personal importance (standardized

Table 3. *Characteristics of emotional future-oriented thoughts associated with positive versus negative anticipated emotions*

Thought property	Positive	Negative	df	t	p
Number of thoughts	6.51 (2.00)	2.82 (1.86)	75	8.64	<.001
Specific thoughts (%)	51 (28)	49 (37)	69	0.37	.71
Nonspecific thoughts (%)	43 (28)	43 (37)	69	0.04	.97
Unclassifiable thoughts (%)	6 (17)	8 (22)	69	0.81	.42
Characteristics (mean ratings)					
Inner speech	3.80 (1.22)	4.34 (1.58)	69	2.81	<.01
Visual images	4.51 (1.28)	3.78 (1.52)	69	3.98	<.001
Intentionality	3.75 (1.29)	3.57 (1.66)	69	0.87	.39
Environment	3.57 (1.89)	3.80 (1.67)	69	0.98	.33
Personal importance	5.01 (0.95)	5.12 (1.27)	69	0.77	.44
Handling	4.06 (1.21)	3.99 (1.61)	69	0.30	.77
Identity	3.27 (1.24)	3.46 (1.27)	69	1.40	.16
Sharing	4.42 (1.29)	3.79 (1.64)	69	3.07	<.01
Recurrence	3.90 (1.00)	3.80 (1.25)	69	0.65	.51
Suppression	1.71 (0.78)	3.91 (1.44)	69	12.45	<.001
Probability	5.77 (0.92)	4.95 (1.23)	69	5.37	<.001
Function (%)					
Decision making	9 (14)	14 (27)	69	1.45	.15
Planning	41 (25)	34 (33)	69	1.39	.17
Intention formation	22 (20)	17 (25)	69	1.38	.17
P.-O. emotional regulation	26 (23)	11 (25)	69	3.58	<.001
F.-O. emotional regulation	11 (15)	35 (37)	69	5.15	<.001
Other	5 (15)	4 (15)	69	0.50	.62
None	19 (20)	24 (34)	69	1.28	.20

Note: Mean ratings; standard deviations in parentheses. $df = 69$ since five participants did not reported any negative anticipatory emotion; P.-O. = present oriented; F.-O. = future oriented.

Table 4. *Intercorrelations between the characteristics of EmoFTs at Level 1 (within participants)*

Thought property	1	2	3	4	5	6	7	8
1. Visual images								
2. Inner speech	−.53*							
3. Intentionality	.01	−.01						
4. Environment	−.01	−.07	.01					
5. Importance	−.04	.09	.02	.01				
6. Identity	.06	.05	−.01	−.03	.40*			
7. Recurrence	−.03	.05	.03	−.08	.24*	.12		
8. Anticipatory emotions	.16*	−.13	−.05	.01	.23*	.15*	.07	
9. Anticipated emotions	.16*	−.10	.03	−.07	.28*	.14*	.11	.47*

Note: EmoFTs = emotional future-oriented thoughts.
*$p < .001$.

coefficient $= 0.23$, $SE = 0.04$, $Z = 5.75$, $p < .001$) and visual imagery (standardized coefficient $= 0.14$, $SE = 0.04$, $Z = 3.32$, $p < .001$) remained significant predictors. Thus, the best and most parsimonious model was to use personal importance and visual imagery to predict the intensity of anticipatory emotions associated with EmoFTs.

Similar analyses were conducted to investigate the independent contribution of different EmoFT properties to the intensity of anticipated emotions. We first looked at the effect of the subjective significance of EmoFTs and found that only personal importance was a significant predictor of anticipated emotions (standardized coefficient = 0.30, $SE = 0.04$, $Z = 7.50$, $p < .001$), when both personal importance and relevance to identity were entered simultaneously in a random intercept model. Adding visual imagery to a model that already included personal importance as predictor resulted in a significantly better fit (LR = 21.02, $df = 1$, $p < .001$); both personal importance (standardized coefficient = 0.31, $SE = 0.04$, $Z = 7.75$, $p < .001$) and visual imagery (standardized coefficient = 0.16, $SE = 0.04$, $Z = 4.62$, $p < .001$) were significant predictors of anticipated emotions in this model. Thus, as for anticipatory emotions, the best and most parsimonious model was to use personal importance and visual imagery to predict the intensity of anticipated emotions associated with EmoFTs.

EmoFTs and mood states

A final goal of this study was to investigate the relationships between the emotional content of future-oriented thoughts and mood states. We assumed that pre-EmoFT mood would impact on both anticipatory and anticipated emotions, which, in turn, would influence post-EmoFT mood. In addition, we hypothesized that the tendency to suppress EmoFTs might exert a detrimental impact on post-EDT mood, given the association between thought suppression and mood (Wenzlaff & Wegner, 2000). The intercorrelations between these variables at Level 1 (i.e., within participants) are shown in Table 5. As expected, both pre- and post-EmoFT mood states were related to anticipatory and anticipated emotions.

To further investigate the independent contribution of the emotional dimensions of future-oriented thoughts on subsequent mood, we conducted a series of multilevel models with post-EmoFT mood as dependent variable. First, we fitted a model in which pre-EmoFT mood was a predictor of post-EmoFT mood, and then we added the ratings of anticipatory emotions accompanying EmoFTs as predictor. This latter model fitted the data significantly better than the simpler model that included only pre-EmoFT mood as predictor

Table 5. *Intercorrelations between mood states and emotional properties of EmoFTs at Level 1 (within participants)*

Variable	1	2	3	4
1. Pre-EmoFT mood				
2. Post-EmoFT mood	.28*			
3. Anticipatory emotions	.23*	.84*		
4. Anticipated emotions	.14*	.71*	.78*	
5. Suppression	−.03	−.58*	−.58*	−.59*

Note: EmoFTs = emotional future-oriented thoughts.
*$p < .001$.

(LR = 811.97, $df = 1$, $p < .001$). Then, we added anticipated emotions to the model, which resulted in a significantly better fit (LR = 29.17, $df = 1$, $p < .001$). Finally, we examined whether the tendency to suppress EmoFTs influenced subsequent mood, once the effects of preceding mood, anticipatory emotions, and anticipated emotions had been taken into account. Adding suppression to the previous model provided a significantly better fit to the data (LR = 22.02, $df = 1$, $p < .001$). Despite the fact that some of the predictor variables were highly intercorrelated (i.e., anticipatory and anticipated emotions), which might introduce bias into the standard errors of coefficients and reduce statistical power (Shieh & Fouladi, 2003), pre-EmoFT mood (standardized coefficient = 0.14, $SE = 0.02$, $Z = 6.43$, $p < .001$), anticipatory emotions (standardized coefficient = 0.64, $SE = 0.03$, $Z = 21.33$, $p < .001$), anticipated emotions (standardized coefficient = 0.11, $SE = 0.03$, $Z = 3.56$, $p < .001$), and suppression (standardized coefficient = −0.12, $SE = 0.03$, $Z = 4.72$, $p < .001$) were all significant predictors of post-EmoFT mood in this model. Together, these four variables accounted for 74% of the within-participants variance in post-EmoFT mood.

Discussion

The present study aimed at examining the frequency, conditions of occurrence, characteristics, and perceived functions of prospections that are accompanied by emotions in natural settings. Our results show that EmoFTs occur quite frequently in daily life, although there is substantial variation across individuals in this respect. This observation complements earlier investigations of future thinking in natural settings that did not specifically target the emotional component of prospection (Berntsen & Jacobsen, 2008; D'Argembeau et al., 2011; Finnbogadóttir & Berntsen, 2013; Klinger & Cox, 1987). For example, when examining any kind of future-oriented thoughts that might occur in daily life, D'Argembeau et al. (2011) found that, on average, around 60 future-oriented thoughts were experienced each day, and the content of about 60% of reported thoughts was judged to have an affective (either positive or negative) tone. However, thoughts about positive or negative futures are not necessarily accompanied by an emotional response (i.e., by an anticipatory emotion; Baumgartner et al., 2008), and the present study indeed suggests that only a subset of prospections experienced in daily life (on average, around 9 thoughts per day) is truly infused with emotion.

We found that EmoFTs occur in various contexts (e.g., at home, at work, in transports) and during a wide range of distinct activities (e.g., working, resting, eating). Around half of the reported EmoFTs referred to a specific event, whereas the other half were more abstract in nature, in line with earlier findings (D'Argembeau et al., 2011). This result is consistent with the idea that prospection involves knowledge structures at different levels of specificity—episodic details (Schacter & Addis, 2007), semantic information (Irish & Piguet, 2013), and conceptual autobiographical knowledge (D'Argembeau, 2015)—which can be flexibly combined to produce various forms of future-oriented thoughts, ranging from detailed episodic simulations of specific events to highly abstract considerations about possible futures (Szpunar et al., 2014). In terms of temporal location, we found that around one fifth of EmoFTs did not refer to any specified time period (except, of course, that they pertained to the future), while most of the EmoFTs targeting a specified time period referred to future events that were reported as likely to occur within the next month; the frequency of EmoFTs decreased

with increasing temporal distance, replicating the temporal gradient previously observed in another diary study of future-oriented thoughts (D'Argembeau et al., 2011).

An important goal of the present study was to investigate valence effects in emotional future thoughts that arise in natural settings. Previous findings have shown that most people tend to have a positive view of their personal future: In laboratory studies, positive events are generated faster and are more detailed than negative events (D'Argembeau & Van der Linden, 2004; De Vito et al., 2015; Newby-Clark & Ross, 2003; Painter & Kring, 2015; Rasmussen & Berntsen, 2013), and diary studies have evidenced that positive future thoughts are more frequent than negative ones in daily life (Berntsen & Jacobsen, 2008; D'Argembeau et al., 2011). A specific contribution of the present study is to distinguish between two types of emotions (anticipatory and anticipated), a distinction that has not been made explicit in previous studies. Interestingly, we found that valence effects in the number of reported thoughts depended on the kind of emotions under consideration: A positivity bias was observed for the anticipated emotional content of EmoFTs (i.e., participants reported positive forecasts more frequently than negative ones), whereas there was no significant difference in the frequency of EmoFTs associated with positive versus negative anticipatory emotions. Thus, the positivity bias in the frequency of future thoughts that has been observed in previous studies seems restricted to affective forecasts, without extending to the emotional impact of future-oriented thoughts in the here and now. When looking at specific emotion categories, we found that fear was more frequently reported as an anticipatory emotion than as an anticipated emotion, while joy, pride, and shame followed the opposite pattern. Overall, these findings highlight the importance of distinguishing between anticipatory and anticipated emotions in characterizing the emotional properties of EmoFTs. In particular, the tendency to conceive one's personal future in a favourable light (Taylor & Brown, 1988) seems to involve anticipated rather than anticipatory emotions.

Notwithstanding these differences between anticipatory and anticipated emotions, the effect of valence on several characteristics of EmoFTs was similar for these two types of emotions. Notably, for both anticipatory and anticipated emotions, positive EmoFTs were more likely to take the form of visual images, whereas negative EmoFTs were more frequently experienced as inner speech, which replicates earlier findings (e. g., D'Argembeau et al., 2011; D'Argembeau & Van der Linden, 2004; Stawarczyk et al., 2013). Positive EmoFTs were also judged as more likely to be shared with others, in line with previous findings (Rasmussen & Berntsen, 2009, 2013), and were perceived as more probable than negative EmoFTs; on the other hand, negative EmoFTs were more likely to be suppressed. Besides these similarities in the effect of valence, some differences were also observed: For anticipatory emotions (but not anticipated emotions), negative EmoFTs were more likely to be triggered by the environment and were judged as more helpful for handling present or future situations. Overall, these findings suggest that although people may be more willing to consider and share positive future events, EmoFTs that are associated with negative anticipatory emotions may nevertheless play important roles in managing present and future situations (see below for further discussion of the perceived functions of EmoFTs).

The present study also sheds light on the determinants of both emotional components of EmoFTs. The intensity of anticipatory and anticipated emotions correlated with multiple characteristics of EmoFTs, including visual imagery, personal importance, relevance to identity, and recurrence. When looking at the independent contribution of these characteristics, we found that visual imagery and personal importance were the best predictors of anticipatory and anticipated emotions. Previous work by Holmes and colleagues has shown that mental images have a particularly powerful affective impact, notably with respect to imagined future events (for review, see Holmes & Mathews, 2010). The present findings provided additional support to this view by showing that the emotions evoked by future-oriented thoughts in natural settings are more intense when people experience them through visual imagery.[2] The second factor that contributed to the prediction of anticipatory and anticipated emotions (i.e., personal relevance) has long been considered an important component of appraisal processes leading to emotion (Ellsworth & Scherer, 2003; Scherer, 2001; see also Brosch & Sander, 2014). The present findings are in line with this componential view of emotion and show that it applies to emotions that are elicited by internally generated thoughts, such as prospections.

Another goal of this study was to investigate the relationships between the emotional components of EmoFTs and mood states. First, we found that anticipatory and anticipated emotions were significantly related to mood states preceding the occurrence of EmoFTs. Second, anticipatory and anticipated emotions were both significant predictors of post-EmoFT mood, even when pre-EmoFT mood was taken into account. These results are in line with recent findings showing similar relationships between mood states and mind-wandering episodes (Poerio et al., 2013). More generally, these observations are in line with the *mood congruency hypothesis*, according to which positive/negative mood states increase the accessibility of material of the same valence (Blaney, 1986; Bower, 1981). Our findings thus extend the extensive literature on mood congruent memory (for review, Sedikides, 1992) by showing that mood states also contributed to determine the affective dimension of prospective thoughts. Emotions associated with EmoFTs in turn

---

[2]Recent evidence has shown that the spatial component of mental imagery is particularly important when constructing future simulations; moreover, when the spatial component of mental imagery is disrupted during future thinking, while the depictive component of mental imagery is spared, fewer emotional details are reported (De Vito, Buonocore, Bonnefon, & Della Sala, 2014). The measure of visual imagery that was used in the present study does not allow us to distinguish between spatial and depictive components, and it would be interesting in future studies to investigate whether these components show specific relationships with the emotional intensity associated with prospection in daily life.

influenced subsequent mood, and, interestingly, the inclination to suppress an EmoFT was also a significant predictor of post-EmoFT mood, with higher suppression leading to worse mood. This detrimental impact of suppression on post-EmoFT mood is consistent with earlier evidence showing an association between thought suppression and negative mood (Wenzlaff & Wegner, 2000). Thought suppression is indeed a dysfunctional emotion regulation strategy (John & Gross, 2004) that has the paradoxical effect of conducing to the state of mind that one is striving to avoid (Wenzlaff & Wegner, 2000). Attempts at suppressing a negative EmoFT might thus actually increase its prominence, leading to a greater degradation of mood. Consequently, an inclination to systematically suppress EmoFTs could play a role in the development or maintenance of some psychopathological states (for a review on psychopathology and thought suppression, see Magee, Harden, & Teachman, 2012): Repeated attempts to suppress negative prospections could induce negative mood states, which in turn would result in an increased accessibility of negative thoughts, thus leading to a vicious cycle of mutually reinforcing EmoFTs and mood states.

Future-oriented emotions are believed to play an important role in motivation and goal pursuit (Baumgartner et al., 2008; Karniol & Ross, 1996; Taylor et al., 1998). Whereas much is known about the functions of autobiographical memories (e.g., Bluck et al., 2005; Harris et al., 2014), little is known about the functions of emotional future thinking in everyday life. The present study contributed to this question by examining participants' beliefs about the functions of their EmoFTs. More than 60% of EmoFTs were judged to be related to planning, intention formation, or decision making, indicating that many EmoFTs occurring in daily life are indeed perceived as highly relevant to goal pursuit. Interestingly, EmoFTs that were associated with negative anticipatory emotions involved decision making to a greater extent than positive EmoFTs and were considered as more helpful for handling present or future situations. These results suggest that negative future thinking is not necessarily maladaptive. All in all, it is likely that

an adequate balance between positive and negative future-oriented thoughts is optimal for successful goal pursuit. It should be emphasized, however, that the present results pertain to participants' beliefs about the functions of their EmoFTs. These beliefs might be partly biased, and it remains to be investigated to what extent positive and negative EmoFTs actually contribute to effective goal pursuit.

While most EmoFTs were perceived as serving goal pursuit, a substantial proportion of EmoFTs were linked to other functions and, notably, emotional regulation. Indeed, around 40% of EmoFTs were perceived as contributing to emotion regulation (Gross, 2002), with either a present-oriented focus (i.e., reassuring oneself or feeling better) or a future-oriented focus (i.e., preparing oneself to deal with an anticipated emotion). Interestingly, for both anticipatory and anticipated emotions, positive EmoFTs were judged as more helpful for present-oriented emotional regulation, whereas negative EmoFTs were judged as more helpful for future-oriented emotional regulation. Again, this finding suggests that negative future-oriented thoughts might convey some adaptive advantages. In particular, envisioning negative future possibilities might contribute to situation selection, situation modification, and attentional deployment, which all constitute antecedent-focused emotion regulation processes (see Gross, 1998).

In summary, the current findings show that emotional prospections are frequent in daily life, occur in various contexts, and take on different representational formats. The positivity bias in the frequency of EmoFTs appears to be restricted to anticipated emotions without extending to anticipatory emotions. Anticipatory and anticipated emotions are both influenced by the visual imagery and personal importance of EmoFTs and have an important impact on ensuing mood states. Finally, emotional future thinking serves a range of important functions related to goal pursuit (i.e., planning, intention formation, and decision making) and emotion regulation (either with a present-oriented focus or with a future-oriented focus).

REFERENCES

Ajzen, I. (1991). The theory of planned behavior. *Organizational Behavior and Human Decision Processes*, *50*(2), 179–211. doi:10.1016/0749-5978(91)90020-t

Atance, C. M., & O'Neill, D. K. (2001). Episodic future thinking. *Trends in Cognitive Sciences*, *5*(12), 533–539. doi:10.1016/s1364-6613(00)01804-0

Baumgartner, H., Pieters, R. & Bagozzi, R. P. (2008). Future-oriented emotions: Conceptualization and behavioral effects. *European Journal of Social Psychology*, 38 (4), 685–696. doi:10.1002/ejsp.467

Bechara, A., & Damasio, A. R. (2005). The somatic marker hypothesis: A neural theory of economic decision. *Games and Economic Behavior*, *52*(2), 336–372. doi:10.1016/j.geb.2004.06.010

Berntsen, D., & Jacobsen, A. S. (2008). Involuntary (spontaneous) mental time travel into the past and future. *Consciousness and Cognition*, *17*(4), 1093–1104. doi:10.1016/j.concog.2008.03.001

Blaney, P. H. (1986). Affect and memory: A review. *Psychological Bulletin*, *99*(2), 229–246. doi:10.1037/0033-2909.99.2.229

Bluck, S., Alea, N., Habermas, T., & Rubin, D. C. (2005). A TALE of three functions: The self-reported uses of autobiographical memory. *Social Cognition*, *23*(1), 91–117. doi:10.1521/soco.23.1.91.59198

Bolger, N., Davis, A., & Rafaeli, E. (2003). Diary methods: Capturing life as it is lived. *Annual Review of Psychology*, *54*(1), 579–616. doi:10.1146/annurev.psych.54.101601.145030

Borkovec, T. D., Ray, W. J., & Stober, J. (1998). Worry: A cognitive phenomenon intimately linked to affective, physiological, and interpersonal behavioral processes. *Cognitive Therapy and Research*, *22*(6), 561–576. doi:10.1023/a:1018790003416

Bower, G. H. (1981). Mood and memory. *American Psychologist*, *36*(2), 129–148. doi:10.1037/0003-066x.36.2.129

Brosch, T., & Sander, D. (2014). Appraising value: The role of universal core values and emotions in decision-making. *Cortex*, *59*, 203–205. doi:10.1016/j.cortex.2014.03.012

Carrera, P., Caballero, A., & Muñoz, D. (2012). Future-oriented emotions in the prediction of binge-drinking intention and expectation: The role of anticipated and anticipatory emotions. *Scandinavian Journal of Psychology*, *53*(3), 273–279. doi:10.1111/j.1467-9450.2012.00948.x

Conway, M. A. (2005). Memory and the self. *Journal of Memory and Language*, *53*(4), 594–628. doi:10.1016/j.jml.2005.08.005

Daselaar, S. M., Rice, H. J., Greenberg, D. L., Cabeza, R., LaBar, K. S., & Rubin, D. C. (2008). The spatio-temporal dynamics of autobiographical memory: Neural correlates of recall, emotional intensity, and reliving. *Cerebral Cortex*, *18*(1), 217–229. doi:10.1093/cercor/bhm048

Davis, F. D., & Warshaw, P. R. (1992). What do intention scales measure? *The Journal of General Psychology*, *119*(4), 391–407. doi:10.1080/00221309.1992.9921181

D'Argembeau, A. (2015). Knowledge structures involved in episodic future thinking. In A. Feeney & V. A. Thompson (Eds.), *Reasoning as memory* (pp. 128–145). Hove: Psychology Press.

D'Argembeau, A., Renaud, O., & Van Der Linden, M. (2011). Frequency, characteristics and functions of future-oriented thoughts in daily life. *Applied Cognitive Psychology*, *25*(1), 96–103. doi:10.1002/acp.1647

D'Argembeau, A., & Van Der Linden, M. (2004). Phenomenal characteristics associated with projecting oneself back into the past and forward into the future: Influence of valence and temporal distance. *Consciousness and Cognition*, *13*(4), 844–858. doi:10.1016/j.concog.2004.07.007

De Vito, S., Buonocore, A., Bonnefon, J-F., & Della Sala, S. (2014). Eye movements disrupt episodic future thinking. *Memory*, 1–10. doi:10.1080/09658211.2014.927888

De Vito, S., Neroni, M. A., Gamboz, N., Della Sala, S., & Brandimonte, M. A. (2015). Desirable and undesirable future thoughts call for different scene construction processes. *The Quarterly Journal of Experimental Psychology*, *68*(1), 75–82. doi:10.1080/17470218.2014.937448

Ellsworth, P., & Scherer, K. R. (2003). Appraisal processes in emotion. In R. J. Davidson, H. H. Goldsmith, & K. R. Scherer (Eds.), *Handbook of affective sciences* (pp. 572–595). Oxford: Oxford University Press.

Finnbogadóttir, H., & Berntsen, D. (2011). Involuntary and voluntary mental time travel in high and low worriers. *Memory*, *19*(6), 625–640. doi:10.1080/09658211.2011.595722

Finnbogadóttir, H., & Berntsen, D. (2013). Involuntary future projections are as frequent as involuntary memories, but more positive. *Consciousness and Cognition*, *22*(1), 272–280. doi:10.1016/j.concog.2012.06.014

Gilbert, D. T., & Wilson, T. D. (2007). Prospection: Experiencing the future. *Science, 317*(5843), 1351–1354. doi:10.1126/science.1144161

Goldstein, H. (2011). *Multilevel statistical models* (4th ed.). Chichester, UK: Wiley. doi:10.1002/9780470973394

Gross, J. J. (1998). The emerging field of emotion regulation: An integrative review. *Review of General Psychology, 2*(3), 271–299. doi:10.1037/1089-2680.2.3.271

Gross, J. J. (2002). Emotion regulation: Affective, cognitive, and social consequences. *Psychophysiology, 39*(3), 281–291. doi:10.1017/s0048577201393198

Harris, A. J. L., & Hahn, U. (2011). Unrealistic optimism about future life events: A cautionary note. *Psychological Review, 118*(1), 135–154. doi:10.1037/a0020997

Harris, C. B., Rasmussen, A. S., & Berntsen, D. (2014). The functions of autobiographical memory: An integrative approach. *Memory, 22*(5), 559–581. doi:10.1080/09658211.2013.806555

Hoerger, M., Chapman, B. P., Epstein, R. M., & Duberstein, P. R. (2012). Emotional intelligence: A theoretical framework for individual differences in affective forecasting. *Emotion, 12*(4), 716–725. doi:10.1037/a0026724

Holmes, E. A., & Mathews, A. (2010). Mental imagery in emotion and emotional disorders. *Clinical Psychology Review, 30*(3), 349–362. doi:10.1016/j.cpr.2010.01.001

Irish, M., & Piguet, O. (2013). The pivotal role of semantic memory in remembering the past and imagining the future. *Frontiers in Behavioral Neuroscience, 7.* doi:10.3389/fnbeh.2013.00027

John, O. P., & Gross, J. J. (2004). Healthy and unhealthy emotion regulation: Personality processes, individual differences, and life span development. *Journal of Personality, 72* (6), 1301–1334. doi:10.1111/j.1467-6494.2004.00298.x

Karniol, R., & Ross, M. (1996). The motivational impact of temporal focus: Thinking about the future and the past. *Annual Review of Psychology, 47*(1), 593–620. doi:10.1146/annurev.psych.47.1.593

Klinger, E., & Cox, W. M. (1987). Dimensions of thought flow in everyday life. *Imagination, Cognition and Personality, 7*(2), 105–128. doi:10.2190/7k24-g343-mtqw-115v

MacLeod, A. K., & Byrne, A. (1996). Anxiety, depression, and the anticipation of future positive and negative experiences. *Journal of Abnormal Psychology, 105* (2), 286–289. doi:10.1037/0021-843x.105.2.286

Magee, J. C., Harden, K. P., & Teachman, B. A. (2012). Psychopathology and thought suppression: A quantitative review. *Clinical Psychology Review, 32* (3), 189–201. doi:10.1016/j.cpr.2012.01.001

Miloyan, B., & Suddendorf, T. (2015). Feelings of the future. *Trends in Cognitive Sciences, 19*(4), 196–200. doi:10.1016/j.tics.2015.01.008

Newby-Clark, I. R., & Ross, M. (2003). Conceiving the past and future. *Personality and Social Psychology Bulletin, 29*(7), 807–818. doi:10.1177/0146167203029007001

Painter, J. M., & Kring, A. M. (2015). Back to the future: Similarities and differences in emotional memories and prospections. *Applied Cognitive Psychology, 29*(2), 271–279. doi:10.1002/acp.3105

Poerio, G. L., Totterdell, P., & Miles, E. (2013). Mind-wandering and negative mood: Does one thing really lead to another? *Consciousness and Cognition, 22*(4), 1412–1421. doi:10.1016/j.concog.2013.09.012

Rasbash, J., Charlton, C., Browne, W. J., Healy, M. & Cameron, B. (2009). *MLwiN Version 2.1.* Bristol: Centre for Multilevel Modelling, University of Bristol. Retrieved from http://www.bristol.ac.uk/cmm/software/mlwin/refs.html

Rasmussen, A. S., & Berntsen, D. (2009). Emotional valence and the functions of autobiographical memories: Positive and negative memories serve different functions. *Memory & Cognition, 37*(4), 477–492. doi:10.3758/mc.37.4.477

Rasmussen, A. S., & Berntsen, D. (2013). The reality of the past versus the ideality of the future: Emotional valence and functional differences between past and future mental time travel. *Memory and Cognition, 41*(2), 187–200. doi:10.3758/s13421-012-0260-y

Schacter, D. L., & Addis, D. R. (2007). The cognitive neuroscience of constructive memory: Remembering the past and imagining the future. *Philosophical Transactions of the Royal Society B: Biological Sciences, 362,* 773–786.

Schacter, D. L., Addis, D. R., Hassabis, D., Martin, V. C., Spreng, R. N., & Szpunar, K. K. (2012). The future of memory: Remembering, imagining, and the brain. *Neuron, 76*(4), 677–694. doi:10.1016/j.neuron.2012.11.001

Scherer, K. R. (2001). Appraisal considered as a process of multi-level sequential checking. In K. R. Scherer, A. Schorr, & T. Johnstone (Eds.), *Appraisal processes in emotion: Theory, methods, research* (pp. 92–120). New York and Oxford: Oxford University Press.

Sedikides, C. (1992). Changes in the valence of the self as a function of mood. *Review of Personality and Social Psychology*, *14*, 271–311.

Seligman, M. E. P., Railton, P., Baumeister, R. F., & Sripada, C. (2013). Navigating into the future or driven by the past. *Perspectives on Psychological Science*, *8*(2), 119–141. doi:10.1177/1745691612474317

Sharot, T., Riccardi, A. M., Raio, C. M., & Phelps, E. A. (2007). Neural mechanisms mediating optimism bias. *Nature*, *450*(7166), 102–105. doi:10.1038/nature06280

Shieh, Y-Y., & Fouladi, R. T. (2003). The effect of multicollinearity on multilevel modeling parameter estimates and standard errors. *Educational and Psychological Measurement*, *63*(6), 951–985. doi:10.1177/0013164403258402

Smallwood, J., Nind, L., & O'Connor, R. C. (2009). When is your head at? An exploration of the factors associated with the temporal focus of the wandering mind. *Consciousness and Cognition*, *18*(1), 118–125. doi:10.1016/j.concog.2008.11.004

Smallwood, J., & Schooler, J. W. (2006). The restless mind. *Psychological Bulletin*, *132*(6), 946–958. doi:10.1037/0033-2909.132.6.946

Song, X., & Wang, X. (2012). Mind wandering in Chinese daily lives – An experience sampling study. *PLoS ONE*, *7*(9), e44423. doi:10.1371/journal.pone.0044423

Stawarczyk, D., Cassol, H., & D'Argembeau, A. (2013). Phenomenology of future-oriented mind-wandering episodes. *Frontiers in Psychology*, *4*. doi:10.3389/fpsyg.2013.00425

Suddendorf, T., & Corballis, M. C. (2007). The evolution of foresight: What is mental time travel and is it unique to humans? *Behavioral and Brain Sciences*, *30*(3). doi:10.1017/s0140525×07001975

Szpunar, K. K., Addis, D. R., & Schacter, D. L. (2012). Memory for emotional simulations: Remembering a rosy future. *Psychological Science*, *23*(1), 24–9. doi:10.1177/0956797611422237

Szpunar, K. K., & Schacter, D. L. (2013). Get real: Effects of repeated simulation and emotion on the perceived plausibility of future experiences. *Journal of Experimental Psychology. General*, *142*(2), 323–7. doi:10.1037/a0028877

Szpunar, K. K., Spreng, R. N., & Schacter, D. L. (2014). A taxonomy of prospection: Introducing an organizational framework for future-oriented cognition. *Proceedings of the National Academy of Sciences*, 201417144. doi:10.1073/pnas.1417144111

Taylor, S. E., & Brown, J. D. (1988). Illusion and well-being: A social psychological perspective on mental health. *Psychological Bulletin*, *103*(2), 193–210. doi:10.1037/0033-2909.103.2.193

Taylor, S. E., Pham, L. B., Rivkin, I. D., & Armor, D. A. (1998). Harnessing the imagination. Mental simulation, self-regulation, and coping. *American Psychologist*, *53*(4), 429–439. doi:10.1037/0003-066x.53.4.429

Van Boven, L., & Ashworth, L. (2007). Looking forward, looking back: Anticipation is more evocative than retrospection. *Journal of Experimental Psychology: General*, *136*(2), 289–300. doi:10.1037/0096-3445.136.2.289

Weinstein, N. D. (1980). Unrealistic optimism about future life events. *Journal of Personality and Social Psychology*, *39*(5), 806–820. doi:10.1037/0022-3514.39.5.806

Wenzlaff, R. M., & Wegner, D. M. (2000). Thought Suppression. *Annual Review of Psychology*, *51*(1), 59–91. doi:10.1146/annurev.psych.51.1.59

Williams, J. M. G., Ellis, N. C., Tyers, C., Healy, H., Rose, G., & Macleod, A. K. (1996). The specificity of autobiographical memory and imageability of the future. *Memory & Cognition*, *24*(1), 116–125. doi:10.3758/bf03197278

Wilson, T., & Gilbert, D. T. (2005). Affective forecasting: Knowing what to want. *Current Directions in Psychological Science*, *14*(3), 131–134. doi:10.1111/j.0963-7214.2005.00355

The degree of disparateness of event details modulates future simulation construction, plausibility, and recall

Valerie van Mulukom[1,2], Daniel L. Schacter[3], Michael C. Corballis[1,2], and Donna Rose Addis[1,2]

[1]School of Psychology, The University of Auckland, Auckland, New Zealand
[2]Centre for Brain Research, The University of Auckland, Auckland, New Zealand
[3]Department of Psychology, Harvard University, Cambridge, MA, USA

Several episodic simulation studies have suggested that the plausibility of future events may be influenced by the disparateness of the details comprising the event. However, no study had directly investigated this idea. In the current study, we designed a novel episodic combination paradigm that varied the disparateness of details through a social sphere manipulation. Participants recalled memory details from three different social spheres. Details were recombined either within spheres or across spheres to create detail sets for which participants imagined future events in a second session. Across-sphere events were rated as significantly less plausible than within-sphere events and were remembered less often. The presented paradigm, which increases control over the disparateness of details in future event simulations, may be useful for future studies concerned with the similarity of the simulations to previous events and its plausibility.

The majority of studies examining future event simulation have focused on events that are likely to occur in the near future; such imagined events are typically very similar to previous events that have occurred (e.g., Addis, Cheng, Roberts, & Schacter, 2011). However, we are also capable of imagining future events that are highly dissimilar to previous events, which allows us to prepare for a greater range of future events. This is an important adaptive feature of event simulation as we never know exactly what the future might bring (see also Bar, 2009). Surprisingly, very little research has investigated differences in unusualness of future events. One study, by Weiler, Suchan, and Daum (2010), investigated differences between imagined future events that were low or high in occurrence probability and found that lower event occurrence probability was accompanied by increases in activity in the right anterior hippocampus (independent of the amount of detail). It was suggested that low probability events are likely to involve more disparate details than high probability events, and that the enhanced hippocampal activation may reflect more extensive recombination processes for these disparate details.

This work was supported by the Marsden Fund [grant number UOA0810], [grant number UOA1210]; Rutherford Discovery Fellowship [grant number RDF-10-UOA-024]; and the National Institutes on Aging [grant number R01 AG08441].

Indeed, future events that are dissimilar to past events probably consist of more dissimilar detail recombinations, involving disparate components gleaned from a variety of distinct episodic memories (Addis & Schacter, 2008). Importantly, whether the disparateness of the details comprising a simulation influences the construction processes or the plausibility of that simulation has not yet been investigated directly. In many studies investigating episodic simulation, such as those utilizing the "episodic recombination paradigm" (Addis, Pan, Vu, Laiser, & Schacter, 2009; Martin, Schacter, Corballis, & Addis, 2011; van Mulukom, Schacter, Corballis, & Addis, 2013), it was not possible to determine the degree of disparateness of newly recombined details. Therefore, we designed a novel version of the recombination paradigm, which incorporates a manipulation of the disparateness of details through *social spheres*. Participants imagined future events incorporating memory details extracted either from the same or from different social spheres. We expected that more extensive detail recombination processes would be required for events composed of details from different spheres than from the same sphere, and that this would be reflected in higher event construction times and lower detail and coherency ratings. In addition, we expected the events with disparate details to be rated as more implausible, following their dissimilarity to previous events. Another aim of this study was to investigate the effect of disparateness on encoding of, and later memory for, future events. We assessed whether recall differs between events with disparate or nondisparate details.

EXPERIMENTAL STUDY

Method

Participants
Twenty-five healthy young adults were recruited via on-campus advertisements and gave written consent to participate in this study, approved by The University of Auckland Human Participants Ethics Committee. All participants were fluent in English, had no history of neurologic or psychiatric conditions or use of psychotropic medications, and had not previously participated in other future simulation studies. Two participants were excluded due to failure to comply with task instructions, and thus data from 23 participants were analysed (9 males, aged 18–32 years, $M = 21.7$ years).

Procedure
We designed a new version of the recombination paradigm (Addis et al., 2009) to include social spheres. The experiment consisted of three sessions: a pre-simulation session where memory details from various social spheres were collected; a simulation session, in which participants imagined future events involving the collected memory details; and a post-simulation session, consisting of a surprise recall test and a post-simulation interview.

Session 1: Recall of memory details. Participants identified three social spheres in their lives, where social spheres were defined as "groups of people who know each other". It was indicated that the selected social spheres should have minimal overlap. Examples of possible social spheres were provided (e.g., university friends, family, work colleagues, or music/rugby friends). Participants recalled 90 episodic details (30 persons, 30 locations, 30 objects) for each of these spheres (resulting in 270 memory details in total), where none of the memory details could be duplicated. In addition, these details were required to be as sphere-specific as possible. Note that unlike previous versions of the recombination paradigm used in our laboratory (Martin et al., 2011; van Mulukom et al., 2013), participants were asked to list episodic details during the pre-simulation session, rather than recall events containing these details. During piloting, participants struggled to recall sufficient numbers of events that had social sphere-specific locations as well as objects, which is crucial for the manipulation in this study. Therefore, we decided to use this listing method for episodic detail generation introduced in a recent study also using the recombination paradigm (Szpunar, Addis, & Schacter, 2012).

The episodic details were used to create cues for two conditions of the simulation session: the *within-sphere* and *across-sphere* conditions. For the

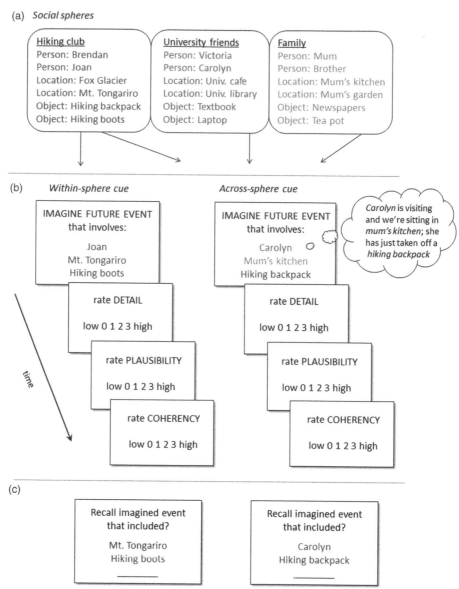

Figure 1. *Social sphere recombination paradigm. (a) The first session: the detail collection session. (b) The second session: the simulation session. Future event cues are created either by combining three memory details within one social sphere (within-sphere, cue left) or by combining three memory details across three social spheres (across-sphere, cue right). (c) The third session: the post-simulation session. Following the simulation session, a surprise cued recall memory test was completed by all participants. Participants were presented with two of the three details from each simulation trial set and were requested to provide the missing detail. In addition, participants completed a post-simulation interview (see the section on Session 3 below for more information). Please note that the colours [online version only] are used here to emphasize the difference between conditions (each colour indicates details from the same sphere) and were not utilized in the experiment. To view this figure in colour, please visit the online version of this Journal.*

within-sphere condition, three memory details (one person, one location, and one object) from the same social sphere were randomly combined. For the across-sphere condition, each memory detail was selected from a different social sphere and was combined into a set of three (see Figures 1a and 1b).

Session 2: Future simulation. Approximately one week later ($M = 7$ days, $SD = 2.4$ days, range = 4–17 days[1]), participants completed the future simulation session. After a practice session involving trials from both conditions, participants were presented 90 future event trials (45 within-sphere, 45 across-sphere), each showing a recombined set of memory details for 8 s (see Figure 1b). Participants were instructed to imagine specific and novel future events that could take place within the next five years, whilst incorporating the three episodic details presented on the screen. Participants made a button press as soon as they had an event in mind; this response, however, did not change the screen, and participants continued imagining until the end of the 8 s of the trial. During this time, participants were encouraged to elaborate and flesh out the event. The event simulation screen was followed by three rating screens, each shown for 3 s. The order of these rating screens was pseudorandomized and counterbalanced. Participants rated each event on a 4-point scale for detail ("0" vague with no or few details, "3" highly detailed and vivid), plausibility ("0" highly implausible, "3" highly plausible), and coherency ("0" a fragmented simulation, the details did not come together well; "3" a fluent simulation, the details came together well). In order to pilot the feasibility of this paradigm for functional magnetic resonance imaging (fMRI), trials were separated by periods of fixation, ranging from 2 to 10 s ($M = 4.22$, $SD = 2.08$ s).

Session 3: Cued recall and post-simulation interview. Each participant was given a 10-minute break after the simulation session, followed by a surprise memory test. In this cued recall test, participants were presented with two of the details from every detail set presented in the simulation session and were required to provide the third missing detail (Martin et al., 2011; see Figure 1c). Participants were encouraged not to guess if they did not know the answer but rather move on to the next question, to ensure that the

participants were confident about their answers. Afterwards, events were classified as successfully or unsuccessfully remembered on the basis of the cued-recall test.

Next, an interview was conducted for all trials that were successfully remembered—we did not probe trials for which the simulations were subsequently forgotten as we could not be sure that these had been successfully constructed. For each trial, participants briefly described the event they had imagined, enabling us to identify trials in which a specific future event (i.e., events that were specific in time and place) was generated; only these trials were included in the analyses. Next, participants rated the likelihood of the co-occurrence of the three details presented in that detail set. This rating provided a measure of the disparateness of the detail recombination by assessing whether these three details naturally occur together in the participant's life, allowing us to test whether our sphere manipulation was successful. In addition, participants made a number of ratings for each of the simulated events. The ratings used were: the difficulty to combine the three presented details into a simulation ("0" not difficult, "3" difficult); the emotionality of the event ("0" not emotional, "3" very emotional); and the similarity of the imagined event to previous thoughts and experiences ("0" not similar at all, never happened before, and "3" very similar, I've imagined this exact event/this event actually occurred). Participants also estimated how far in the future the events might take place. Finally, all participants were asked whether they were aware there was going to be a memory test, and all responded that they were not.

Results

Trials were excluded if there was no button press (signalling lack of event construction) or a reaction time of less than 500 ms (to exclude accidental button presses). Also excluded were trials with a similarity rating of "3", indicating that the imagined

---

[1]The delay duration between detail recollection and future event simulations did not correlate with the percentage of future events recalled ($r = .04$, $p = .85$).

future event was identical to previously experienced or imagined events (and therefore noncompliant with task instructions to imagine novel events). These criteria resulted in 5.9% of trials being excluded.

Behavioural data for across-sphere and within-sphere events are provided in Table 1. Results in this section were analysed through a series of paired-sample t-tests. To correct for multiple comparisons, we computed a Bonferroni-corrected alpha threshold of $p = .004$, derived by dividing our original alpha criterion ($p = .05$) by the number of t-tests we ran (12 in total). Note that unadjusted p-values are provided.

First, to confirm that our sphere manipulation had worked, we examined post-simulation ratings to determine whether the details in the across-sphere detail sets were indeed more disparate than the details in the within-sphere set. As expected, we found that the likelihood of co-occurrence of the details comprising the details sets was significantly lower for across-sphere events than for within-sphere events, $t(22) = -12.50$, $p < .001$, $d_z = -2.61$. Furthermore, across-sphere events were rated as less similar to previous thoughts, $t(22) = -29$, $p < .001$, $d_z = -6.05$, and experiences, $t(22) = -10.58$, $p < .001$, $d_z = -2.21$, than within-sphere events (noting, though, that events from both sphere conditions were still considerably far removed from being similar to previous events). Together, these findings confirm that the sphere manipulation affected the disparateness of the details in an imagined event.

Next, we were interested in whether differences in the disparateness of the details influenced the plausibility of events. We found that, as predicted, across-sphere events were rated during simulation as significantly less plausible than within-sphere events, $t(22) = -12.43$, $p < .001$, $d_z = -2.59$.

Furthermore, the disparateness of details also affected the construction of future events. A series of paired-samples t-tests demonstrated that within-sphere events were faster, $t(22) = 4.99$, $p < .001$, $d_z = 1.04$, and less difficult, $t(22) = 8.96$, $p < .001$, $d_z = 1.87$, to construct, and they were rated during simulation as more coherent, $t(22) = -6.47$, $p < .001$, $d_z = -1.35$, and more detailed, $t(22) = -5.65$, $p < .001$, $d_z = -1.18$, than across-sphere events. We also tested whether post-simulation ratings of emotionality and temporal distance differed between the sphere conditions. While the temporal distance exceeded the alpha level of $p = .05$, neither effect exceeded the Bonferroni-corrected alpha level of $p = .004$ [temporal distance, $t(22) = 2.23$, $p = .04$, $d_z = 0.46$; emotionality, $t(22) = -0.98$, $p = .34$, $d_z = -0.20$].[2]

Finally, we were interested in whether the disparateness of details influenced recall. We analysed whether the percentage of recalled events of total events differed between across-sphere and within-sphere conditions.[3] A paired-sample t-test revealed that across-sphere events ($M = 59.48\%$, $SD = 19.50\%$) were remembered significantly less often than within-sphere events ($M = 72.20\%$, $SD = 14.48\%$), $t(22) = -4.33$, $p < .001$, $d_z = -0.90$.

Discussion

The main objectives of this study were to investigate whether the disparateness of details comprising imagined future events modulates event plausibility and event recall. To this end, we designed a novel paradigm in which simulation cues incorporated details extracted from the same or from various social spheres. Consistent with the idea that the integration of disparate details requires more extensive processing, the disparateness of details affected construction-related

---

[2]Given that the nonsignificant difference of temporal distance between the sphere conditions was nevertheless a medium-sized effect ($d_z = 0.46$), we explored whether temporal distance correlated with key dependent variables. These correlations were generally weak (plausibility: $r = -.25$, $p = .26$; detail: $r = -.14$, $p = .54$; coherency: $r = -.22$, $p = .32$; difficulty: $r = .14$, $p = .52$; emotion: $r = .10$, $p = .64$), suggesting that differences in temporal distance on the order of a few weeks are not likely to affect phenomenology of details that were imagined to occur more than a year in the future.

[3]We calculated a percentage of the total events per sphere condition rather than comparing the number of recalled trials directly as the number of total trials could slightly differ between the sphere conditions due to the exclusion criteria (see the beginning of the *Results* section).

Table 1. *Behavioural data for across-sphere and within-sphere events*

Ratings	Mean (SD)	
	Across	Within
Simulation ratings and RT		
Coherency***	1.36 (0.47)	1.90 (0.40)
Detail***	1.67 (0.48)	2.04 (0.40)
Plausibility***	0.56 (0.41)	1.52 (0.54)
RT (s)***	4.92 (0.91)	4.41 (0.86)
Post-simulation ratings[a]		
Difficulty***	1.53 (0.47)	0.79 (0.30)
Emotionality	0.43 (0.39)	0.49 (0.37)
Likelihood of co-occurrence***	0.32 (0.21)	1.29 (0.39)
Similarity of event to previous experiences***	0.34 (0.21)	0.98 (0.38)
Similarity of event to previous thoughts***	0.14 (0.17)	0.35 (0.39)
Temporal distance of event (years)	1.71 (0.81)	1.47 (0.42)

Note: All participant ratings were made using a 4-point rating scale, ranging from 0 (low) to 3 (high), except for temporal distance (in years).
[a]These ratings were only collected for events that were successfully recalled.
***$p < .001$.

processes: Across-sphere events took significantly longer to generate, were less coherent, were rated as more difficult to construct, and contained less detail than within-sphere events. Furthermore, our analyses also confirmed previous assumptions that events containing disparate details from different social realms of one's life were rated as less plausible than events containing details that are more closely connected.

Little is known, however, about the psychological mechanisms by which people evaluate the plausibility of imagined future events (Szpunar & Schacter, 2013). If plausible and implausible future events are distinguished in the way that we also distinguish between real (i.e., past) and imagined events, then according to the *reality monitoring theory*, the distinction is made on the basis of the events' phenomenology, and more particularly on the basis of the amount of sensory and perceptual detail present (Johnson & Raye, 1981; Johnson, Suengas, Foley, & Raye, 1988), where more detailed events may be considered more plausible. This idea is supported by a number of findings that real events are typically associated with greater sensory and perceptual detail than imagined future events (e.g., Addis et al., 2009;

D'Argembeau & Van der Linden, 2004; Weiler, Suchan, & Daum, 2011). From the perspective of the *availability heuristic*, on the other hand, the fluency with which the event is imagined influences plausibility estimations (Bernstein, Godfrey, & Loftus, 2009; Tversky & Kahneman, 1973; Whittlesea & Leboe, 2003). Thus an unexpected fluency of imagining for a novel event may lead individuals to mistake this fluency for familiarity with the event, thus inflating the belief that the event (or part of the event) is present in episodic memory, resulting in a higher plausibility rating (Bernstein et al., 2009). This idea is consistent with the suggestions that a crucial difference between imagined future events and recalled real past event lies in the effort that is required to construct them (Hassabis, Kumaran, & Maguire, 2007; McDonough & Gallo, 2010). Unfortunately, our data could not adjudicate between these two hypotheses because more plausible within-sphere events were both more fluently constructed and more detailed. Therefore, future studies that can differentiate the reality monitoring theory and availability heuristic are needed.

Another aim of this study was to investigate the effect of novelty (i.e., dissimilarity to past events and unusualness of event details occurring together) on the recall of future simulations. A long-held view maintains that novelty enhances encoding (Tulving, Markowitsch, Craik, Habib, & Houle, 1996), following the idea that information is encoded to the extent it is novel (see also, Knight, 1996). Consistent with this idea is the finding that anterior hippocampus supports memory for novel but not repeated stimuli (Poppenk, McIntosh, Craik, & Moscovitch, 2010) and the finding that medial temporal responses to novel stimuli are correlated with subsequent memory for these stimuli (Kirchhoff, Wagner, Maril, & Stern, 2000). Typically, these studies have examined recall effects for novel relative to repeated stimuli; however, increased rates of remembering are also evident for novel or *uncommon* stimuli over common stimuli. This encoding advantage for uncommon stimuli has been called the "bizarreness effect" and has been documented over a range of studies including bizarre images and word-pairs (for reviews see, Einstein, Lackey, & McDaniel, 1989; Hirshman, Whelley, & Palij, 1989). In contrast, a recent theory suggests that novelty may be a disadvantage for later recall, in particular with regards to episodic events. Poppenk and Norman (2012) propose in their *scaffolding hypothesis* that the similarity of novel stimuli to previous experiences facilitates the binding of new information as the previous experience provides a "scaffold" to which the new information can be attached. Consistent with this hypothesis are findings that familiarity with the stimuli enhances encoding (Klein, Robertson, Delton, & Lax, 2012; Poppenk, Köhler, & Moscovitch, 2010). Thus, it would be expected that for future events that are less similar to previous experiences, encoding may be less efficient, leading to decreased recall rates for events with more disparate details relative to events with less disparate details.

Our results demonstrated that within-sphere events were remembered significantly more often than across-sphere events. Thus, it would appear that novelty (manifested in the across-sphere condition as the inclusion of disparate details that normally do not occur together) impairs the encoding and recall of imagined future events, while heightened similarity to previous experiences

(as evident for the within-sphere condition) enhances encoding. This pattern of findings therefore suggests, in line with the scaffolding hypothesis (Poppenk & Norman, 2012), that factors that scaffold the event into memory—such as preexisting memories of similar content—probably enhance encoding. This finding is in line with recent work focused on the key role played by encoding of integrated memory representations (Shohamy & Wagner, 2008). Future work that directly evaluates this prediction should provide a basis for evaluating the suggestions laid out here.

The finding that novelty does not seem to enhance encoding and may even disrupt it was somewhat surprising given previous findings that bizarre information is often better encoded than common information (Hirshman et al., 1989; McDaniel & Einstein, 1986). It may be, however, that this pattern of results is related to the type of memory test that was employed in studies that found the bizarreness effect (free recall) versus the current study (cued recall; Einstein et al., 1989). Free-recall tests measure how strong an entire memory is represented in the mind and how well this memory can be accessed. Cued recall, on the other hand, tests how strong the connections are between the components that were cued (Einstein et al., 1989). It has previously been suggested that increased remembering rates for bizarre relative to common stimuli (e.g., imagery) are due to increased access to the images rather than enhanced retrieval of the components of the images (McDaniel & Einstein, 1986). Accordingly, the bizarreness effect is found with free recall, but not with cued recall (Einstein et al., 1989). This idea can also explain why our participants frequently mentioned during the post-simulation session that they remembered the "bizarre" events in particular (i.e., an instance of free recall), even though their performance on the cued-recall test showed the contrary. For future research, it would be interesting to do an experiment with the same social sphere paradigm as that used in the present study, but using both free- and cued-recall tests to further explore this hypothesis. Although our piloting of the standard recombination paradigm (e.g., Martin et al., 2011) has demonstrated that participants remember very few events through the free-recall method, it is possible that the sphere manipulation might increase rates of free recall, and thus this may be an interesting venture for future research.

It is important to note that across-sphere events were rated as occurring slightly further into the future than within-sphere events, and this medium effect approached significance. This observation suggests that the sphere differences in temporal distance may have influenced some of the results observed in this study. Indeed, close future events (within a year) tend to contain more sensory and contextual details and be associated with stronger feelings of "preexperience" than distant events (e.g., 5–10 years into the future; D'Argembeau & Van der Linden, 2004). However, our finding that temporal distance was not significantly associated with construction difficulty and event phenomenology speaks against this interpretation and suggests that differences in temporal distance on the order of a few weeks are not likely to affect the phenomenology of events imagined to occur more than a year into the future.

In summary, this study demonstrates that future events composed of more disparate details were not only more dissimilar from previous events and less plausible but also required more extensive constructive processes. Moreover, these future events were recalled less frequently than events with similarities to preexisting memories, highlighting the importance of scaffolding in successful encoding. Further investigation of event novelty and plausibility should provide important insights in the nature and function of future simulation.

REFERENCES

Addis, D. R., Cheng, T., Roberts, R. P., & Schacter, D. L. (2011). Hippocampal contributions to the episodic simulation of specific and general future events. *Hippocampus, 21,* 1045–1052.

Addis, D. R., Pan, L., Vu, M., Laiser, N., & Schacter, D. L. (2009). Constructive episodic simulation of the future and the past: Distinct subsystems of a core brain network mediate imagining and remembering. *Neuropsychologia, 47,* 2222–2238.

Addis, D. R., & Schacter, D. L. (2008). Constructive episodic simulation: Temporal distance and detail of past and future events modulate hippocampal engagement. *Hippocampus, 18,* 227–237.

Bar, M. (2009). The proactive brain: Memory for predictions. *Philosophical Transactions of the Royal Society B: Biological Sciences, 364*(1521), 1235–1243. doi:10.1098/rstb.2008.0310

Bernstein, D. M., Godfrey, R. D., & Loftus, E. F. (2009). False memories: The role of plausibility and autobiographical belief. In K. D. Markman, W. M. P. Klein, & J. A. Suhr (Eds.), *Handbook of imagination and mental simulation* (pp. 89–102). New York, NY: Taylor & Francis.

D'Argembeau, A., & Van der Linden, M. (2004). Phenomenal characteristics associated with projecting oneself back into the past and forward into the future: Influence of valence and temporal distance. *Consciousness and Cognition, 13,* 844–858.

Einstein, G. O., Lackey, S., & McDaniel, M. A. (1989). Bizarre imagery, interference, and distinctiveness. *Journal of Experimental Psychology: Learning Memory and Cognition, 15*(1), 137–146. doi:10.1037//0278-7393.15.1.137

Hassabis, D., Kumaran, D., & Maguire, E. A. (2007). Using imagination to understand the neural basis of episodic memory. *Journal of Neuroscience, 27*(52), 14365–14374.

Hirshman, E., Whelley, M. M., & Palij, M. (1989). An investigation of paradoxical memory effects. *Journal of Memory and Language, 28*(5), 594–609. doi:10.1016/0749-596x(89)90015-6

Johnson, M. K., & Raye, C. L. (1981). Reality monitoring. *Psychological Review, 88*(1), 67–85. doi:10.1037//0033-295x.88.1.67

Johnson, M. K., Suengas, A. G., Foley, M. A., & Raye, C. L. (1988). Phenomenal characteristics of memories for perceived and imagined autobiographical events. *Journal of Experimental Psychology: General, 117*(4), 371–376. doi:10.1037//0096-3445.117.4.371

Kirchhoff, B. A., Wagner, A. D., Maril, A., & Stern, C. E. (2000). Prefrontal-temporal circuitry for episodic encoding and subsequent memory. *Journal of Neuroscience, 20*(16), 6173–6180.

Klein, S. B., Robertson, T. E., Delton, A. W., & Lax, M. L. (2012). Familiarity and personal experience as mediators of recall when planning for future contingencies. *Journal of Experimental Psychology: Learning Memory and Cognition, 38*(1), 240–245. doi:10.1037/a0025200

Knight, R. T. (1996). Contribution of human hippocampal region to novelty detection. *Nature, 383*, 256–259.

Martin, V. C., Schacter, D. L., Corballis, M. C., & Addis, D. R. (2011). A role for the hippocampus in encoding simulations of future events. *Proceedings of the National Academy of Sciences of the United States of America, 108*(33), 13858–13863. doi:10.1073/pnas.1105816108

McDaniel, M. A., & Einstein, G. O. (1986). Bizarre imagery as an effective memory aid: The importance of distinctiveness. *Journal of experimental psychology: Learning, memory, and cognition, 12*(1), 54–65.

McDonough, I. M., & Gallo, D. A. (2010). Separating past and future autobiographical events in memory: Evidence for a reality monitoring asymmetry. *Memory & Cognition, 38*(1), 3–12. doi:10.3758/mc.38.1.3

Poppenk, J., Köhler, S., & Moscovitch, M. (2010). Revisiting the novelty effect: When familiarity, not novelty, enhances memory. *Journal of Experimental Psychology: Learning Memory and Cognition, 36*(5), 1321–1330. doi:10.1037/a0019900

Poppenk, J., McIntosh, A. R., Craik, F. I. M., & Moscovitch, M. (2010). Past experience modulates the neural mechanisms of episodic memory formation. *Journal of Neuroscience, 30*(13), 4707–4716. doi:10.1523/jneurosci.5466-09.2010

Poppenk, J., & Norman, K. A. (2012). Mechanisms supporting superior source memory for familiar items: A multi-voxel pattern analysis study. *Neuropsychologia, 50*, 3015–3026.

Shohamy, D., & Wagner, A. D. (2008). Integrating memories in the human brain: Hippocampal-midbrain encoding of overlapping events. *Neuron, 60*(2), 378–389. doi:10.1016/j.neuron.2008.09.023

Szpunar, K. K., Addis, D. R., & Schacter, D. L. (2012). Memory for emotional simulations: Remembering a rosy future. *Psychological Science, 23*(1), 24–29. doi:10.1177/0956797611422237

Tulving, E., Markowitsch, H. J., Craik, F. I. M., Habib, R., & Houle, S. (1996). Novelty and familiarity activations in PET studies of memory encoding and retrieval. *Cerebral Cortex, 6*(1), 71–79.

Tversky, A., & Kahneman, D. (1973). Availability: A heuristic for judging frequency and probability. *Cognitive Psychology, 5*(2), 207–232.

van Mulukom, V., Schacter, D. L., Corballis, M. C., & Addis, D. R. (2013). Re-imagining the future: Repetition decreases hippocampal involvement in future simulation. *PLoS One, 8*(7), e69596. Retrieved from http://journals.plos.org/plosone/article?id=10.1371/journal.pone.0069596

Weiler, J. A., Suchan, B., & Daum, I. (2010). Foreseeing the future: Occurrence probability of imagined future events modulates hippocampal activation. *Hippocampus, 20*(6), 685–690. doi:10.1002/hipo.20695

Weiler, J. A., Suchan, B., & Daum, I. (2011). What comes first? Electrophysiological differences in the temporal course of memory and future thinking. *European Journal of Neuroscience, 33*(9), 1742–1750. doi:10.1111/j.1460-9568.2011.07630.x

Whittlesea, B. W. A., & Leboe, J. P. (2003). Two fluency heuristics (and how to tell them apart). *Journal of Memory and Language, 49*(1), 62–79. doi:10.1016/s0749-596x(03)00009-3

Visual perspective in remembering and episodic future thought

Kathleen B. McDermott[1], Cynthia L. Wooldridge[2], Heather J. Rice[1], Jeffrey J. Berg[1], and Karl K. Szpunar[3]

[1]Department of Psychology, Washington University in St. Louis, St Louis, MO, USA
[2]Department of Psychology, Washburn University, Topeka KS, USA
[3]Department of Psychology, University of Illinois at Chicago, Chicago, IL, USA

According to the constructive episodic simulation hypothesis, remembering and episodic future thinking are supported by a common set of constructive processes. In the present study, we directly addressed this assertion in the context of third-person perspectives that arise during remembering and episodic future thought. Specifically, we examined the frequency with which participants remembered past events or imagined future events from third-person perspectives. We also examined the different viewpoints from which third-person perspective events were remembered or imagined. Although future events were somewhat more likely to be imagined from a third-person perspective, the spatial viewpoint distributions of third-person perspectives characterizing remembered and imagined events were highly similar. These results suggest that a similar constructive mechanism may be at work when people remember events from a perspective that could not have been experienced in the past and when they imagine events from a perspective that could not be experienced in the future. The findings are discussed in terms of their consistency with—and as extensions of—the constructive episodic simulation hypothesis.

Episodic future thought refers to a type of self- and future-oriented imagery that is centred on a distinct episode in time (Atance & O'Neill, 2001; Szpunar, 2010). For example, envisioning oneself taking an upcoming driving test or imagining oneself passing through a busy intersection on a cycling outing that is to take place later in the day would both be considered episodic future thinking. This type of thought is tied closely to remembering, in terms of both definition (it is, in essence, the flip side of remembering) and its characteristics (see Szpunar, 2010, for a review).

For example, people with medial temporal lobe amnesia experience not only profound breakdowns in the ability to remember episodes from their personal past, but also impaired ability to envision events in their personal futures (Hassabis, Kumaran, Vann, & Maguire, 2007; Klein, Loftus, & Kihlstrom, 2002; Tulving, 1985). Moreover, conditions that lead to more subtle memory impairments, such as schizophrenia and depression, also produce an accompanying deficit in episodic future thought (D'Argembeau, Raffard, & Van der Linden, 2008; Williams et al., 1996, for schizophrenia and depression, respectively). Both capacities develop at about four years of age (Busby & Suddendorf, 2005). Furthermore, neuroimaging studies have shown similarities in the neural signature of remembering and episodic future thought (e.g., Addis, Wong, & Schacter, 2007; Szpunar, Watson, & McDermott, 2007; for a recent review, see Schacter et al., 2012).

These similarities do not, however, tell the complete story. Clearly, there is a phenomenological difference between remembering and envisioning the future—the two subjective experiences are not typically confused. For instance, remembering leads to higher ratings of sensorial details and clarity of location than does episodic future thought (Arnold, McDermott, & Szpunar, 2011; D'Argembeau & Van der Linden, 2004). A network of brain regions in parahippocampal cortex and retrosplenial complex activates more during remembering than during episodic future thinking (Gilmore, Nelson, & McDermott, 2014), a finding that has been interpreted in light of the observation that contextual associations are more readily available for episodes being remembered than for those being constructed from scratch. That is, parahippocampal cortex and retrosplenial complex activate more when contextual associations are plentiful (Bar & Aminoff, 2003) and activate more for remembering than for episodic future thought (Gilmore et al., 2014).

Notwithstanding the differences, the similarities between these two processes have been integrated in a hypothesis termed the *constructive episodic simulation hypothesis* (Schacter & Addis, 2007). This hypothesis holds that memory's constructive nature may exist in order to allow individuals to flexibly recombine prior experiences in order to envision potential future scenarios. As such, "memory" regions are engaged when envisioning the future so as to merge elements from memory in the creation of a plausible future scenario. This hypothesis, of course, also fits well with the aforementioned neuropsychological findings that impairments of memory in hippocampal amnesia are accompanied by deficits in episodic future thought (Addis & Schacter, 2012).

Visual perspective in remembering

A central distinction in the autobiographical memory literature has been made between field memories (i.e., memories experienced from the first-person perspective, as originally experienced) and observer memories (i.e., those recollected from the third person perspective, as if an observer watching oneself participate in the event; Nigro & Neisser, 1983). Yet Rice and Rubin (2011) noted that although numerous studies have asked subjects to classify individual memories as having occurred from the first- or third-person perspective, none had directly queried third-person perspectives to examine the location(s) from which these memories take place. That is, third-person perspectives had been bundled together as an undifferentiated category, despite the possibility that third-person perspectives could originate from disparate spatial locations.

Using a novel methodology, Rice and Rubin (2011) demonstrated that third-person perspectives are common—indeed, more common than previously observed. More importantly, however, the authors also showed that the third-person perspective is quite variable with respect to spatial position (i.e., where the observer is located) and that the typical third-person perspective varies across event type. For example, when remembering a group performance, people tended to remember the event from a perspective in front of their location during the original event. When remembering running from a threat, conversely, individuals tended to remember the event from a perspective behind their location during the original event.

Goals of the present study

In the present study, we applied the approach of Rice and Rubin (2011) to both memory *and* episodic future thought. Our study addressed the following questions: Do first-person (or third-person) perspectives predominate in episodic future thought? How does the distribution of spatial perspectives in future thought compare to that of remembering? To the extent that third-person perspectives are adopted in episodic future thought, do they vary (as do third-person perspective memories), or do they occur from a canonical perspective? To this end, a direct comparison of first- and third-person perspectives for memory and episodic future thought was conducted.

An ancillary goal was to contrast phenomenological characteristics of remembering and episodic future thought (Arnold et al., 2011; D'Argembeau & Van der Linden, 2004; Szpunar & McDermott, 2008). How do they differ with respect to difficulty, vividness, and other characteristics? Further, can the first-person/third-person classification shed light on these phenomenological differences between memories and future thoughts?

EXPERIMENTAL STUDY

Method

Participants
Sixty undergraduate volunteers from Washington University in St. Louis participated in partial fulfilment of a course requirement.

Materials
Ten event cues, previously employed by Rice and Rubin (2011), were utilized: (a) *being in a group performance*; (b) *demonstrating a skilled act to a child or a friend*; (c) *giving an individual public presentation*; (d) *having a face-to-face conversation*; (e) *in an accident or near an accident*; (f) *running for exercise*; (g) *studying*; (h) *swimming*; (i) *walking or running from a threatening situation*; and (j) *watching the news on television*.[1] Each event cue was paired at random with a time cue, which instructed participants either to remember a personal event (*past*) or to imagine a plausible personal future event (*future*). Each participant received a different randomized order of event cues, and counterbalancing ensured that each event cue was paired with the past and future orientations equally often across participants. Stimuli and instructions were presented using E-prime software (Psychology Software Tools, Pittsburgh, PA).

Procedure
Participants were informed that they would be given a series of cues and asked to, for each cue, recall a specific related memory or imagine a specific related event that could plausibly occur in the future. They then completed three temporally contiguous phases (see Figure 1 for a summary), adapted from Rice and Rubin (2011). The entire session lasted approximately 45 minutes.

Phase 1. Participants were presented with a series of event–time cue pairs (e.g., future–*watching the news*). For each pair, the participants were asked to think of a related specific future event or memory, as indicated by the cue. After an individual event–time cue pair had been imagined or remembered, participants were to type in a two-to-three-word description of the event that could help them retrieve their mental image at a later time. This phase was self-paced with no time limits.

Phase 2. Participants were informed that visual perspectives can originate from various spatial positions within an image, then were subsequently given examples of different positions and how each one could be described in terms of *height* (e.g., from about ceiling height), *location* (e.g., directly behind myself), and *distance* (e.g., approximately 10 feet away). These instructions included a first-person perspective position, as well as several third-person perspective positions. Subsequently, participants were informed that they would be provided their two-to-three-word descriptions from Phase 1, be asked to reenvision the events, and—finally—describe the locus of each perspective in relation to its position within the scene. For example, if a participant saw the scene from a position floating in front and above, he or she might respond, "directly in front of myself, from above, about 10 feet away". These instructions were provided *after* Phase 1 to avoid introducing any bias into the original mental images.

After typing their perspective description, participants answered 15 questions about the

---

[1]These cues had been adapted from Nigro and Neisser's (1983) study.

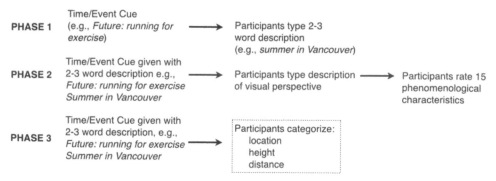

Figure 1. *Schematic of the steps involved in each of the three temporally contiguous phases.*

phenomenological characteristics of the remembered or imagined event (e.g., "The relative spatial arrangement of people in my image of the event is. . . . 1 = vague to 7 = clear"), which were adapted from the Memory Characteristics Questionnaire (Johnson, Foley, Suengas, & Raye, 1988). This process (cue, description, ratings) was repeated for all 10 event–time cue pairs.

Phase 3. Each event–time cue pair and its accompanying two-to-three-word description were presented once again. Participants were asked to categorize their perspectives from among sets of options within the *height*, *location*, and *distance* dimensions (as opposed to the self-generated depictions in Phase 2). For example, the height of a given visual perspective could be classified as: "own eye level," "slightly above own head," "from waist height," "from ceiling height," and so forth. These categorizations were queried in a separate phase so that participants would not simply adopt them in their Phase 2 descriptions of visual perspective (please see the Appendix for descriptions and response options for each visual perspective dimension).

Results

The primary question of interest was whether, and how, the visual perspective that participants adopted differed as a function of engaging in episodic future thought or remembering, which is described first. The secondary question of how

episodic future thought and memory differ on other phenomenological characteristics is then described. The significance threshold was set at $p < .05$, with corrections for multiple comparisons where appropriate (described below).

Visual perspective
Visual perspective was considered in two ways, both of which derive from the data obtained at the end of Phase 3 (see Figure 1). We consider first whether the perspective adopted was from a first- or a third-person point of view. Subsequently, we explore more deeply the nuances of the third-person perspective.

First or third person. A perspective was coded as being first person if the participant chose "from my own eyes" for both the distance and location questions. In any event in which there was disagreement between these two questions (33 of 600 perspectives, or 5.5%), the descriptions of perspective from Phase 2 (see Figure 1) were examined by the authors, and the data were coded according to that description. The height dimension was not used to code perspective because there was no "from my own eyes" option, and a perspective "from the level of my eyes" could represent a first- or third-person perspective.

As can be seen in Figure 2, there were more third- than first-person perspectives in both episodic future thought and memory. Further, the figure suggests that third-person perspectives occurred more often in episodic future thought than in

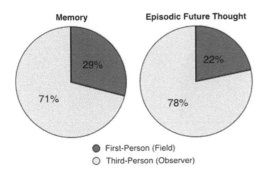

Memory   Episodic Future Thought

29%   22%

71%   78%

● First-Person (Field)
○ Third-Person (Observer)

Figure 2. *Percentage of first- and third-person perspectives in memory and episodic future thought (data from Phase 3).*

memory. Indeed, a Wilcoxon signed-rank test (computed with SPSS, v.22) comparing the frequency with which future thoughts were experienced in the third-person perspective to the frequency with which memories were experienced in the third-person perspective showed that participants reported significantly more third-person perspectives—and therefore fewer first-person perspectives—for episodic future thoughts (78.3%, $SEM = 3.6\%$) than memories (70.7%, $SEM = 4.0\%$), $z(N = 32) = 2.60,^2$ $p = .009$, $r = .24$. The correlation between the number of third-person memories and the number of third-person future thoughts across people was $r = .70$, $p < .001$, indicating that people who were more likely to generate third-person memories were also more likely to generate third-person future thoughts.

Characteristics of the third-person perspective. To answer whether or not episodic future thought and memory differed with respect to the spatial location of third-person perspectives, perspective locations were categorized into four dimensions taken from their Phase 3 categorizations: (a) *height* (below eye level, eye level, above eye level); (b) *distance* (less than six feet, greater than six feet); (c) *location—front/back* (behind, alongside, in front); and (d) *location—side/body* (right, centre, left). To the extent that the constructive nature of memory is responsible for the occurrence

of remembered third-person perspectives, one might predict little difference between memories and future thoughts given that episodic future thought is inherently constructive.

Indeed, as can be seen in Figure 3, perspective locations were fairly similar for memories and future thoughts when collapsing across the event cues. In particular, there was no difference between memories and future thoughts in the distance, $\chi^2(2, N = 436) = 0.3$, $p = .581$, $\phi_c = .03$, front/back, $\chi^2(3, N = 416) = 1.68$, $p = .431$, $\phi_c = .06$, and side/body dimensions, $\chi^2(3, N = 416) = 1.30$, $p = .522$, $\phi_c = .06$. Within the height dimension, however, memories and future thoughts differed significantly, $\chi^2(3, N = 426) = 6.96$, $p = .031$, $\phi_c = .13$, as future thoughts (vs. memories) were more often viewed from below eye level (27.1% vs. 19.0%, respectively) and less often viewed from eye level (30.3% vs. 41.5%, respectively), with similar proportions of above eye level perspectives across future and past episodes (42.5% vs. 39.5%, respectively).

Furthermore, because each participant responded to 10 different event cues and therefore could have contributed multiple responses to any one cell in Figure 3, we chose to analyse the data for each event cue separately. That is, memory and episodic future thought were contrasted for each of the 10 events in each of the four dimensions. Within the height and side/body dimensions, none of the event cues demonstrated significant differences in perspective. Moreover, only one event cue within the distance dimension was significant at the $p < .05$ level: Imagining running for exercise in the future was associated with more perspectives from closer than six feet than a memory of the same event, $\chi^2(2, N = 45) = 4.98$, $p = .026$, $\phi_c = .33$. Lastly, two of the 10 event cues featured significant differences within the front/back dimension: Imagining being in a group performance, $\chi^2(3, N = 47) = 7.85$, $p = .02$, $\phi_c = .41$, and watching the news on television, $\chi^2(3, N = 38) = 7.92$, $p = .019$, $\phi_c = .46$, in the future were both associated with more perspectives from behind than memories of

---

^{2}The sample size is reduced as a result of tied observations.

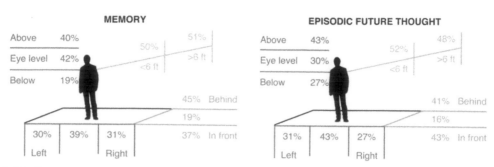

Figure 3. *Distributions of third-person perspective locations across memory and episodic future thought. Data from Phase 3. Cases in which the sum of percentages do not equal 100 are due to rounding error. To view this figure in color, please visit the online version of this Journal.*

the same events. When correcting for multiple comparisons (10 events; corrected $\alpha = .05/10 = .005$), however, memory and episodic future thought did not differ on any of the four dimensions for any event. Of course, this Bonferroni correction is conservative, so the appropriate conclusion from these suggestive results awaits future research.

Phenomenological characteristics

Here we consider the phenomenological characteristics that were rated at the end of Phase 2 (see Figure 1). Of primary interest was the comparison between memories and future thoughts. Therefore, phenomenological ratings were compared across memories and episodic future thoughts, collapsing across first- and third-person perspectives, as well as across event cues. We then averaged each subject's rating per characteristic for memory and future thought and conducted paired t tests. Because there were 15 characteristics, we corrected for multiple comparisons by setting $\alpha = .05/15 = .003$. Prior work in this area (e.g., Arnold et al., 2011; D'Argembeau & Van der Linden, 2004) led to the prediction that memories would be accompanied by more vivid ratings than future thought.

As can be seen in Figure 4, memories were rated higher than future thoughts on 10 of the 15 phenomenological characteristics, all of which survived the multiple comparisons correction (corrected $\alpha = .003$): (p)reexperiencing ($M_P = 4.91$, $M_F = 4.50$; $d = 0.45$); location clarity ($M_P = 6.20$, $M_F = 4.94$; $d = 1.1$); movement

clarity ($M_P = 4.69$, $M_F = 4.25$; $d = 0.48$); object clarity ($M_P = 5.32$, $M_F = 4.57$; $d = 0.68$); people clarity ($M_P = 5.21$, $M_F = 4.48$; $d = 0.79$); sound details ($M_P = 3.89$, $M_F = 3.31$; $d = 0.55$); visual details ($M_P = 5.45$, $M_F = 4.89$; $d = 0.63$); whether the event formed a coherent story ($M_P = 4.82$, $M_F = 3.64$; $d = 0.86$); time clarity ($M_P = 5.39$, $M_F = 4.35$; $d = 0.87$); and the feeling of traveling in time to the event ($M_P = 4.88$, $M_F = 4.05$; $d = 0.84$) [all ts(59) > 3.5, ps ≤ .001].

Memories were rated as less difficult to produce than were future thoughts ($M_P = 3.14$, $M_F = 3.57$), $t(59) = -2.7$, $p = .009$, $d = -0.35$, although this difference did not survive the correction for multiple comparisons (corrected $\alpha = .05/15 = .003$). The remaining four characteristics (emotion intensity, positive emotion, negative emotion, and taste/smell details) did not differ between memory and episodic future thought [all ts(59) ≤ 2.07, ps ≥ .043] when subjected to the multiple comparisons corrections. An examination of these data broken down by solely first-person or solely third-person perspective produced no appreciable differences.

To the extent that memory and episodic future thought represent two manifestations of a common set of processes, one might expect individual variation in the clarity of memory to correlate with that for episodic future thought. For example, do subjects who tend to report high clarity of visual details when remembering also report high clarity of visual details for their episodic future thoughts? To this end, we again collapsed

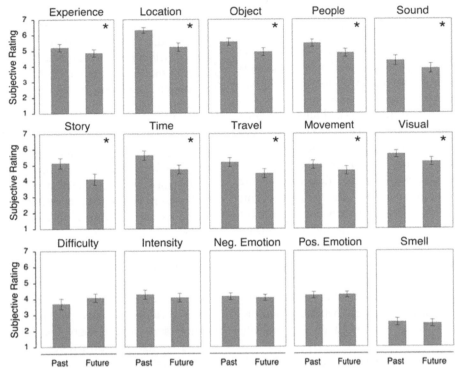

Figure 4. *Mean ratings for phenomenological characteristics for memory and episodic future thought. Error bars represents standard errors of the mean.*
Note: * = comparisons achieving statistical significance after correcting for multiple comparisons $(p < .003)$.

across perspectives and across event cues, averaged each subject's rating (per characteristic) for memory and future thought, and found the correlation (across subjects) between the average memory and average future thought ratings for each phenomenal characteristic. In particular, subjects' average ratings for memories and future thoughts were significantly correlated on each phenomenological characteristic (all $rs \geq .40$, $ps < .001$) except for negative emotion ($r = .18$, $p = .177$).

Are first-person perspectives more vivid than third-person perspectives? The data reported here revealed some evidence of such a difference (cf. Nigro & Neisser, 1983; Robinson & Swanson, 1993). For each phenomenal characteristic, paired t tests were conducted on each subject's average rating for first- and third-person perspectives (collapsing across memory and future thought). Four phenomenal characteristics

demonstrated differences between first- and third-person perspectives: Feelings of (p)reexperiencing, people clarity, and negative emotion all were rated higher when subjects maintained first-person perspectives, whereas positive emotion was greater when subjects took third-person perspectives. After correcting for multiple comparisons (corrected $\alpha = .05/15 = .003$), however, only feelings of (p)reexperiencing and people clarity differed, with negative emotion, $t(41) = 2.65$, $p = .011$, $d = 0.41$, and positive emotion, $t(41) = -2.23$, $p = .031$, $d = -0.34$, failing to meet the correction. Specifically, feelings of re- (or pre) experiencing were greater for first-person ($M_{1st} = 5.09$) than for third-person ($M_{3rd} = 4.52$) perspectives, $t(41) = 3.43$, $p = .001$, $d = 0.53$, and the clarity of people in the image was also greater for first-person ($M_{1st} = 5.21$) than for third-person ($M_{3rd} = 4.60$) perspectives, $t(41) = 3.27$, $p = .002$, $d = 0.5$.

41

Discussion

In this study, episodic future thought and remembering both tended to be experienced from the third-person perspective, but this tendency was greater for future thought than for remembering. An examination of third-person perspective events suggests that memories and future thoughts are experienced from similar vantage points and that these vantage points vary as a function of the specific event being remembered or imagined. Further, we observed individual differences in the experiences of remembering and future thought; people who experience remembering in a vivid way tend also to imagine vividly. And finally, remembering and episodic future thought tend to covary, but with the tendency for remembering to be more vividly experienced. For example, both remembering and future thinking elicited low vividness ratings for smell, whereas location, objects, and people were rated as being vividly experienced.

The results reported replicate Rice and Rubin's (2011) findings that third-person perspectives can be common in remembering and extend this conclusion to future thought. Further, these results replicate their finding that third-person perspectives are not a single, canonical visual perspective but instead vary from event to event. Interestingly, this variability in third-person perspective locations extends to future thought. In addition, people who tended to experience third-person perspectives when remembering also tended to experience them during future thought, suggesting that this may be a stable individual difference across people.

The results extend our understanding of the constructive episodic simulation hypothesis, which states that constructive processes support remembering and episodic future thought. One corollary of this hypothesis is that episodic future thought is inherently more constructive than remembering because future events have not yet taken place. Our data support this view in that future thoughts were more likely to be imagined from a third-person perspective than memories. Notably, there is no indication in the literature as

to whether common or distinct constructive processes are at work in the context of remembering and episodic future thought. The highly overlapping distributions of third-person perspectives observed for remembered and simulated events in the present study point to the influence of a common constructive process.

Specific experimental factors may have contributed to the high rate of third-person perspectives reported in our study and the accompanying overlap in distribution of third-person perspectives. For instance, it is possible that the high level of third-person perspectives observed in this study may be an artefact of the particular cues employed (e.g., remember/imagine having a face-to-face conversation). In another study, D'Argembeau and Van der Linden (2004) used a more open-ended cueing technique (e.g., remember/imagine an event within the past/next year) and found a higher incidence of first-person perspectives in memory and future thought than what we report here. Importantly, instructions about perspective in that study were presented prior to simulation. Moreover, studies that find high rates of first-person perspectives often explicitly instruct participants to generate memories and future events from a first-person (as opposed to a third-person) perspective (e.g., Addis et al., 2007). Clearly, additional work that takes into consideration the role of cue type and nature of instructions about visual perspective is needed to tease apart experimental influences on the occurrence of first- and third-person perspectives in memory and future thought. Such work should also take into consideration the role of temporal distance, as prior research has shown that temporally distant memories and future events are more likely to be experienced from a third-person perspective (D'Argembeau & Van der Linden, 2004).

The above considerations aside, it was still the case that the distribution of third-person perspectives in our study was highly similar across remembered and simulated events. Along these lines, we note that participants were required to remember simulated events prior to categorizing those events in terms of perspective. Although the extent to which such short-term reactivation of a simulated event might change its mental representation is unclear at this point, additional work that more stringently controls for the presence/absence of prior simulation would be valuable. Finally, it will be important for future work to discriminate between the influence of episodic and semantic memory in terms of the constructive process(es) that support remembering and future thought. For example, it is possible that third-person perspectives of remembered and simulated events can be based on either elements extracted from relevant episodic memories (Szpunar & McDermott, 2008) or schematic biases extracted from repeated experiences or perhaps repeated viewings of movies and television shows (e.g., a speech is typically seen from the perspective of an audience, and such schematic biases may influence the formation of third-person perspectives in memory and future thought).

Our findings provide further insights into the emerging literature on individual differences in the phenomenology of episodic future thought and remembering. For example, Arnold et al. (2011) demonstrated that one's proclivities in thinking about time (as measured by the Zimbardo Time Perspective Inventory; Zimbardo & Boyd, 1999) predict the phenomenological experiences of remembering and future thought. D'Argembeau and Van der Linden (2006) demonstrated that people who are low in expressive suppression (i.e., who are open to experiencing feelings) tend to have high levels of re- and preexperiencing events. Here we show that across visual perspective, people who tend to report vivid experiences in remembering tend also to report vivid episodic future thinking.

In summary, the present data extend the observations that episodic future thought and remembering share phenomenological characteristics. As has often been observed, the phenomenology is not identical across the two (nor should it be, as we can typically tell the difference between imagining and remembering). Commonalities in third-person perspective across remembering and episodic future thought add to the literature demonstrating similarities in the two processes. These similarities have been proposed to arise from constructive processes (Schacter & Addis,

2007), a hypothesis that fits well with the data observed here. In short, visual perspective (first or third person, and the specific vantage point within third person) is highly similar for remembering and future thinking.

REFERENCES

Addis, D. R., & Schacter, D. L. (2012). The hippocampus and imagining the future: Where do we stand? *Frontiers in Human Neuroscience, 5*, 173. doi:10.3389/fnhum.2011.00173

Addis, D. R., Wong, A. T., & Schacter, D. L. (2007). Remembering the past and imagine the future: Common and distinct neutral substrates during event construction and elaboration. *Neuropsychologica, 45*, 1363–1377. doi:10.1016/j.neuropsychologia.2006.10.016

Arnold, K. M., McDermott, K. B., & Szpunar, K. K. (2011). Individual differences in time perspective predict autonoetic experience. *Consciousness and Cognition, 20*, 712–719. doi:10.1016/j.concog.2011.03.006

Atance, C. M., & O'Neill, D. K. (2001). Episodic future thinking. *Trends in Cognitive Sciences, 5*, 533–539. doi:10.1016/S1364-6613(00)01804-0

Bar, M., & Aminoff, E. (2003). Cortical analysis of visual context. *Neuron, 38*, 347–358. doi:10.1016/S0896-6273(03)00167-3

Busby, J., & Suddendorf, T. (2005). Recalling yesterday and predicting tomorrow. *Cognitive Development, 20*, 362–372. doi:10.1016/j.cogdev.2005.05.002

D'Argembeau, A., Raffard, S., & Van der Linden, M. (2008). Remembering the past and imagining the future in schizophrenia. *Journal of Abnormal Psychology, 117*, 247–251. doi:10.1037/0021-843X.117.1.247

D'Argembeau, A., & Van der Linden, M. (2004). Phenomenal characteristics associated with projecting oneself back into the past and forward into the future: Influence of valence and temporal distance. *Consciousness and Cognition, 13*, 844–858. doi:10.1016/j.concog.2004.07.007

D'Argembeau, A., & Van der Linden, M. (2006). Individual differences in the phenomenology of mental time travel: The effect of vivid visual imagery and emotion regulation strategies. *Consciousness and Cognition, 15*, 342–350. doi:10.1016/j.concog.2005.09.001

Gilmore, A. W., Nelson, S. M., & McDermott, K. B. (2014). The contextual association network activates more for remembered than for imagined events. *Cerebral Cortex*. Advance online publication. doi:10.1093/cercor/bhu223

Hassabis, D., Kumaran, D., Vann, S. D., & Maguire, E. A. (2007). Patients with hippocampal amnesia cannot imagine new experiences. *Proceedings of the National Academy of Sciences of the United States of America, 104*, 1726–1731. doi:10.1073/pnas.0610561104

Johnson, M. K., Foley, M. A., Suengas, A. G., & Raye, C. L. (1988). Phenomenal characteristics of memories for perceived and imagined autobiographical events. *Journal of Experimental Psychology: General, 117*, 371–376. doi:10.1037/0096-3445.117.4.371

Klein, S. B., Loftus, J., & Kihlstrom, J. F. (2002). Memory and temporal experience: The effects of episodic memory loss on an amnesic patient's ability to remember the past and imagine the future. *Social Cognition, 20*, 353–379. doi:10.1521/soco.20.5.353.21125

Nigro, G., & Neisser, U. (1983). Point of view in personal memories. *Cognitive Psychology, 15*, 467–482. doi:10.1016/0010-0285(83)90016-6

Rice, H. J., & Rubin, D. C. (2011). Remembering from any angle: The flexibility of visual perspective during retrieval. *Consciousness and Cognition, 20*, 568–577. doi:10.1016/j.concog.2010.10.013

Robinson, J. A., & Swanson, K. L. (1993). Field and observer modes of remembering. *Memory, 1*, 169–184. doi:10.1080/09658219308258230

Schacter, D. L., & Addis, D. R. (2007). The cognitive neuroscience of constructive memory: Remembering the past and imagining the future. *Philosophical Transactions of the Royal Society of London B: Biological Sciences, 362*, 773–786. doi:10.1098/rstb.2007.2087

Schacter, D. L., Addis, D. R., Hassabis, D., Martin, V. C., Spreng, R. N., & Szpunar, K. K. (2012). The future of memory: Remembering, imagining, and the brain. *Neuron, 76*, 677–694. doi:10.1016/j.neuron.2012.11.001

Szpunar, K. K. (2010). Episodic future thought: An emerging concept. *Perspectives on Psychological Science, 5*, 142–162. doi:10.1177/1745691610362350

Szpunar, K. K., & McDermott, K. B. (2008). Episodic future thought and its relation to remembering: Evidence from ratings of subjective experience. *Consciousness and Cognition, 17*(1), 330–334. doi:10.1016/j.concog.2007.04.006

Szpunar, K. K., Watson, J. M., & McDermott, K. B. (2007). Neural substrates of envisioning the future. *Proceedings of the National Academy of Sciences, 104*, 642–647. doi:10.1073/pnas.0610082104

Tulving, E. (1985). Memory and consciousness. *Canadian Psychology, 26*, 1–12. doi:10.1037/h0080017

Williams, J. M. G., Ellis, N. C., Tyers, C., Healy, H., Rose, G., & MacLeod, A. K. (1996). The specificity of autobiographical memory and imageability of the future. *Memory & Cognition, 24*, 116–125. doi:10.3758/BF03197278

Zimbardo, P. G., & Boyd, J. N. (1999). Putting time in perspective: A valid, reliable individual-differences metric. *Journal of Personality and Social Psychology, 77*, 1271–1288. doi:10.1037/0022-3514.77.6.1271

APPENDIX

Descriptions and response options for each visual perspective dimension in Phase 3

Height

When referring to the height, we are interested in the height of the origin of your visual perspective. That is, does it seem that you are floating above the scene or lying on the floor looking up at the scene? We are interested in the ORIGIN of the perspective, NOT where you are looking.

1: Own eye level
2: Slightly above own head

3: From waist height
4: From ceiling height
5: From above ceiling height
6: From level of floor
7: Other

Location

When referring to the spatial location, we are again interested in the location of the origin of your visual perspective. That is, does it seem that the origin is directly to the left of your location during the event or is it in front and to the right of your location during the event? We are interested in the ORIGIN of the perspective, NOT where you are looking.

1: Directly in front of yourself, facing yourself

2: Directly behind yourself
3: To the left and behind you
4: To the right and behind you
5: To your left and in front of you
6: To your right and in front of you
7: Directly to your left
8: Directly to your right
9: From your own eyes
10: Other

Distance

When referring to the distance, we are again interested in the distance of the origin of your visual perspective from yourself.

That is, does it seem that you are approximately a foot away from your location during the event or 10 feet away from your location?

1: From your own eyes
2: 3 feet away or closer (arm's length is approximately 2–3 feet)
3: 3–6 feet away (6 feet is the wingspan of a 6-foot-tall person)
4: 6–20 feet away (20 feet is approximately the distance from the top of the key on the basketball court to the basket)
5: 20–100 feet away (100 feet is approximately the length of a college basketball court)
6: 100 feet away or more

Prevalence and determinants of direct and generative modes of production of episodic future thoughts in the word cueing paradigm

Olivier Jeunehomme and Arnaud D'Argembeau

Department of Psychology, University of Liège, Liège, Belgium

Recent research suggests that episodic future thoughts can be formed through the same dual mechanisms, direct and generative, as autobiographical memories. However, the prevalence and determinants of the direct production of future event representations remain unclear. Here, we addressed this issue by collecting self-reports of production modes, response times (RTs), and verbal protocols for the production past and future events in the word cueing paradigm. Across three experiments, we found that both past and future events were frequently reported to come directly to mind in response to the cue, and RTs confirmed that events were produced faster for direct than for generative responses. When looking at the determinants of direct responses, we found that most past and future events that were directly produced had already been thought of on a previous occasion, and the frequency of previous thoughts predicted the occurrence of direct access. The direct production of autobiographical thoughts was also more frequent for past and future events that were judged important and emotionally intense. Collectively, these findings provide novel evidence that the direct production of episodic future thoughts is frequent in the word cueing paradigm and often involves the activation of personally significant "memories of the future."

Much of our actions are guided by the events we anticipate, their envisioned consequences, and our goals and plans in attaining or avoiding imagined states of affairs. Elucidating the cognitive systems and processes that allow us to consider possible futures is, therefore, an important step in understanding human behaviour. In the past few years, substantial evidence—from cognitive, neuroimaging, developmental, and patient studies—has accumulated to show that future scenarios are founded on information stored in episodic and semantic memory (for review, see D'Argembeau, 2012; Schacter, Addis, Hassabis, Martin, Spreng, & Szpunar, 2012; Szpunar, 2010). Memory for specific happenings in one's personal past provides an important source of details (e.g., about people, objects, locations, and actions) for imagining events that might lie ahead (Schacter & Addis, 2007; Suddendorf & Corballis, 2007), and semantic knowledge contributes the framework or scaffolding into which episodic details can be integrated (D'Argembeau & Mathy, 2011; Irish, Addis, Hodges, & Piguet, 2012). Despite these recent advances in understanding the sources of information supporting episodic future thinking, the mechanisms underlying the production of future event representations are not fully understood. Here, we examine the prevalence and determinants of two possible modes of production of episodic future thoughts: direct and generative.

Direct and generative retrieval in autobiographical memory

A prominent theoretical account of autobiographical memory contends that memories for past experiences are transitory mental constructions generated from a hierarchically structured knowledge base that includes general knowledge about one's life (lifetime periods and general events) as well as specific episodic details (Conway, 2005, 2009; Conway & Pleydell-Pearce, 2000). The construction of specific memories from this system can occur through two mechanisms: generative and direct retrieval. Generative retrieval refers to an effortful process involving iterative searches in the knowledge base: Cues are first elaborated, then memory is searched, the outputs from memory are evaluated, and, if required, these outputs are elaborated further, and another search is undertaken (Burgess & Shallice, 1996; Conway & Pleydell-Pearce, 2000). During this process, knowledge access typically proceeds from the abstract to the specific, with general autobiographical knowledge (i.e., a lifetime period and/or an associated general event) being accessed first and then used to access episodic details (Haque & Conway, 2001). Direct retrieval, on the other hand, refers to an automatic and effortless associative process whereby internal or external cues activate episodic details such that a coherent event representation is formed without the need of elaborating retrieval cues to search memory. The main difference between the two modes of retrieval is that the search process is modulated by control processes in generative retrieval but not, or not so extensively, in direct retrieval (Conway & Pleydell-Pearce, 2000).

While it has been generally assumed that generative retrieval is the dominant mode of formation of autobiographical memories (Conway & Pleydell-Pearce, 2000; Haque & Conway, 2001),

recent findings suggest that direct retrieval is more frequent than previously thought. First, there is evidence that involuntary memories—memories that come to mind with no preceding attempt at retrieval (Berntsen, 1996)—are common in everyday life, perhaps even more frequent than effortfully generated memories (Rasmussen & Berntsen, 2011). Although the concepts of involuntary retrieval and direct access are not equivalent (i.e., direct access can potentially occur whether or not memories are retrieved voluntarily), such studies nevertheless suggest that memories are frequently accessed with no search effort. Second, a recent study by Uzer, Lee, and Brown (2012) has shown that direct retrieval also occurs frequently when people voluntarily recall memories in experimental tasks such as the Crovitz word cueing paradigm—one of the most commonly used method to elicit autobiographical memories. In a first experiment, participants were cued with objects (e.g., *bag*) or emotions (e.g., *happy*) words and were instructed to say aloud everything that came to mind until they had a specific past event in mind. Once a suitable memory was recalled, participants were asked to report whether the memory was retrieved directly and without apparent effort (i.e., the memory was immediately triggered by the cue word) or whether they had to make a conscious effort by searching memory and using other information in order to recall a suitable memory. The verbal protocols and self-reports of retrieval strategy converged to show that the direct retrieval of memories was more frequent than generative retrieval (the overall prevalence of self-reported direct retrieval was 60%). Furthermore, response times (RTs) indicated, as predicted, that direct retrieval was much faster than generative retrieval. The authors replicated these results in a second experiment in which participants performed the same cueing task silently and alone without providing verbal protocols. Finally, converging results were obtained in a third experiment in which direct retrieval was assessed in terms of information use during retrieval (i.e., whether or not participants used information about their life to help them recall the memory). Taken together, these studies suggest that direct retrieval in autobiographical memory is at least as common as generative retrieval, even in experimental cueing paradigms that were previously thought to rely mainly on generative processes.

Modes of production of episodic future thoughts

According to the constructive episodic simulation hypothesis (Schacter & Addis, 2007; Schacter, Addis, & Buckner, 2008), episodic future thoughts are formed by extracting and flexibly recombining details from past experiences into coherent representations of realistic, yet novel, future events. This constructive process is presumably effortful and relies on executive control processes to search for, monitor, and combine relevant memory elements. The contribution of control processes has indeed been evidenced in several studies that examined the production of episodic future thoughts in the word cueing paradigm. In particular, it has been found that individual differences in executive resources and working memory capacity correlate with the ability to imagine specific future events (D'Argembeau, Ortoleva, Jumentier, & Van der Linden, 2010; Hill & Emery, 2013) and that the experimental depletion of executive resources by means of a concurrent task results in increased RTs and error rates, at least when future thoughts are cued with low-imageability words (Anderson, Dewhurst, & Nash, 2012). Furthermore, there is evidence that the imagination of future events in response to word cues often involves a protracted generative process in which general information about one's life is retrieved and used for constructing a specific episode (D'Argembeau & Mathy, 2011).

Other evidence suggests, however, that episodic future thoughts are sometimes produced without the intervention of controlled processes. Berntsen and colleagues (Berntsen & Jacobsen, 2008; Finnbogadottir & Berntsen, 2013) have shown that episodic future thoughts can occur involuntarily (i.e., with no conscious attempt to generate these representations), suggesting that the production of future event representations does not necessarily require search effort. Other recent findings suggest that voluntary prospection can also entail the direct production of episodic future thoughts. Anderson et al. (2012) found that participants were faster in imagining specific future events when they were cued with high-imageability words (e.g., *butterfly*) than with low-imageability words (e.g., *attitude*). Insofar as decreases in the time needed to produce a specific event to high- versus low-imageability words may reflect greater reliance on direct retrieval (Williams, Healy, & Ellis, 1999), these findings suggest that episodic future thoughts may sometimes be produced directly. This evidence is somewhat indirect, however, because the modes of formation of future event representations were inferred from differences in RTs and were not explicitly assessed. As noted by Uzer et al. (2012), differences in RTs could either reflect the existence of two distinct processes (direct vs. generative) or could be due to variations in the difficulty of a single generative process. More straightforward evidence for the direct production of episodic future thoughts during voluntary prospection comes from a study that used a think-aloud procedure to determine the types of information accessed when future event representations were formed. D'Argembeau and Mathy (2011) asked participants to report everything that came to their mind when they attempted to imagine specific future events in response to cue words. They found that on most trials, participants reported general information about their life before a specific event was constructed, thus showing that they relied on generative processes. Interestingly, however, episodic future thoughts were sometimes produced rapidly (as indicated by RT data), and the verbal protocols suggested that the event representations came directly to mind (i.e., without using general information about one's life for constructing the event).

The present research

Overall, the available evidence suggests that episodic future thoughts can be produced through the same dual mechanisms, direct and generative, as autobiographical memories. The prevalence of the direct formation of future event representations during voluntary prospection remains unclear, however, because the modes of production of future thoughts have not been directly assessed. Different possibilities can be considered. On the one hand, given the necessity to flexibly recombine appropriate memorial information to form coherent future event representations (Schacter & Addis, 2007), one would expect that generative processes would be the dominant mode of episodic future thinking. On the other hand, considering the adaptive value of prospective thought (Boyer, 2008; Schacter, 2012), it is conceivable that mechanisms allowing rapid anticipation and planning of future contingencies based on memory had been designed by natural selection (Klein, 2013). Furthermore, some imagined future events may be encoded in memory—as "memories of the future" (Ingvar, 1985; Szpunar, Addis, McLelland, & Schacter, 2013)—such that they could later be directly accessed in response to relevant cues. In everyday life, people indeed frequently and repeatedly think about some future events (D'Argembeau, Renaud, & Van der Linden, 2011; Watkins, 2008), which might lead to readily accessible memories of the future.

To shed light on these possibilities, we investigated the prevalence and determinants of the direct production of episodic future thoughts, using an adaptation of the strategy reports created by Uzer et al. (2012) to gauge the frequency of directly retrieved memories in the word cueing paradigm. In Experiment 1, we assessed the frequency of the direct mode of production of past and future events and further investigated whether certain characteristics of memories and future thoughts (such as previous recall/imagination, emotional value, and personal importance) predicted direct access. Experiment 2 was then conducted to ensure that directly accessed future thoughts indeed referred to anticipated future events and did not simply consist in directly retrieved past events that would be recast as future events. Finally, in Experiment 3, we sought to delve deeper into the nature of the direct mode of production of memories and future thoughts by distinguishing between two dimensions of direct access: search effort and information use.

EXPERIMENT 1

The aim of Experiment 1 was threefold. First, we sought to assess the frequency of the direct production of episodic future thoughts in the word cueing paradigm and to compare it to the frequency of directly retrieved memories (Uzer et al., 2012). Second, we investigated whether the frequency of directly produced memories and future thoughts is modulated by the temporal distance of events. Considering that the quantity and quality of memories and future thoughts decrease with increasing temporal distance (D'Argembeau & Van der Linden, 2004; Spreng & Levine, 2006), we expected that the direct production of event representations would be less frequent for the distant past and future than for closer time periods. Third, we aimed to explore possible determinants of the direct mode of production of autobiographical representations. Frequently retrieved memories tend to be more accessible than less frequently retrieved memories (Thompson, Skowronski, Larsen, & Betz, 1996), and, therefore, one would expect that direct retrieval would be more common for the former kind of memories. In a similar vein, as suggested above, future events that are frequently thought about might lead to readily accessible memories of the future that could be directly accessed in response to relevant cues. Therefore, we predicted that the frequency of previous thoughts about an event would predict the occurrence of direct access, both for memories and for future thoughts. Finally, in addition to frequency of thoughts, we also explored the possible impact of other dimensions that have been shown to influence the accessibility of memories and future thoughts, such as emotional value and personal importance (e.g., D'Argembeau & Van der Linden, 2004; Newby-Clark & Ross, 2003).

Method

Participants

Thirty-two undergraduate students (21 females and 11 males) aged between 18 and 27 years ($M = 21.5$ years, $SD = 2.5$ years) participated in this study. They all provided written informed consent for their participation and were tested individually.

Materials and procedure

Participants were asked to recall specific events that happened in their personal past and to imagine specific events that might reasonably happen to them in the future, in response to a series of cue words. Twenty cue words referring to common places, objects, persons, and feelings (e.g., *friend, school, garden, restaurant, book, love*) were selected from previous studies of autobiographical memory and future thinking (Conway, Pleydell-Pearce, & Whitecross, 2001; D'Argembeau & Mathy, 2011). These words were divided into four lists of five cues that were matched for word length (number of letters), imageability, and frequency of use (Desrochers & Bergeron, 2000). These four lists served as stimuli for the four conditions (near past, distant past, near future, distant future), and the assignment of lists to conditions was counterbalanced across participants. Within each condition, the five cue words were presented in random order.

The tasks were programmed and presented using E-Prime 2.0 software (Psychology Software Tools, Inc.). Depending on the condition, an instruction slide first informed participants that their task was to recall events that happened during the last week (near past condition), to recall events that happened more than one year ago (distant past condition), to imagine events that might happen during the next week (near future condition), or to imagine events that might happen in more than a year from now (distant future condition). It was emphasized that the produced events should be specific (i.e., unique events taking place in a specific place at a specific time and lasting a few minutes or hours but not more than a day) and refer to personal experiences; for future events, it was also mentioned that the events should be plausible (i.e., something that might reasonably happen).

Following the instruction slide, the five trials of the corresponding condition were presented. Each trial started with the presentation of the signal "READY" during 1.5 s, followed by an empty screen for 1 s. Then, a cue word was presented, and participants were instructed to retrieve or imagine a specific event, as quickly as possible, in response to the cue. As soon as an appropriate event came to mind, they were instructed to press the spacebar (which recorded RT) and to briefly describe the recalled or imagined event on a sheet of paper (if no appropriate event came to mind within 90 s of the word cue being presented, the computer terminated the trial automatically, and the participant was requested to initiate a new trial). Then, participants were asked to judge whether the event came to their mind directly or whether some active search was necessary. Definitions of direct and generative access (which were adapted from Uzer et al., 2012) were provided before the experiment: Participants were told that sometimes memories or imagined future events can automatically come to mind (i.e., with little or no effort), whereas at other times memories or imagined future events have to be actively searched. It was explained that during the task, some memories or imagined future events might be immediately triggered by the cue word, but sometimes the participant would need to actively search and reflect in order to find a suitable event. For each trial, participants were instructed to press the key *1* if the event came directly to mind or the key *2* if the production of the event required an active search effort. Finally, participants were invited to write a brief title for the event, which was used to identify events in the next phase of the experiment.

Immediately after the five trials of a given condition, participants were instructed to come back to each of the five events they had recalled or imagined, and, for each event, they estimated its temporal distance (in hours, days, weeks, months, and years), they indicated whether they already thought about this event on a previous occasion (by answering "yes" or "no"; if they responded "yes", they were also asked to rate the extent to which they previously thought about this event, from 1 = very rarely, to 7 = very often), and they rated the personal importance of the event (from 1 = not at all important, to 7 = extremely important) and its affective valence (−3 = very negative, 0 = neutral, +3 = very positive). For future events, participants also estimated the probability that the imagined event would actually happen in the future ($1 = $ extremely low, $7 = $ extremely high). After making these judgements for each of the five events, participants proceeded to the next condition. The order of presentation of the four conditions was counterbalanced across participants.

Before starting the experimental trials, participants received two practice trials (with different cue words) that were then discussed with the experimenter to ensure that participants correctly understood all instructions.

Results

The reported estimations of temporal distance confirmed that participants produced events falling in the required categories; the mean temporal distance was, in days, 1550 in the distant past condition, 1071 in the distant future condition, 4 in the recent past condition, and 5 in the near future condition.

The mean proportion of events that directly came to mind in response to the cues (as estimated by participants) is shown in Figure 1, as a function of temporal orientation (past vs. future) and temporal distance (near vs. distant). As can be seen, the majority of events were produced directly, not only for past events but also for future events. A 2

(temporal orientation) $\times$ 2 (temporal distance) repeated measures analysis of variance (ANOVA) showed a significant effect of temporal distance, $F(1, 31) = 5.08$, $p = .03$, $\eta_p^2 = 0.14$, indicating that the direct mode of production was more frequent when recalling or imagining events that refer to a temporally close time period. There was no significant effect of temporal orientation, $F(1, 31) = 2.59$, $p = .12$, $\eta_p^2 = 0.08$, and no interaction between temporal orientation and temporal distance, $F(1, 31) = 0.38$, $p = .54$, $\eta_p^2 = 0.01$.

Next, we examined whether RTs differed as a function of the mode of production of events. For each participant, we computed median RTs for direct and generative responses in the past and future event conditions (four participants were excluded from this analysis because they did not provide any generative response in the past or future event conditions). The means across participants are shown in Figure 2, as a function of the mode of production of events and their temporal orientation. A 2 (mode) $\times$ 2 (temporal orientation) ANOVA yielded a significant effect of mode, $F(1, 27) = 49.67$, $p < .001$, $\eta_p^2 = 0.65$, showing that RTs were faster for direct than generative responses. There was no significant effect of temporal orientation, $F(1, 27) = 2.72$, $p = .11$, $\eta_p^2 = 0.09$, but the interaction between temporal

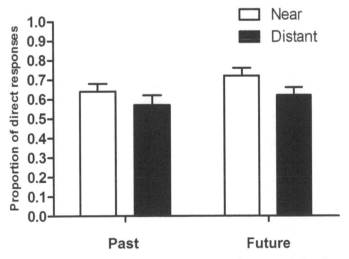

Figure 1. *Mean proportion of direct responses reported for near and distant past and future events in Experiment 1. Error bars represent the standard error of the mean.*

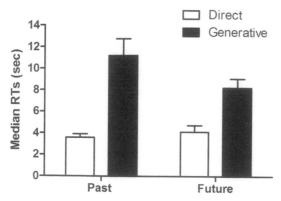

Figure 2. Median response times (RTs) as a function of the mode of production (direct vs. generative) and temporal orientation (past vs. future) of events in Experiment 1. Error bars represent the standard error of the mean.

orientation and mode was significant, $F(1, 27) = 6.03$, $p = .02$, $\eta_p^2 = 0.18$. Follow-up t-contrasts showed that there was no significant difference in RTs between past and future events when the events were produced directly, $t(27) = 0.95$, $p = .35$, $d = 0.18$, but future events were produced faster than past events in the generative mode, $t(27) = 2.21$, $p = .04$, $d = 0.42$.

This latter finding was somewhat unexpected, and we further investigated whether these faster RTs could be related to the extent to which participants had previously thought about the reported events.[1] We indeed found that, for events reported in the generative mode, the proportion of events that had been thought of previously was significantly higher for future events than for past events, $t(27) = 2.05$, $p = .049$, $d = 0.39$, such that the above-mentioned difference in RTs could in part be due to differences in previous thoughts about the events. Therefore, we examined whether RTs in the generative mode would differ between past and future events that had not been thought of previously. When looking specifically at this subset of events, there was no significant difference in RTs between past and future events, $t(18) = 1.24$, $p = .23$, $d = 0.29$, although numerically the difference was still in the same direction (means of median RTs were 12,573 ms, $SEM = 1720$ ms, and 9938 ms,

$SEM = 1231$ ms, for past and future events, respectively). This analysis should be taken with caution, however, because only 19 participants could be included (i.e., participants who reported at least one event that had not been thought of previously, both in the past and in the future conditions), and many of the data points (12 out of 38) were computed on the basis of only one event (e.g., because the participant reported only one past event that had not been thought of previously).

An important goal of this study was to determine whether the direct mode of production of events depended on having previously thought about these events, as well as on their personal importance and affective value. To examine these questions, we conducted a series of logistic regressions to investigate the effect of each variable on the odds of the direct mode of production of events. To account for the hierarchical structure of the data (i.e., events are nested within participants and are thus not independent), we used multilevel modelling (Goldstein, 2011) with events as Level 1 units and participants as Level 2 units. These analyses were performed using MLwiN and second-order penalized quasilikelihood as estimation method (Rasbash, Charlton, Browne, Healy, & Cameron, 2011).

For past events, we found that the percentage of direct responses was significantly higher for events that had been thought of previously (67%) than for events that had not been thought of previously (50%; coefficient = 0.73, $SE = 0.25$, $Z = 2.92$, $p = .003$). A similar trend was observed for future events, with the percentage of direct responses being, respectively, 71% and 60% for events that had been thought of and events that had not been thought of previously (coefficient = 0.52, $SE = 0.27$, $Z = 1.93$, $p = .054$). Furthermore, among past and future events that had been thought of previously, the odds of direct response increased with the frequency of previous thoughts about the events (past events: coefficient = 0.22, $SE = 0.11$, $Z = 2.00$, $p = .04$; future events: coefficient = 0.34, $SE = 0.10$, $Z = 3.45$, $p < .001$). Thus, for both the past and future, having previously thought about an event seems an important

---

[1]We thank an anonymous reviewer for calling our attention to this possibility.

determinant of direct production. In fact, the majority of events that were produced directly had been thought of on a previous occasion, with no significant difference between past and future events in this respect (69% of past events and 73% of future events that were produced directly had been thought of on a previous occasion; coefficient = 0.18, $SE = 0.23$, $Z = 0.78$, $p = .43$).

For both past and future events, we also found that the odds of direct response increased with the importance attributed to the events (past events: coefficient = 0.41, $SE = 0.08$, $Z = 5.13$, $p < .001$; future events: coefficient = 0.20, $SE = 0.08$, $Z = 2.40$, $p = .02$). With regard to affective valence, an initial inspection of the percentage of direct responses across levels of the 7-point rating scale showed a V-shaped relation between the ratings and the frequency of direct responses, suggesting that the occurrence of direct responses increased with the affective intensity of events. Therefore, we created a new variable reflecting the affective intensity of events by taking the absolute value of affective ratings and investigated whether this variable predicted the odds of direct response. For past events, the odds of direct response significantly increased with the affective intensity of events (coefficient = 0.42, $SE = 0.12$, $Z = 3.50$, $p < .001$); a similar trend was observed for future events, although the effect failed to reach statistical significance (coefficient = 0.22, $SE = 0.12$, $Z = 1.83$, $p = .066$). Finally, we found that for future events the odds of direct response significantly increased with the perceived probability that the imagined events would actually happen in the future (coefficient = 0.40, $SE = 0.10$, $Z = 4.08$, $p < .001$).

All the effects on the odds of direct response reported above remained unchanged when temporal distance was entered in the model (coded as a categorical variable: close vs. distant), and there was no interaction between the variable of interest and temporal distance, with the following exception: For future events, a significant interaction term indicated that the effect of importance on the odds of direct response was more pronounced for the distant future than for the near future (coefficient = 0.40, $SE = 0.17$, $Z = 2.35$, $p = .02$).

Discussion

The results of Experiment 1 replicated the findings of Uzer et al. (2012) that directly retrieved memories are common in the cue-word paradigm, with the percentage of directly retrieved memories being comparable between the two studies (i.e., around 60%). Also consistent with the results of Uzer et al., we found that directly retrieved memories were formed much faster than generated memories. Our main aim was then to investigate whether the direct production of episodic future thoughts is also frequent in the same word cueing conditions. We found that the direct production of event representations was as frequent and as fast for episodic future thoughts as it was for memories. This finding provides novel and more straightforward evidence that future event representations are often produced immediately and with no apparent search effort in the word cueing task.

Another important contribution of Experiment 1 is to shed light on the determinants of the direct mode of production of events. As expected, we found that most events that were produced directly had already been thought of on previous occasions. This was the case not only for past events but also for future events, suggesting that most episodic future thoughts that were produced directly were memories of the future rather than newly imagined future events. Besides previous thoughts, our results also showed that the frequency of direct access increased as a function of the personal significance of events (as reflected by ratings of importance and emotional value).

While the direct production of autobiographical representations was common whatever the contemplated time period, a significant effect of temporal distance was observed, showing that direct access was more frequent for the recent past and near future than for the distant past and future. This suggests that memories and future thoughts are more readily accessible when they are closely related to the present. Such highly accessible memories for recent past experiences and anticipated experiences might function to keep us tightly connected to our current goals and plans (Conway, 2009).

An unexpected finding was that, when event representations were not produced directly, people took more time for generating memories than future thoughts. This seems in contradiction with some previous studies that used a similar cueing paradigm and found that RTs either did not differ between past and future events or were faster for past than for future events (e.g., Anderson et al., 2012). However, these previous studies did not assess the modes of production of event representations, and they probably involved a mix of direct and generative processes, which renders a direct comparison with our data difficult. One possible explanation for the present finding is that this difference in RTs was, at least in part, due to previous thoughts about the events: Indeed, the proportion of events produced in the generative mode that had been thought of previously was significantly higher for future events than for past events, and, when only the subset of events that had not been thought of previously were analysed, the difference between past and future events in RTs was no longer statistically significant (although numerically, the difference was still in the same direction). Another, not necessarily mutually exclusive, explanation would capitalize on the idea that remembering the past is more constrained by reality concerns than is imagining the future (Van Boven, Kane, & McGraw, 2009): In the former case, one has to produce an event that actually happened (or at least that one believes actually happened), thus adding a search parameter that could slow down the generative process.

EXPERIMENT 2

The results of Experiment 1 show that the direct production of episodic future thoughts is common in the word cueing paradigm and is predicted by the frequency of previous thoughts about the events. In Experiment 2, we aimed to replicate these findings while excluding several contaminating factors that could have contributed to the high frequency of direct responses in Experiment 1.

First, in Experiment 1, the specificity of future event representations could not be meaningfully checked (because participants only provided a brief title for each event they imagined), and it could thus be the case that some event representations that were directly formed were in fact not specific. Therefore, in Experiment 2, we asked participants to describe imagined events in more detail, such that we could estimate the prevalence of direct access for trials involving the imagination of a specific event. Second, it could be that some events participants reported as directly produced were in fact not truly anticipated experiences: Another possibility would be that the cue word directly triggered the memory of a past event, which would then be simply recast as a future event. To investigate this "recasting hypothesis" (Addis, Pan, Vu, Laiser, & Schacter, 2009; Gamboz, Brandimonte, & De Vito, 2010), we asked participants to evaluate the similarity of each imagined future event to past experiences and to indicate whether they used past events to imagine this future event.

Method

Participants

Twenty undergraduate students (10 females and 10 males) aged between 18 and 30 years ($M = 21.3$ years, $SD = 2.8$ years) participated in this study. A power analysis using the MLPowSim Software Package (Browne, Golalizadeh, & Parker, 2009) indicated that this sample size at Level 2 (participants) and a sample size of 15 at Level 1 (events; see below) yielded an estimated power of above .95 to detect a significant effect (with an alpha of .05, two-tailed) of the size observed in Experiment 1 for the effect of frequency of previous thoughts on the odds of direct responses for future events. All participants provided written informed consent and were tested individually.

Materials and procedure

The procedure and materials were similar to those in Experiment 1, with the following modifications. First, only cue words referring to objects (e.g., *book*, *bottle*) or locations (e.g., *hotel*, *restaurant*) were used in this experiment; feeling words were removed because previous research suggests that they may involve distinct retrieval processes in the word cueing task (e.g., Conway et al., 2001). Second, participants were only instructed to imagine future events (i.e., there was no past event condition), and the number of trials was increased to 15. Third, there was no constraint regarding the temporal distance of imagined events. Fourth, after each event had been produced, participants were invited to describe the event on a sheet of paper, and they were asked to describe it with sufficient details so that someone could understand that it referred to a specific event. We used these instructions so that we could later check that each reported event was specific.

Immediately after the 15 trials, participants were instructed to come back to each future event they imagined, and, as in Experiment 1, they were invited to date each event, to indicate whether they previously thought about this event (by answering "yes" or "no"; if they responded "yes", they were also asked to rate the extent to which they previously thought about this event, from 1 = very rarely, to 7 = very often), and to rate its personal importance (from 1 = not at all important, to 7 = extremely important). Furthermore, in the present experiment, participants were also asked to rate the similarity of the imagined event to previously experienced events (from 1 = the exact same event was experienced previously, to 5 = the event is completely novel; Addis, Musicaro, Pan, & Schacter, 2010) and to indicate whether they thought about one or more past events during the imagination of the future event (by answering "yes" or "no"; if they responded "yes", they were also asked to indicate whether they used this event or these events for imagining the future event, by answering "yes or "no"). Finally, participants were asked to determine whether the imagined future event was linked to one or more future events produced on a previous trial (by answering "yes" or "no").

Scoring

All events were scored as specific or nonspecific by the first author. Events were considered specific if they referred to events happening on a particular occasion (i.e., in a specific place at a specific time) and lasting no longer than a day (Williams et al., 1996). A random selection of 20% of events was also independently scored by the second author. The coefficient of raw agreement was $ra = .95$; we did not assess the degree of rater agreement using Cohen's k because of the marginal dependency of k for extreme marginal distributions (see von Eye & von Eye, 2008; in the present case, the marginal distributions were not uniform, with the cell frequency corresponding to a rating of nonspecificity by both raters being 0).

Results

In total, the 20 participants reported 300 future events. However, 17 of these events (from 12 participants) did not refer to a specific happening; these events were excluded from the analyses, thus leaving 283 specific future events. As in Experiment 1, participants reported that the majority of future events they imagined came directly to mind (mean proportion = .63, $SEM = .03$), and a paired-sample t-test showed that RTs were significantly faster for direct than generative responses (means of median RTs were 4943 ms, $SEM = 571$ ms, and 17,352 ms, $SEM = 1912$ ms, for direct and generative responses, respectively), $t(19) = 8.19$, $p < .001$, $d = 1.83$.

In line with Experiment 1, we found that the direct mode of production of future events depended on having previously thought about these events: The percentage of direct responses was significantly higher for events that had been thought of previously (71%) than for events that had not been thought of previously (41%; coefficient = 1.34, $SE = 0.29$, $Z = 4.62$, $p < .001$); among future events that had been thought of previously, the odds of direct response increased with the frequency of previous thoughts about the events (coefficient = 0.41, $SE = 0.15$, $Z = 2.73$, $p = .006$). As in Experiment 1, the majority of future events that were accessed directly (84%) had been thought of on a previous occasion. We also replicated the finding of Experiment 1 that the odds of direct response increased with the importance attributed to the events (coefficient = 0.27, $SE = 0.08$, $Z = 3.23$, $p = .001$).

Next, we investigated whether the direct mode production of future events could be explained by the recasting hypothesis (i.e., participants may directly retrieve a past event and then simply recast it as a future event). As a first test of this hypothesis, we examined whether future events that were produced directly were more similar to previously experienced events than future events that were produced using generative processes. We found that similarity with past experiences did not significantly predict the odds of direct response (coefficient = -0.08, $SE = 0.08$, $Z = 1.06$, $p = .29$). Furthermore, reports of having thought about one or more past events during the imagination process were not significantly related to the production of direct responses (coefficient = -0.34, $SE = 0.25$, $Z = 1.33$, $p = .18$). Finally, reports of having used one or more past events for imagining the future event were also not significantly related to the production of direct responses (coefficient = 0.44, $SE = 0.40$, $Z = 1.10$, $p = .27$).

Discussion

The results of Experiment 2 replicated the findings of Experiment 1 that the direct production of episodic future thoughts is common and fast, while ensuring that the future events that were directly produced were specific and did not simply consist in remembered past events recast as future events. In line with Experiment 1, we also found that having previously thought about events was a significant predictor of direct access and that the large majority of future events that were directly produced had been thought of on at least one previous occasion. Finally, the personal importance of events was also a significant predictor of direct access, again replicating the results of Experiment 1.

EXPERIMENT 3

The results of Experiments 1 and 2 converged to show that when they attempted to produce future events in response to cue words, participants frequently reported that a future event representation was directly triggered by the cue. In Experiment 3, we sought to delve deeper into the nature of the direct formation of episodic future thoughts by distinguishing between two dimensions that could characterize direct responses: search effort and information use (Uzer et al., 2012). Experiments 1 and 2 characterized the distinction between direct and generative responses in terms of search effort: Directly produced memories and future thoughts were defined as being formed with no search effort, whereas generated memories and future thoughts were defined as involving an active search process. Another important feature of generative retrieval is the use of general information about one's life (e.g., people we know, activities we engage in, or places we frequent) that can serve as cues for retrieving a specific event (Conway & Pleydell-Pearce, 2000).

Using this definition of generative retrieval in terms of information use during retrieval, Uzer et al. (2012, Experiment 3) found that direct responses were still common, occurring in around 50% of trials. It remains unknown whether this is also the case for episodic future thoughts. In fact, a previous study that used a think-aloud method to assess the kinds of information that were accessed when producing episodic future thoughts found that people most often accessed general information about their life before a specific event was produced (D'Argembeau & Mathy, 2011). It thus remains possible that the production of episodic future thoughts is frequently direct in the sense that it requires no or little search effort (as shown by Experiments 1 and 2), but may still involve the activation of general information about one's life (as shown by D'Argembeau & Mathy, 2011). For instance, a cue could automatically (i.e., with no search effort) trigger knowledge about a familiar person or location, which then could automatically trigger the representation of an associated future event. To investigate this possibility in Experiment 3, we used two different questions to assess direct production not only in terms of search effort, but also in terms of the type of information accessed. In addition, we used a think-aloud method, as in D'Argembeau and Mathy (2011) and Uzer et al. (2012), to independently assess the activation of general information during the production of events.

Method

Participants

Twenty undergraduate students (11 females and 9 males) aged between 18 and 29 years ($M = 23$ years, $SD = 2.64$ years) participated in this study. In addition to determining the proportion of direct responses as defined by search effort and information use, we also sought to investigate whether RTs differed as a function of these two dimensions of event production. A power analysis using G*Power 3 (Faul, Erdfelder, Lang, & Buchner, 2007) indicated that a sample size of 20 participants yielded a power of above .95 to detect a significant effect (with an alpha of .05, two-tailed) of the size observed in Experiment 1 for the influence of the mode of production of events on RTs. All participants provided written informed consent and were tested individually.

Materials and procedure

Twenty cue words referring to common objects and locations were divided into two lists of 10 cues that were matched for word length (number of letters), imageability, and frequency of use (Desrochers & Bergeron, 2000). Participants were instructed to retrieve 10 past events and to imagine 10 future events in response to these cues, with no constraint in terms of temporal distance. While recalling or imagining an event, participants were instructed to say aloud everything that came into their mind from the time they read the cue word to the time they had a specific event in mind. They were asked to report every thought or image they experienced, even if it had apparently nothing to do with the cue word. A digital audio recorder was used to record all vocalizations during this phase. As soon as an appropriate event came to mind, participants were instructed to press the spacebar, which recorded RT. Contrary to Experiments 1 and 2, we did not impose a time limit for producing an event because we expected that the recall or

imagination process would be slowed down by the think-aloud procedure (Fox, Ericsson, & Best, 2011).

We sought to distinguish between two dimensions that may characterize the direct mode of production of autobiographical events: the type of information accessed versus the effort deployed for producing an event. These two dimensions were assessed using two different questions. Once they pressed the space bar, participants were first asked to assess whether the event came immediately to mind following the presentation of the cue word or whether some general information about their lives came to their mind before they accessed a specific event. Before the experiment, it was explained that in some cases, a specific event can directly come to mind following the presentation of a cue, while in other cases, a specific event may not come immediately to mind, and other general information about one's life, such as information about an extended period of life (e.g., when I was child) or event categories (e.g., football games), is activated before a specific event is produced. Participants were instructed to press either the 1 key to indicate that a specific event immediately came to mind or the 2 key to indicate that some general information about their lives came to mind before they accessed a specific event. Next, participants were asked to evaluate whether or not the retrieval of the past event or the imagination of the future event required search effort. Before the experiment, it was explained that in some cases, a memory or an imagined event can come to mind automatically, effortlessly, while in other cases, an active search effort is required. Participants were instructed to press either the 1 key to indicate that the event was retrieved or imagined with no search effort, or the 2 key to indicate that the event was retrieved or imagined after an active search effort. Before starting the experiment, the experimenter made sure that the two dimensions were correctly understood, and it was further specified that they represent potentially orthogonal dimensions (e.g., an event could be imagined using some general information about one's life, yet without the need to engage in an active search to access such information).

Scoring

Event specificity was scored using the same method as that used in Experiment 2. The coefficient of raw agreement was $ra = .94$; again we did use Cohen's k to express the degree of rater agreement because the marginal distributions were not uniform (von Eye & von Eye, 2008).

Results

In total, the 20 participants reported 200 past events and 200 future events. However, eight past events and 15 future events were not specific. These nonspecific events were discarded, and thus the reported analyses were based on 192 past events and 185 future events. Although the two dimensions used to define direct access (i.e., search effort and information use) were related, they were not entirely equivalent (26% of events were classified as direct for one dimension and non-direct for the other dimension); thus we analysed each dimension separately.

First, we investigated the frequency of direct access as defined by search effort. The mean proportion of events that were produced with no apparent search effort is shown in Figure 3a, as a function of temporal orientation (past vs. future). In line with Experiments 1 and 2, the majority of events were produced without search effort, and there was no significant differences between past and future events in this respect, $t(19) = 0.90$, $p = .38$, $d = 0.20$. Next, we examined the frequency of direct access as defined by information use. The

mean proportion of self-reports of events that immediately came to mind in response to the cue (i.e., without accessing more general information first) is shown in Figure 3b, as a function of temporal orientation. As can be seen, participants judged that the majority of events came immediately to mind, not only for past events but also for future events; a paired sample t-test showed no significant difference between past and future events in terms of immediacy of access, $t(19) = 1.04$, $p = .31$, $d = 0.23$.

We also examined whether the verbal protocols obtained while participants produced the events were consistent with their answers to the information use question. The first author listened to each verbal report and judged whether or not some general information was reported before a specific event was described, while blind to how participants had answered the information use question. We then compared the degree of concordance between the coder's judgements and the participants' answers to the information use question. The strength of agreement was good ($k = .74$, 95% confidence interval, CI: [.67, .81]), thus providing support for the reliability of participants' judgements of information use.

Next, we examined whether RTs differed as a function of the two dimensions of event production that were investigated here. First, we computed, for each participant, the median RT for past and future events, as a function of whether or not the production of the event required search effort. Means across participants are shown in Figure 4a. A 2

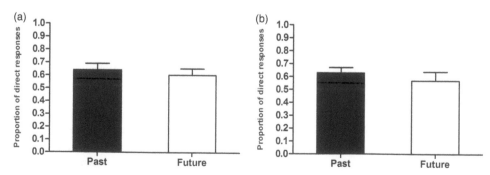

Figure 3. *Mean proportion of past and future events that (a) were produced with no search effort and (b) came immediately to mind (without accessing general information) in Experiment 3. Error bars represent the standard error of the mean.*

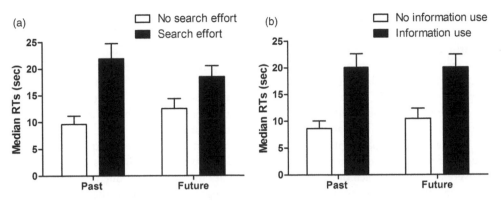

Figure 4. *Median response times (RTs) for past and future events as a function of (a) search effort and (b) information use in Experiment 3. Error bars represent the standard error of the mean.*

(search effort) × 2 (temporal orientation) ANOVA revealed a main effect of search effort, $F(1, 19) = 23.87$, $p < .001$, $\eta_p^2 = 0.56$, indicating that participants took on average more time to produce a specific event when it required a search effort than when it required no search effort. There was no main effect of temporal orientation, $F(1, 19) = 0.08$, $p = .78$, $\eta_p^2 = 0.004$, but there was a significant interaction between temporal orientation and search effort, $F(1, 19) = 5.14$, $p = .04$, $\eta_p^2 = 0.21$. This interaction indicated that RTs tended to be faster for past events than for future events when the events were produced with no search effort, $t(19) = -1.97$, $p = .06$, $d = 0.44$, whereas the difference tended to be in the opposite direction when events were produced with search effort, $t(19) = 1.81$, $p = .09$, $d = 0.40$.

Finally, we computed, for each participant, the median RT for past and future events, as a function of whether or not the event immediately came to mind. Four participants were excluded from this analysis because one cell was empty (e.g., they did not produce any future event without accessing more general information first). Means across participants are shown in Figure 4b. A 2 (mode of access) × 2 (temporal orientation) ANOVA showed a significant main effect of mode, $F(1, 15) = 34.56$, $p < .001$, $\eta_p^2 = 0.69$, showing that RTs were faster when the events came immediately to mind. There was no significant effect of temporal orientation $F(1, 15) = 0.56$, $p = .46$,

$\eta_p^2 = 0.03$, and no interaction between temporal orientation and mode, $F(1, 15) = 0.87$, $p = .36$, $\eta_p^2 = 0.05$.

Discussion

In Experiment 3, we sought to delve deeper into the nature of the direct production of memories and future thoughts by distinguishing between two dimensions: search effort and information use. We found that participants reported not only that memories and future thoughts were frequently produced with no search effort (in line with Experiments 1 and 2), but also that they came immediately to mind (i.e., without accessing more general information first). Moreover, for both dimensions characterizing direct production, we found that RTs were faster for direct than for generative responses. We also asked participants to report everything that went into their minds while producing the events, and the analysis of their verbal protocols showed that the information described was in good agreement with participants' own judgements of information use.

In the present experiment, the percentage of episodic future thoughts that were produced without accessing general information first (57%) was much higher than the percentage we observed in a previous study (16%) in which direct and generative modes of formation were assessed using a think-aloud method (D'Argembeau & Mathy,

2011, Study 1). A key difference between the two studies is that D'Argembeau and Mathy (2011) required participants to imagine novel events (i.e., events that had not been previously experienced or thought of), whereas in the current experiment the novelty of events was left unspecified. As shown by the results of Experiments 1 and 2, the large majority of future events that were directly produced in the current paradigm had already been thought of on at least one previous occasion (and thus were not novel), which probably explains why the frequency of direct access is substantially higher than in D'Argembeau and Mathy (this point is discussed further in the General Discussion).

GENERAL DISCUSSION

The present research aimed to unravel the modes of production of episodic future thoughts in the word cueing paradigm. Across three experiments, we found that future event representations were produced through the same dual mechanisms, direct and generative, as autobiographical memories (Uzer et al., 2012). On the majority (i.e., around 60%) of trials, participants reported that a future event directly came to mind in response to the cue, and the prevalence of such direct production of episodic future thoughts was comparable to the prevalence of directly retrieved memories for past events; RT data confirmed that event representations were produced much faster for direct than for generative responses, both for the past and for the future. These results were observed not only when direct access was conceptualized in terms of search effort, but also in terms of information use (Experiment 3). When looking at the determinants of direct responses, we found that most past and future events that were directly produced had already been thought of on a previous occasion, and the frequency of previous thoughts predicted the occurrence of direct access (Experiments 1 and 2). Importantly, however, the future thoughts that were directly produced did not simply consist in remembered past events recast as future events (Experiment 2) and could more

appropriately be conceptualized as "memories of the future"—that is, future events that had been envisioned on a previous occasion (Ingvar, 1985; Szpunar et al., 2013). The personal relevance of events was also a significant predictor of the mode of production of autobiographical thoughts, with direct access being more frequent for past and future events that were important and emotionally intense (Experiments 1 and 2). Collectively, these findings provide novel evidence that the direct production of episodic future thoughts is frequent in the word cueing paradigm and often involves the activation of personally significant memories of the future.

Recent theoretical and empirical work on episodic future thinking has emphasized the role of constructive and generative processes in the production of future event representations (e.g., D'Argembeau & Mathy, 2011; Schacter et al., 2012; Suddendorf & Corballis, 2007). While not downplaying the importance of such processes, the present research shows that episodic future thoughts are not necessarily effortfully generated, but instead can come to mind rapidly, with no search effort and information manipulation, in response to a cue (see also Anderson et al., 2012; Berntsen & Jacobsen, 2008; D'Argembeau & Mathy, 2011; Finnbogadottir & Berntsen, 2013, for related observations). Most importantly, however, our data suggest that the direct production of episodic future thoughts mainly occur for events that have already been contemplated on a previous occasion, rather than novel events (i.e., events that have not been previously imagined or experienced). Thus, the mechanisms involved in the formation of episodic future thoughts seem to critically depend on the types of future events that are produced: Effortful constructive processes (e.g., the extraction and flexible recombination of details from past experiences) may mainly be required for simulating *novel* future events (Schacter et al., 2008, 2012), and indeed our previous work suggests that for newly imagined events, the use of generative processes is the dominant mode of production of episodic future thoughts (D'Argembeau & Mathy, 2011). On the other hand, as shown by the present research, when the constraint to imagine a novel event is removed, episodic future thoughts frequently consist in prestored representations of previously imagined events that are accessed directly.

While it is true that most of the episodic future thoughts that were produced directly referred to previously imagined events, it is worth noting that some future thoughts were formed directly, and yet participants reported that they had not previously thought about the corresponding events; this occurred for 27% and 16% of directly produced future thoughts in Experiments 1 and 2, respectively. These figures align with the previous observation that when participants were explicitly instructed to imagine novel future events, 16% of episodic future thoughts appeared to be formed directly (D'Argembeau & Mathy, 2011, Study 1). Thus, while the notion of memories of the future can account for most instances of direct production of episodic future thoughts, some thoughts do not easily conform to this explanation. This is an intriguing finding, and the exact nature of the mechanisms allowing the direct formation of novel episodic future thoughts should be further investigated in future studies. Perhaps some personal goals or concerns are so salient in a person's mind that anticipations and plans related to these goals are formed rapidly and automatically (Klinger, 2013). The finding that providing cues referring to personal goals increases the frequency of direct formation of episodic future thoughts (D'Argembeau & Mathy, Study 3) might be taken as supporting this idea. Another possibility would be that all events that are produced directly have in fact been constructed on a previous occasion, but in some cases the individual does not remember or was not aware of the previous act of imagination. For example, future thoughts are frequent during mind-wandering episodes (e.g., Stawarczyk, Cassol, & D'Argembeau, 2013) but people do not necessarily take explicit note of these thoughts (Schooler et al., 2011), such that they could erroneously be considered novel when subsequently reactivated.

Be it as it may, from a methodological point of view, the present results highlight the importance of instructions in determining the processes by which episodic future thoughts are produced. In some previous studies of episodic future thinking, participants were explicitly instructed to produce novel future events (e.g., Addis, Wong, & Schacter, 2007), whereas novelty was left unspecified in other studies (e.g., D'Argembeau & Van der Linden, 2004). Taken together, the present research and our previous work (D'Argembeau & Mathy, 2011) suggest that these two situations differ in the relative contribution of direct and generative processes to the production of future event representations: Generative processes are dominant when event novelty is emphasized, whereas direct access is more frequent when event novelty is left unspecified. This may be an important point to consider when interpreting deficits in the production of episodic future thoughts that are observed in various clinical populations (e.g., Addis, Sacchetti, Ally, Budson, & Schacter, 2009; D'Argembeau, Raffard, & Van der Linden, 2008; Williams et al., 1996).

It is also worth mentioning that, in the present experiments, participants were asked to bring specific future events to mind but did not have to construct detailed mental simulations of these events (e.g., by visualizing the location, persons, objects, and actions involved). Referring to the proposed distinction between construction and elaboration phases of episodic future thought (Addis et al., 2007), our results mainly pertain to the former phase (i.e., bringing a specific event to mind), and additional work should therefore be conducted to determine whether and to what extent detailed mental simulations of future events can also be formed directly, with no search effort.

The implications of the present findings for hierarchical models of autobiographical memory and future thinking warrant further discussion. According to such models (see e.g., Conway, 2009; Conway & Pleydell-Pearce, 2000; D'Argembeau, 2015), different types of knowledge structures varying in levels of abstraction contribute to past and future thoughts, with knowledge structures at higher levels of abstraction (e.g., broad goals, lifetime periods, and general events) providing information that contextualizes and locates specific event representations. What the present results show is that the production of specific memories and future thoughts does not necessarily involve a top-down search through such a hierarchical knowledge base—in which more abstract representations would be used for constructing specific event representations (Conway & Pleydell-Pearce, 2000). Top-down search processes may be involved in the formation of novel or infrequently considered event representations, but for events that are more frequently part of one's mental landscape, the present data suggest the existence of prestored representations that can be directly accessed in response to relevant cues. This does not imply, however, that these event representations are not part of a hierarchical autobiographical knowledge base. In other words, while the present findings argue against the necessity of a hierarchical *search process* in the production of specific event representations, they are neutral with respect to the assumed hierarchical *organization* of autobiographical knowledge. In fact, many past and future event representations are structured in higher level clusters that organize sequences of causally or thematically related events (Brown, 2005; Brown & Schopflocher, 1998; D'Argembeau & Demblon, 2012; Demblon & D'Argembeau, 2014), and such higher order organization could actually promote the direct production of autobiographical thoughts (i.e., the activation of one event within a cluster may tend to automatically trigger other related events). In line with this view, it has been found that pairs of associated events are produced faster when the events are part of the same cluster (Brown & Schopflocher, 1998).

Finally, we note that the direct and rapid access to previously imagined events may be an important factor in the adaptive value of prospection. Episodic future thought allows the anticipation and simulation of potential goal-relevant events and actions (i.e., events and actions that are conductive or obstructive to reaching personal goals), which in turn can inform decisions and plans and, ultimately, guide behaviour (D'Argembeau & Mathy, 2011; Schacter, 2012; Suddendorf & Corballis, 2007; Taylor, Pham, Rivkin, & Armor, 1998). Successful goal pursuit may depend, in part, on

the ability to remember the content of anticipated events, plans, and outcomes (Ingvar, 1985; Szpunar et al., 2013), and the rapid access to such representations may help guide behaviour more efficiently and effectively. In other words, successful goal pursuit may benefit from mechanisms that make goal-related future thoughts highly accessible, such that they can be automatically triggered when a relevant (internal or external) cue is encountered. The present research provides preliminary support for the existence of such mechanisms, by showing that episodic future thoughts that are considered more important (thus presumably involving goal-related contents) are more frequently formed in a fast and direct way.

REFERENCES

Addis, D. R., Musicaro, R., Pan, L., & Schacter, D. L. (2010). Episodic simulation of past and future events in older adults: Evidence from an experimental recombination task. *Psychology and Aging, 25*, 369–376.

Addis, D. R., Pan, L., Vu, M. A., Laiser, N., & Schacter, D. L. (2009). Constructive episodic simulation of the future and the past: Distinct subsystems of a core brain network mediate imagining and remembering. *Neuropsychologia, 47*, 2222–2238.

Addis, D. R., Sacchetti, D. C., Ally, B. A., Budson, A. E., & Schacter, D. L. (2009). Episodic simulation of future events is impaired in mild Alzheimer's disease. *Neuropsychologia, 47*, 2660–2671.

Addis, D. R., Wong, A. T., & Schacter, D. L. (2007). Remembering the past and imagining the future: Common and distinct neural substrates during event construction and elaboration. *Neuropsychologia, 45*, 1363–1377.

Anderson, R. J., Dewhurst, S. A., & Nash, R. A. (2012). Shared cognitive processes underlying past and future thinking: The impact of imagery and concurrent task demands on event specificity. *Journal of Experimental Psychology: Learning, Memory, and Cognition, 38*, 356–365.

Berntsen, D. (1996). Involuntary autobiographical memories. *Applied Cognitive Psychology, 10*, 435–454.

Berntsen, D., & Jacobsen, A. S. (2008). Involuntary (spontaneous) mental time travel into the past and future. *Consciousness and Cognition, 17*, 1093–1104.

Boyer, P. (2008). Evolutionary economics of mental time travel?. *Trends in Cognitive Sciences, 12*, 219–224.

Brown, N. R. (2005). On the prevalence of event clusters in autobiographical memory. *Social Cognition, 23*, 35–69.

Brown, N. R., & Schopflocher, D. (1998). Event clusters: An organization of personal events in autobiographical memory. *Psychological Science, 9*, 470–475.

Browne, W. J., Golalizadeh, M., & Parker, R. M. A. (2009). *A guide to sample size calculations for random effect models via simulation and the MLPowSim software package.* University of Bristol: Centre for Multilevel Modelling.

Burgess, P. W., & Shallice, T. (1996). Confabulation and the control of recollection. *Memory, 4*, 359–411.

Conway, M. A. (2005). Memory and the self. *Journal of Memory and Language, 53*, 594–628.

Conway, M. A. (2009). Episodic memories. *Neuropsychologia, 47*, 2305–2313.

Conway, M. A., & Pleydell-Pearce, C. W. (2000). The construction of autobiographical memories in the self-memory system. *Psychological Review, 107*, 261–288.

Conway, M. A., Pleydell-Pearce, C. W., & Whitecross, S. E. (2001). The neuroanatomy of autobiographical memory: A slow cortical potential study of autobiographical memory retrieval. *Journal of Memory and Language, 45*, 493–524.

D'Argembeau, A. (2012). Autobiographical memory and future thinking. In D. Berntsen & D. C. Rubin (Eds.), *Understanding autobiographical memory: Theories and approaches* (pp. 311–330). New York: Cambridge University Press.

D'Argembeau, A. (2015). Knolwedge structures involved in episodic future thinking. In A. Feeney & V. A. Thompson (Eds.), *Reasoning as memory* (pp. 128–145). Psychology Press.

D'Argembeau, A., & Demblon, J. (2012). On the representational systems underlying prospection: Evidence from the event-cueing paradigm. *Cognition, 125*, 160–167.

D'Argembeau, A., & Mathy, A. (2011). Tracking the construction of episodic future thoughts. *Journal of Experimental Psychology: General, 140*, 258–271.

D'Argembeau, A., Ortoleva, C., Jumentier, S., & Van der Linden, M. (2010). Component processes underlying future thinking. *Memory & Cognition, 38*, 809–819.

D'Argembeau, A., Raffard, S., & Van der Linden, M. (2008). Remembering the past and imagining the future in schizophrenia. *Journal of Abnormal Psychology, 117*, 247–251.

D'Argembeau, A., Renaud, O., & Van der Linden, M. (2011). Frequency, characteristics, and functions of future-oriented thoughts in daily life. *Applied Cognitive Psychology, 25*, 96–103.

D'Argembeau, A., & Van der Linden, M. (2004). Phenomenal characteristics associated with projecting oneself back into the past and forward into the future: Influence of valence and temporal distance. *Consciousness and Cognition, 13*(4), 844–858.

Demblon, J., & D'Argembeau, A. (2014). The organization of prospective thinking: Evidence of event clusters in freely generated future thoughts. *Consciousness and Cognition, 24*, 75–83.

Desrochers, A., & Bergeron, M. (2000). Valeurs de fréquence subjective et d'imagerie pour un échantillon de 1916 substantifs de la langue française. *Revue Canadienne de Psychologie Expérimentale, 54*, 274–325.

Faul, F., Erdfelder, E., Lang, A.-G. & Buchner, A. (2007). G*Power 3: A flexible statistical power analysis program for the social, behavioral, and biomedical sciences. *Behavior Research Methods, 39*, 175–191.

Finnbogadottir, H., & Berntsen, D. (2013). Involuntary future projections are as frequent as involuntary memories, but more positive. *Consciousness and Cognition, 22*, 272–280.

Fox, M. C., Ericsson, K. A., & Best, R. (2011). Do procedures for verbal reporting of thinking have to be reactive?. A meta-analysis and recommendations for best reporting methods. *Psychological Bulletin, 137*, 316–344.

Gamboz, N., Brandimonte, M. A., & De Vito, S. (2010). The role of past in the simulation of autobiographical future episodes. *Experimental Psychology, 57*, 419–428.

Goldstein, H. (2011). *Multilevel statistical models* (4th ed.). Chichester, UK: Wiley.

Haque, S., & Conway, M. A. (2001). Sampling the process of autobiographical memory construction. *European Journal of Cognitive Psychology, 13*, 529–547.

Hill, P. F., & Emery, L. J. (2013). Episodic future thought: Contributions from working memory. *Consciousness and Cognition, 22*, 677–683.

Ingvar, D. H. (1985). "Memory of the future": An essay on the temporal organization of conscious awareness. *Human Neurobiology, 4*, 127–136.

Irish, M., Addis, D. R., Hodges, J. R., & Piguet, O. (2012). Considering the role of semantic memory in episodic future thinking: Evidence from semantic dementia. *Brain, 135*, 2178–2191.

Klein, S. B. (2013). The temporal orientation of memory: It's time for a change of direction. *Journal of Applied Research in Memory and Cognition, 2,* 222–234.

Klinger, E. (2013). Goal Commitments and the content of thoughts and dreams: Basic principles. *Frontiers in Psychology, 4,* 415.

Newby-Clark, I. R., & Ross, M. (2003). Conceiving the past and future. *Personality and Social Psychology Bulletin, 29,* 807–818.

Rasbash, J., Charlton, C., Browne, W. J., Healy, M., & Cameron, B. (2011). *MLwiN Version 2.24.* University of Bristol: Centre for Multilevel Modelling.

Rasmussen, A. S., & Berntsen, D. (2011). The unpredictable past: Spontaneous autobiographical memories outnumber autobiographical memories retrieved strategically. *Consciousness and Cognition, 20,* 1842–1846.

Schacter, D. L. (2012). Adaptive constructive processes and the future of memory. *American Psychologist, 67,* 603–613.

Schacter, D. L., & Addis, D. R. (2007). The cognitive neuroscience of constructive memory: Remembering the past and imagining the future. *Philosophical Transactions of the Royal Society B Biological Sciences, 362,* 773–786.

Schacter, D. L., Addis, D. R., & Buckner, R. L. (2008). Episodic simulation of future events: Concepts, data, and applications. *Annals of the New York Academy of Sciences, 1124,* 39–60.

Schacter, D. L., Addis, D. R., Hassabis, D., Martin, V. C., Spreng, R. N., & Szpunar, K. K. (2012). The future of memory: Remembering, imagining, and the brain. *Neuron, 76,* 677–694.

Schooler, J. W., Smallwood, J., Christoff, K., Handy, T. C., Reichle, E. D., & Sayette, M. A. (2011). Meta-awareness, perceptual decoupling and the wandering mind. *Trends in Cognitive Sciences, 15,* 319–326.

Spreng, R. N., & Levine, B. (2006). The temporal distribution of past and future autobiographical events across the lifespan. *Memory & Cognition, 34,* 1644–1651.

Stawarczyk, D., Cassol, H., & D'Argembeau, A. (2013). Phenomenology of future-oriented mind-wandering episodes. *Frontiers in Psychology, 4.* doi:10.3389/fpsyg.2013.00425.

Suddendorf, T., & Corballis, M. C. (2007). The evolution of foresight: What is mental time travel and is it unique to humans?. *Behavioral and Brain Sciences, 30,* 299–351.

Szpunar, K. K. (2010). Episodic future thought: An emerging concept. *Perspectives on Psychological Science, 5,* 142–162.

Szpunar, K. K., Addis, D. R., McLelland, V. C., & Schacter, D. L. (2013). Memories of the future: New insights into the adaptive value of episodic memory. *Frontiers in Behavioral Neuroscience, 7,* 47.

Taylor, S. E., Pham, L. B., Rivkin, I. D., & Armor, D. A. (1998). Harnessing the imagination: Mental simulation, self-regulation, and coping. *American Psychologist, 53,* 429–439.

Thompson, C. P., Skowronski, J. J., Larsen, S. F., & Betz, A. L. (1996). *Autobiographical memory: Remembering waht and remembering when.* Mahwah, NJ: Lawrence Erlbaum.

Uzer, T., Lee, P. J., & Brown, N. R. (2012). On the prevalence of directly retrieved autobiographical memories. *Journal of Experimental Psychology: Learning, Memory, and Cognition, 38,* 1296–1308.

Van Boven, L., Kane, J. M., & McGraw, A. P. (2009). Temporally asymmetric constraints on mental simulation: Retrospection is more constrained than prospection. In K. D. Markman, W. M. P. Klein, & J. A. Suhr (Eds.), *Handbook of imagination and mental simulation* (pp. 131–147). New York: Psychology Press.

von Eye, A., & von Eye, M. (2008). On the marginal dependency of Cohen's k. *European Psychologist, 13,* 305–315.

Watkins, E. R. (2008). Constructive and unconstructive repetitive thought. *Psychological Bulletin, 134,* 163–206.

Williams, J. M. G., Ellis, N. C., Tyers, C., Healy, H., Rose, G., & MacLeod, A. K. (1996). The specificity of autobiographical memory and imageability of the future. *Memory & Cognition, 24,* 116–125.

Williams, J. M. G., Healy, H. G., & Ellis, N. C. (1999). The effect of imageability and predicability of cues in autobiographical memory. *The Quaterly Journal of Experimental Psychology, 52A,* 555–579.

Do future thoughts reflect personal goals? Current concerns and mental time travel into the past and future

Scott N. Cole and Dorthe Berntsen

Department of Psychology and Behavioral Sciences, Center on Autobiographical Memory Research (CON AMORE), Aarhus University, Aarhus, Denmark

Our overriding hypothesis was that future thinking would be linked with goals to a greater extent than memories; conceptualizing goals as current concerns (i.e., uncompleted personal goals). We also hypothesized that current-concern-related events would differ from non-current-concern-related events on a set of phenomenological characteristics. We report novel data from a study examining involuntary and voluntary mental time travel using an adapted laboratory paradigm. Specifically, after autobiographical memories or future thoughts were elicited (between participants) in an involuntary and voluntary retrieval mode (within participants), participants self-generated five current concerns and decided whether each event was relevant or not to their current concerns. Consistent with our hypothesis, compared with memories, a larger percentage of involuntary and voluntary future thoughts reflected current concerns. Furthermore, events related to current concerns differed from non-concern-related events on a range of cognitive, representational, and affective phenomenological measures. These effects were consistent across temporal direction. In general, our results agree with the proposition that involuntary and voluntary future thinking is important for goal-directed cognition and behaviour.

Recent theoretical and empirical work indicates that autobiographical memories often have relevance to our current goals (e.g., Conway, 2005, for a review; Johannessen & Berntsen, 2009, 2010). In addition to recollecting their past, humans can vividly imagine possible self-referential future scenarios (i.e., episodic future thinking; Atance & O'Neill, 2001). Reexperiencing the past and preexperiencing the future are inherently linked, cognitively and neurologically, and can be subsumed under one capacity: *mental time travel* (MTT; Wheeler, Stuss, & Tulving, 1997). Importantly, there is a growing focus on how *future* MTT might underlie or influence important

The authors would like to thank Søren Staugaard, Daniella Villadsen, Thorbjørn Larsen, and Marie Kirk for assisting in conducting the study. We also extend our gratitude to Lia Kvavilashvili who kindly provided the materials for the adapted paradigm. The authors declare no conflict of interest. These results were presented at the inaugural International Convention of Psychological Science in Amsterdam (March, 2015) as part of a symposium on the "Functions of Future Thoughts".

This work was supported by the Danish National Research Foundation [grant number DNRF93].

goal-oriented human functions, such as intention, planning, decision making, and goal attainment (e.g., Klein, 2013; Schacter, 2012; Seligman, Railton, Baumeister, & Sripada, 2013; Suddendorf & Corballis, 2007; Szpunar & Jing, 2013), which may garner beneficial outcomes for one's future. Other authors argue that *spontaneous* thoughts also garner benefits toward future goals (Antrobus, Singer, & Greenberg, 1966; Baird, Smallwood, & Schooler, 2011). This study examines the relation between personal goals and past and future MTT with the expectation that future thought will reflect current goals to a greater extent than memories and that goal-related past and future MTT will differ from their non-goal-related counterparts in similar ways. We here use the term past and future MTT as a reference to remembering past events and imagining possible future events, whilst acknowledging that these vary with regard to their spatiotemporal specificity (Anderson & Dewhurst, 2009; Berntsen & Jacobsen, 2008; Klein, 2013).

Johannessen and Berntsen (2010) examined the relation between goals and involuntary (spontaneously arising) versus voluntary (strategically retrieved) autobiographical memories by utilizing the concept of current concerns (Klinger, 1975). Current concerns refer to personal goals that have a specific onset (goal commitment) and offset (goal achievement or disengagement) and which have observable effects upon thoughts, perception, and behaviour until they are completed or discarded (Klinger, 1975). These generally refer to higher order goals (e.g., obtaining one's preferred job, having children)—which remain sensitive to current-concern-related cues (Klinger, 1975)—rather than drives (e.g., hunger) or specific action plans (e.g., implementation intentions; see Gollwitzer, 1993). In their diary study, Johannessen and Berntsen (2010) found that approximately half of the recorded autobiographical memories (both involuntary and voluntary) were judged by participants to be related to one or more of their current concerns. They also found that memories related to current concerns were more rehearsed, important for self-identity and life story, and closer to the present, supporting

the idea that goal-related memories may have a cognitive and representational status akin to "self-defining memories" (Singer & Salovey, 1993). Interestingly, Johannessen and Berntsen (2010) also found that participants judged that current-concern-related *memories* would have a greater effect upon their *future* life. However, in their study, Johannessen and Berntsen (2010) did not pursue the role of current concerns for future thinking.

In the present paper, we similarly operationalize goals with the construct of current concerns (Klinger, 1975). Our main aim was to examine whether current concerns would be represented more frequently in future MTT than past MTT, thus filling a critical gap in the existing literature on MTT and its relation to current concerns. Another important aim was to examine whether current-concern-related events would be distinguishable from non-current-concern-related events on key phenomenological variables, and whether this pattern of differences would be similar for both future and past events.

Consistent with Johannessen and Berntsen (2010), we examine both involuntary and voluntary MTT, and, consistent with their findings, we expect concern-related events to be equally frequent for the involuntary and voluntary conditions. However, our study differs from previous work by including a future condition and by using an adapted laboratory paradigm (Cole, Staugaard, & Berntsen, 2014; Schlagman & Kvavilashvili, 2008) to elicit the events. In this paradigm, the participants are asked to conduct a vigilance task, while reporting memories and future thoughts that potentially may arise spontaneously in response to subtly presented word phrase cues. The external validity of this paradigm was verified by finding consistent results with diary studies (e.g., Berntsen & Jacobsen, 2008; Cole et al., 2014). Involuntary future projections are a particularly important phenomenon to investigate in the laboratory as, although we know that they are experienced in daily life (Berntsen & Jacobsen), very little is known about how and why they occur. Following Johannessen and Berntsen's (2010) findings for past events, we predicted that concern-related future thoughts would be more important to self-identity and life story, more rehearsed, and be dated more closely to the present than events judged to be unrelated to current concerns.

Summary and hypotheses of the present study

We examined both involuntary and voluntary MTT, including future MTT. We used a recently validated laboratory paradigm to elicit involuntary and voluntary MTT. This method was originally developed to measure involuntary autobiographical memories in a laboratory setting (Schlagman & Kvavilashvili, 2008), but has recently been adapted to also measure involuntary future thinking (Cole et al., 2014; see Method). Considering the unique contribution of future MTT to goal-related cognition and behaviour, we predicted that, compared with memories, a larger proportion of future thoughts would be current-concern related. Based on aforementioned theoretical and empirical work assigning voluntary and involuntary future thoughts a role in representing personal goals, as with memories, we did not predict that one retrieval mode would be especially important for goal representation (Johannessen & Berntsen, 2010). Also, we did not expect our findings regarding involuntary MTT to necessarily replicate findings in the mind-wandering literature. This is because involuntary past and future MTT is conceptually distinct from the notion of mind wandering, by the former being clearly autobiographical, typically cue dependent, short-lived, and not necessarily off-task thinking (see Berntsen, 2009, for an extended discussion).

In addition, we predicted that, like memories, current-concern-related future thoughts have a privileged cognitive, emotional, and representational status. Specifically, we expected concern-related representations to be rated higher on a number of phenomenological measures related to rehearsal frequency, temporal distance, and self-relevance. More speculatively, if self-defining memories (Singer & Salovey, 1993) and self-defining future projections (D'Argembeau, Lardi, & Van der Linden, 2012) conceptually overlap with current-concern-related representations, then current-

concern-related MTT should be rated higher on vividness and affective characteristics, in agreement with findings from past events (Johannessen & Berntsen, 2010). We expected this advantage of goal-related events to be similar for future and past MTT, consistent with evidence of both temporal directions being supported by many of the same neurocognitive substrates, and responding similarly to a number of experimental manipulations (see D'Argembeau, 2012; Szpunar, 2010a, for reviews).

EXPERIMENTAL STUDY

Method

The data presented here derive from a more extensive study examining differences between involuntary and voluntary mental time travel using an adapted laboratory paradigm (Cole et al., 2014; Schlagman & Kvavilashvili, 2008) in which the participants were asked to report autobiographical memories or future thoughts (depending on group assignment) that occur during and after a vigilance task. The vigilance task was presented to participants as their "primary task" and was implemented to simulate the moderately demanding tasks in which involuntary future mental time travel occurs in daily life (e.g., washing the dishes, see Berntsen & Jacobsen, 2008). If participants believed their main role was to generate past/future thoughts, they may have contaminated the involuntary condition by using voluntary self-generation processes. In a different part of this study, we obtained data on their current concerns and whether their future thoughts (or memories) were related (or unrelated) to their reported current concerns. Here we report these previously unpublished data on the role of current concerns in the frequencies and qualities of the reported future thoughts and autobiographical memories.

Participants

From the initial 64 participants who were recruited for the study, data from 55 Danish-speaking participants were included here (reasons for exclusion were: psychological illness, $n = 2$; an absence of involuntary representations, $n = 1$; and noncompliance with task instructions, $n = 6$). The included participants, who also participated in Cole et al. (2014), were randomly assigned to report memories from their past ($n = 28$) or imagined events in their future ($n = 27$).[1] The two groups were alike regarding age (*past*: $M = 24.29$ years, $SD = 6.19$; *future*: $M = 24.33$ years, $SD = 6.93$) and male:female ratio (*past* = 5:23; *future* = 7:20) and were psychologically and neurologically healthy. All tested participants received two cinema tickets as recompense.

Design

For analyses addressing the frequencies of current-concern-related memories and future thoughts, we employed a 2 (future, past; between-participants) by 2 (involuntary, voluntary; within-participants) mixed design. For analyses comparing phenomenological characteristics of current-concern- and non-current-concern-related past and future event representations, only voluntarily retrieved representations were included in the analyses. The involuntary conditions were not analysed here because participants varied greatly in the frequencies of involuntary representations, and several lacked sufficient numbers of representations related and not related to current concerns to generate participant averages and render the analyses meaningful. (This variability is a natural consequence of having the number of current-concern-related events as a dependent variable for the first part of our analyses.) Also, previous work examining involuntary versus voluntary memories that were related versus unrelated to current concerns found no interactions between these two factors (Johannessen & Berntsen, 2010). This part therefore utilized a 2 (future, past; between-participants) × 2 (concern-related,

concern-unrelated; within-participants) mixed design.

Materials

All instructions and measures were presented in Danish (see Cole et al., 2014, for details on translations). The Current Concerns Questionnaire was presented after all involuntary and voluntary past/future thoughts were elicited. Whereas some questions of the Autobiographical Characteristics Questionnaire were presented immediately after elicitation because they required immediate recording (Part 1), most were administered retrospectively (Part 2).

Current Concerns Questionnaire. A questionnaire was administered in which participants were asked to provide five current concerns.[2] It included a written part, which described current concerns as something you would like to have, achieve, or complete (i.e., *positive*) or something that you might want to get rid of, prevent, or avoid (i.e., *negative*). Participants were also provided with two examples ("devote more time to my hobbies—especially singing" or "avoid getting into debt with the bank") for clarity. Participants were free to choose their own current concerns and were not prompted or cued. This represented a shortened version of the instrument used by Sellen, Murran, Cox, Theodosi, and Klinger (2006) and Johannessen and Berntsen (2010).

Autobiographical Characteristics Questionnaire. For each recorded future or past representation, participants completed the structured Autobiographical Characteristics Questionnaire, consisting of two parts (represented on the same page). In Part 1, participants provided a short description of the representation, followed by a vividness rating (1 to 7; 1 = vague, almost no image; 7 = very vivid,

---

[1]Participants were taken from an original pool of 64 participants (32 for each temporal direction condition). Four participants were excluded from the past condition analyses due to noncompliance with concentration task instructions ($n = 1$; e.g., confusion over button press to identify targets), an absence of any involuntary memories ($n = 1$), and self-reported mental illness ($n = 2$), and five were excluded from the future condition analyses due to noncompliance with concentration task instructions ($n = 4$) and reporting only involuntary memories or images not concerning the future ($n = 1$).

[2]As part of the standardized questionnaire, participants also rated the importance of each on an 11-point scale (0 = *not important*; 10 = *very important*).

almost like normal vision).[3] In Part 2, participants completed a more extensive description followed by indicating whether the representation was a specific event (binary; specific, not specific), their age within the past/future representation, how often the representation had been thought of before (1 to 5; 1 = never, 5 = very often), and the emotional valence (−2 to +2; −2 = negative, 0 = neutral, +2 = positive) and emotional intensity (1 to 5; 1 = no intensity, 5 = very intense) of the representation. They also rated the impact of the representation on current mood (−1 = negative, 0 = neutral, +1 = positive) and the extent to which the representation was/will be a central part of one's life story and was/will be a part of one's personal identity (1 to 5; 1 = totally disagree, 5 = totally agree; both items). Finally, participants were asked to refer back to current concerns and document which, if any, were related to the particular memory or future thought being reported (consistent with Johannessen & Berntsen, 2010). Items were presented in the above order.

Involuntary and voluntary MTT session equipment. Both involuntary and voluntary conditions were presented on E-Prime Professional Version 2.0 on desktop computers. Cue phrases (e.g., "coffee jar", "lucky find") served as the stimuli presented in both conditions, presented in the centre of the screen (18-point Arial font). Each slide consisted of a cue phrase embedded in line arrays distributed on a white background. For the involuntary condition, 600 slides were presented in the context of a vigilance task in which participants were required to identify targets (1.5 s/slide). Targets were line arrays presented vertically ($N = 11$, presented every 40–60 slides). All others were horizontal. For the voluntary condition, slides were formatted similarly except that all line arrays were presented horizontally, and 12 cue phrases were presented (maximum = 60 s/slide). Different cue phrases were assigned to each retrieval mode condition and were consistent across past/future conditions. The implementation of these experimental materials was based upon a paradigm that successfully elicited involuntary autobiographical memories (Schlagman & Kvavilashvili, 2008). Here, they were utilized to elicit both involuntary memories and future thoughts.

Procedure
Upon entering the laboratory, each participant completed informed consent procedures. Each participant completed all tasks individually in workstations, consisting of a desktop computer and questionnaire booklet.

For the involuntary condition, on-screen instructions introduced a vigilance task in which participants had to press a button (spacebar) each time a target (vertical lines) was identified. No response was required for nontargets (horizontal lines). Participants were also informed that they would see phrases, but they were to ignore these as these would be detected by participants in another condition (actually, no such condition existed) to maintain the impression that successful performance on the vigilance task was paramount. Thereafter, participants completed a one-minute practice session consisting of 40 trials (three targets).

Following the practice vigilance task, screen instructions varied depending on past/future group assignment. In the future condition, participants were initially informed that since the vigilance task was monotonous, they may have other thoughts, including goals, daydreams, and memories (the last example was "imagined future events" in the past condition), which was normal. Instructions highlighted that participants might experience imagined future events that "pop" into their mind spontaneously. Participants were told that future MTT could be temporally near or far, refer to a specific event that referred to a particular day in the future, or be a more general scene with no reference to a specific day (we allowed variable temporal distances to be consistent with Schlagman & Kvavilashvili, 2008, and we allowed different levels of specificity because both

---

[3]Two additional items were included for involuntary representations; participants were asked to describe the event's trigger or cue, if known, and their level of concentration when the representation came to mind.

autobiographical memories and future thoughts can be specific or general; see, for example, Anderson & Dewhurst, 2009; Berntsen & Jacobsen, 2008; Klein, 2013). In addition to the "primary" vigilance task, participants were asked to press the left mouse button when involuntary future MTT occurred. Once pressed, the vigilance task was paused, and text instructed participants to complete Part 1 of the Autobiographical Characteristics Questionnaire, then press enter to return to the task. Overall, involuntary session duration depended upon amount and length of pauses. Participants then completed the voluntary condition, which differed from the involuntary condition in the following ways: (a) There was no parallel vigilance task, (b) participants were asked to consciously imagine future events associated with 12 different cue phrases (although see below), and (c) if a representation was not recorded in 60 s (by pressing left mouse button), the next cue phrase was presented. Cue phrases used in involuntary and voluntary conditions were selected from the same pool of standardized cue phrases as those used in the respective conditions in Schlagman and Kvavliashvili (2008), matched for imagery and concreteness. The order of the involuntary and voluntary conditions was fixed as one important aim of this study was to compare our findings with those of the original involuntary memory paradigm (Schlagman & Kvavilashvili, 2008). Also having the involuntary condition before the voluntary condition was important in order not to disclose the actual purpose of the experiment and thus potentially contaminate the involuntary condition with strategic search for future and past events.

After a short break, participants were given three tasks in the following order: the Current Concerns Questionnaire, Part 2 of the Autobiographical

Characteristics Questionnaire for each representation, and the Consideration of Future Consequences Scale (Strathman, Gleicher, Boninger, & Edwards, 1994).[4] Part 2 contained an extensive series of phenomenological rating scales (see Materials) and was completed for each Part 1 entry. Part 2 items were only revealed when participants removed an adhesive piece of paper that had covered them up to this point. No time limit was imposed for these measures. The rationale for not administering the Current Concerns Questionnaire before the past/future elicitation phase was that it was likely that it would have affected the main phenomena of interest (i.e., past and future thoughts) by priming personal goals. Priming has been shown to affect both past and future thoughts (Mace, 2005; Szpunar, 2010b; Wang, 2008). Overall testing time per participant was approximately 100 minutes. In the memory group, instructions were identical except references to temporal direction.

Results

Descriptive data

Due to the nature of the tasks, the total number of recorded events depended on retrieval mode and temporal direction conditions (past involuntary = 239, past voluntary = 307, future involuntary = 154, and future voluntary = 267). When the mean participant frequencies were entered in a 2 × 2 mixed analysis of variance (ANOVA; within-participant factor was retrieval mode; between-participant factor was temporal direction), voluntary representations significantly outnumbered involuntary ones (in line with Schlagman & Kvavilashvili, 2008), and there were significantly fewer future versus past

---

[4]The Consideration of Future Consequences (CFC) scale was administered to assess whether this general disposition was related to frequency of goal-related future thoughts. Two subscales were used due to recent research showing a two-factor structure (see Joireman, Balliet, Sprott, Spangenberg, & Schultz, 2008). No relationship was evident between the mean CFC scores and proportion of involuntary ($r = -.08$, $p = .69$) or voluntary ($r = -.05$, $p = .79$) current-concern-related future thoughts. When immediate (involuntary $r = .11$, $p = .57$, voluntary $r = .16$, $p = .43$) and future (involuntary $r = .004$, $p = .99$, voluntary $r = .15$, $p = .45$) subscales of the CFC were correlated against proportion of current-concern-related future thoughts, this lack of a correlation remained. It remains an open question whether other individual differences (e.g., self-consciousness) moderate the CFC future-thinking relationship and/or whether the CFC correlates with other aspects of future thought (e.g., objectively coded detail).

Table 1. *Frequencies of involuntary and voluntary past and future MTT*

| Variable | Past | | | | Future | | | | Main effects and interaction | | | | | |
| | Involuntary | | Voluntary | | Involuntary | | Voluntary | | Past vs. future | | Involuntary vs. voluntary | | Interaction | |
	M	SD	M	SD	M	SD	M	SD	F	η_p^2	F	η_p^2	F	η_p^2
Total frequency	8.54	3.85	10.96	1.60	5.70	4.23	9.89	1.74	11.62*	.18	29.76*	.36	2.10	.04
CC-related (frequencies)	3.43	2.30	3.14	2.56	3.48	2.78	4.67	2.47	1.77	.03	1.11	.02	2.96	.05
CC-related (proportions)	.41	.30	.29	.24	.65	.32	.47	.24	11.55*	.18	12.01*	.19	0.46	.01

Note: Main effects and interactions indicated by analysis of variance (ANOVA) statistics. All effect sizes are partial eta-squared (η_p^2). MTT = mental time travel; CC = current concerns.
*$p < .005$.

representations.[5] See Table 1 for relevant means, statistics, and effect sizes.

The frequency of current-concern-related MTT elicited involuntarily and voluntarily

The frequencies and percentages of future and past events that were perceived as related to one or more of participant's five stated current concerns are presented in Table 1. As can be observed from proportional data, on average, future representations were more frequently related to current concerns than those directed toward the past. A mixed ANOVA (retrieval mode, within-participants; temporal direction, between-participants; see Table 1 for F-values, p-values, and effect sizes) using current-concern-related-to-total proportions demonstrated a main effect of temporal direction, with individuals having a higher proportion of current-concern-related future than past representations ($M = .56$, 95% CI [.47, .65], where CI = confidence interval, versus $M = .35$, 95% CI [.27, .44]). In contrast to predictions, there was a main effect of retrieval mode whereby involuntary conditions contained a higher proportion of current-concern-related representations than voluntary conditions ($M = .53$, 95% CI [.45, .61], versus $M = .38$, 95% CI [.32, .45], respectively). There was no interaction.

Given the fact that more past than future events were recorded, as well as the fact that more voluntary than involuntary representations were recorded, we conducted the same analysis on the basis of the means of the raw frequencies (cf. Table 1). For this analysis, there was no main effect of temporal direction or retrieval mode. Nor was there a significant interaction, but a trend was seen ($p = .09$). Given this trend, and given the fact that an inspection of the numbers in Table 1 suggested a difference between the frequencies of past- and future-concern-related events in the voluntary condition (but less so in the involuntary condition), we conducted a t-test following up on the numerically greater frequency of voluntary future than of voluntary past representations that were concern related. This test confirmed a reliable difference between these conditions, $t(53) = 2.25$, $p < .05$. Thus, whereas proportional data showed increased goal relatedness of future representations in general, when analysing frequencies, only the voluntary future representations were more frequently goal related (when compared with its contrasting voluntary past condition).

Phenomenological characteristics as a function of temporal direction and current-concern relatedness

The means, standard deviations, and ANOVA statistics of all phenomenological characteristics are reported in Table 2. Note that no significant

---

[5]These frequencies differed slightly from those of Cole et al. (2014), as two representations were excluded here due to not having current-concerns relatedness data.

Table 2. *Descriptive and ANOVA statistics for phenomenological characteristics of MTT as a function of temporal direction and current-concern relatedness*

| | Past | | | | Future | | | | Main Effects and Interaction | | | | | |
| | NCC | | CC | | NCC | | CC | | Past vs. future | | NCC vs. CC | | Interaction | |
Measure	M	SD	M	SD	M	SD	M	SD	F	η_p^2	F	η_p^2	F	η_p^2
Specificity	.51	.26	.67	.33	.44	.31	.52	.37	2.34	.05	5.68*	.10	0.68	.01
Vividness	4.51	1.02	5.02	1.46	3.87	1.20	4.13	1.14	6.72*	.12	5.20*	.10	0.59	.01
Rehearsal	2.28	0.58	2.56	1.02	2.04	0.81	2.86	1.01	0.03	.00	13.37**	.21	3.14	.06
Life story	1.85	0.65	2.41	1.21	1.88	0.66	2.44	0.99	0.02	.00	14.39**	.23	0.00	.00
Identity	1.82	0.65	2.33	1.26	1.91	0.68	2.37	0.99	0.08	.00	9.80**	.17	0.03	.00
Valence	0.38	0.38	0.70	0.92	0.46	0.57	0.80	0.65	0.49	.01	6.57*	.12	0.01	.00
Intensity	1.94	0.70	2.50	0.99	2.30	0.77	2.91	0.91	3.99	.08	17.15**	.26	0.03	.00
Pos. mood impact	.37	.24	.58	.38	.35	.24	.57	.29	0.04	.00	18.29**	.24	0.00	.00
Neg. mood impact	.07	.11	.16	.25	.15	.18	.17	.17	1.01	.02	3.02	.06	1.14	.02
No mood impact	.56	.26	.26	.33	.51	.33	.26	.29	0.33	.00	29.45**	.38	0.33	.01
Temporal distance	4.90	3.42	3.49	3.85	2.87	3.85	2.80	3.76	2.66	.05	1.34	.03	1.08	.02

Note: CC = current-concern-related; NCC = non-current-concern-related; ANOVA = analysis of variance; MTT = mental time travel; pos. = positive; neg. = negative. Most dependent measures are scale averages, except specificity and mood impact measures, which are denoted by mean proportions. See Materials for scale anchor points. Temporal distance = difference between current age and age-in-event (in years).

*p < .05. **p < .005.

interactions were found for any autobiographical characteristic (see Table 2).

Main effects of current-concern relatedness. Overall our findings are consistent with our prediction that there would be increases on several autobiographical characteristics for past/future representations related (versus not related) to current concerns. First, in line with predictions, on average, representations associated with current concerns were rehearsed more frequently ($M = 2.71$ versus $M = 2.16$), and were more important to one's life story ($M = 2.43$ versus $M = 1.86$) and self-identity ($M = 2.35$ versus $M = 1.85$) than non-current-concern-related representations (see Table 2 for statistical analyses). Current-concern-related representations were, on average, more vivid ($M = 4.58$ versus $M = 4.19$) and more frequently referred to specific spatiotemporal events ($M = .59$ versus $M = .47$). They also differed with regard to affective characteristics; current-concern-related representations were more emotionally intense ($M = 2.71$ versus $M = 2.12$), were more emotionally

positive ($M = .75$ versus $M = .42$), and more frequently had impact on current mood ($M = .26$ versus $M = .53$, see "no mood impact", Table 2), with the most pronounced differences in positive ($M = .58$ versus $M = .37$) rather than negative ($M = .17$ versus $M = .11$) mood impact (see Table 2).

Main effects of temporal direction. Comparing past and future representations showed that, in line with previous studies (see D'Argembeau, 2012, for a review), past representations were more vivid than projections into the future ($M = 4.77$ versus $M = 4.00$). Except for a marginally significant difference indicating that future representations had greater emotional intensity than past representations ($M = 2.61$ versus $M = 2.22$), no other main effects emerged.

Discussion

In this study, we utilized a recently validated laboratory-based paradigm (Cole et al., 2014) to elicit involuntary and voluntary memories and future thoughts and asked participants to indicate which were related to their personal goals—operationalized here as current concerns (Klinger, 1975). Several novel results were found. First, in line with our principal hypothesis that future MTT is more important for goal-oriented cognition than past MTT, analysis of proportions and frequencies indicated that current-concern-related representations are more prevalent for future than for past MTT, at least when generated voluntarily through a top-down strategic manner. Additionally, when comparing the two retrieval modes, a higher proportion of current-concern-related representation was found in the involuntary (i.e., nonstrategic) condition. However, this difference was not present in analysis based on the frequencies, perhaps indicating a selective bias toward the goal relatedness of involuntary MTT experiences due to their relatively scarce nature in the present experiment (see below for further details). Because the participants retrieved more voluntary representations overall, they may have reported more events that were of less goal relevance and personal significance. Secondly, across almost all phenomenological characteristics, representations related to at least one current concern could be reliably distinguished from those unrelated to any. Consequently, this study was the first to demonstrate that MTT future and past representations related to our current goals have a privileged status across cognitive, representational, and affective dimensions, building upon prior memory research (Johannessen & Berntsen, 2010). The implications and limitations of these findings are discussed below.

Here, we found supporting evidence for the hypothesis that future-directed MTT would be more goal related than past MTT, at least when sampled through a voluntary retrieval task. This finding builds on findings from cognitive studies showing, for example, that individuals perceive a role for their own future thoughts in planning and goal attainment (D'Argembeau, Renaud, & Van der Linden, 2011; Rasmussen & Berntsen, 2013) and that episodic future thoughts are elicited more fluently when cued by personal goals

(D'Argembeau & Mathy, 2011, Study 3). The current study extends the latter result by showing that goals increase the phenomenological prominence of future thoughts.

This study also found that MTT related to current concerns differed from non-current-concern-related MTT in a similar way on various phenomenological characteristics, regardless of the temporal direction. First, current-concern-related representations were significantly more rehearsed. This result agrees with how self-defining *memories* are characterized (Johannessen & Berntsen, 2010; Singer & Salovey, 1993) and extends these to future projections suggesting that goal-related representations have a history of being brought to mind. Second, current-concern-related representations were rated higher on sensory–perceptual vividness and were more frequently classed as being spatiotemporally specific. The latter result was in line with previous findings from future thinking (D'Argembeau & Mathy, 2011; but not memory research, see Johannessen & Berntsen, 2010). Several authors have argued that representing specific, rather than general, future events may be especially important for goal planning (Atance & O'Neill, 2001; Szpunar, 2010a, see also Gollwitzer, 1993). Third, current-concern-related representations had greater relevance to life story and identity. This can be seen as consistent with D'Argembeau et al. (2010) who found that goal-related future thoughts had greater perceived personal import (see also Johannessen & Berntsen, 2010, for similar results concerning memories). Fourth, the present results established several differences concerning emotion: Current-concern-related MTT was more emotionally positive and intense and garnered a greater positive impact upon present mood. In contrast to previous findings in memory (Johannessen & Berntsen, 2010), concern-related MTT was not closer to the present. Finally, the overall lack of interaction in phenomenological characteristics analyses supports the prevailing theoretical view that episodic past and future thinking relies on shared cognitive and neuropsychological processes (Schacter et al., 2012, for a review).

In this study, goal-related MTT was more prevalent in an involuntary mode than in a voluntary mode when analysed proportionally. This contrasted with our expectation based on previous work that retrieval mode should not affect the proportion of goal-related representations, for past (see Johannessen & Berntsen, 2010) and future MTT. As shown by the frequencies in Table 1, the number of voluntary event representations greatly exceeded the number of involuntary ones in the present study, whereas the numbers of involuntary and voluntary memories were kept similar in Johannessen and Berntsen's (2010) diary study. Given that relatively fewer involuntary representations were recorded in the present study, these may have been more selective and thus perceived as more frequently referring to current concerns than their (more frequent) voluntary counterparts. Analyses of the raw frequencies of current-concern-related representations indeed showed no differences between involuntary and voluntary MTT in terms of current-concern relatedness. To clarify this issue, future studies investigating involuntary and voluntary MTT may benefit from equating frequencies of representations across conditions.

Directions for future research

Although this study supports the view that many future thoughts serve goal functions (see Schacter, 2012), some important questions remain unexplored. Research has demonstrated that brief goal-related future thought interventions induce behaviour change (e.g., Pham & Taylor, 1999). However, do people who spontaneously experience more goal-related future thoughts complete plans more often and more effectively? Furthermore, how might *involuntary* future projections be involved in goal-directed cognition and behaviour? Also, there is evidence that future thinking sometimes may become dysfunctional (Schacter, 2012), such as in terms of worry (e.g., Borkovec, Robinson, Pruzinsky, & DePree, 1983), but little is known as to the underlying mechanisms of adaptive versus maladaptive forms of future thinking. Future investigations using the present and related paradigms could help uncover how goal-

related involuntary and voluntary future thoughts contribute to goal-directed behaviour and related functions.

Summary

The current study assessed the relation between goals and MTT, conceptualizing goals as current concerns —a self-selected set of uncompleted personal goals. The past MTT data complement recent data and theory about autobiographical memory (Johannessen & Berntsen, 2010; see also Conway, 2005) indicating that goals are related to a sizeable proportion of memories. Analysis of phenomenological characteristics uncovered that being goal related affected representational, cognitive, and affective aspects of past and future MTT. Crucially, the frequency data indicate that, in comparison to the past, future thought has a tighter relation with one's goals. Overall, this finding corresponds with recent reviews (e.g., Seligman et al., 2013) indicating that controlled and spontaneous thoughts about the future have an important role in optimizing goal-directed cognition and behaviour.

REFERENCES

Anderson, R. J., & Dewhurst, S. A. (2009). Remembering the past and imagining the future: Differences in event specificity of spontaneously generated thought. *Memory, 17*, 367–373.

Antrobus, J. S., Singer, J. L., & Greenberg, S. (1966). Studies in the stream of consciousness: Experimental enhancement and suppression of spontaneous cognitive processes. *Perceptual and Motor Skills, 23*(2), 399–417.

Atance, C. M., & O'Neill, D. K. (2001). Episodic future thinking. *Trends in Cognitive Sciences, 5*(12), 533–539.

Baird, B., Smallwood, J., & Schooler, J. W. (2011). Back to the future: Autobiographical planning and the functionality of mind-wandering. *Consciousness and Cognition, 20*, 1604–1611. doi:10.1016/j.concog.2011.08.007

Berntsen, D. (2009). *Involuntary autobiographical memories: An introduction to the unbidden past.* Cambridge: Cambridge University Press.

Berntsen, D., & Jacobsen, A. S. (2008). Involuntary (spontaneous) mental time travel into the past and future. *Consciousness and Cognition, 17*, 1093–1104.

Borkovec, T. D., Robinson, E., Pruzinsky, T., & DePree, J. A. (1983). Preliminary exploration of worry: Some characteristics and processes. *Behaviour Research and Therapy, 21*, 9–16.

Conway, M. A. (2005). Memory and the self. *Journal of Memory and Language, 53*, 594–628.

Cole, S. N., Staugaard, S., & Berntsen, D. (2014). Inducing involuntary and voluntary mental time travel using a laboratory paradigm. Under review.

D'Argembeau, A. (2012). Autobiographical memory and future thinking. In D. Berntsen & D. C. Rubin (Eds.), *Understanding autobiographical memory: Theories and approaches* (pp. 311–330). Cambridge: Cambridge University Press.

D'Argembeau, A., Lardi, C., & Van der Linden, M. (2012). Self-defining future projections: Exploring the identity function of thinking about the future. *Memory, 20*, 110–120. doi:10.1080/09658211.2011.647697

D'Argembeau, A., & Mathy, A. (2011). Tracking the construction of episodic future thoughts. *Journal of Experimental Psychology: General, 140*, 258–271. doi:10.1037/a0022581.

D'Argembeau, A., Renaud, O., & Van der Linden, M. (2011). Frequency, characteristics and functions of future-oriented thoughts in daily life. *Applied Cognitive Psychology, 25*, 96–103. doi:10.1002/acp.1647

D'Argembeau, A., Stawarczyk, D., Majerus, S., Collette, F., Van der Linden, M., Feyers, D., … Salmon, E. (2010). The neural basis of personal goal processing when envisioning future events. *Journal of Cognitive Neuroscience, 22*, 1701–1713. doi:10.1162/jocn.2009.21314

Gollwitzer, P. M. (1993). Goal achievement: The role of intentions. In W. Stroebe, & M. Hewstone (Eds.), *European review of social psychology* (Vol. 4, pp. 141–185). Chichester, England: Wiley.

Johannessen, K. B., & Berntsen, D. (2009). Motivation for weight loss affects recall from autobiographical memory in dieters. *Memory, 17*, 69–83. doi:0.1080/09658210802555616

Johannessen, K. B., & Berntsen, D. (2010). Current concerns in involuntary and voluntary autobiographical memories. *Consciousness and Cognition, 19*, 847–860. doi:10.1016/j.concog.2010.01.009

Joireman, J., Balliet, D., Sprott, D., Spangenberg, E., & Schultz, J. (2008). Consideration of future consequences, ego-depletion, and self-control: Support for distinguishing between CFC-Immediate and CFC-Future sub-scales. *Personality and Individual Differences, 45*, 15–21. doi:10.1016/j.paid.2008.02.011

Klein, S. B. (2013). The complex act of projecting oneself into the future. *Wiley Interdisciplinary Reviews: Cognitive Science, 4*, 63–79. doi:10.1002/wcs.1210

Klinger, E. (1975). Consequences of commitment to and disengagement from incentives. *Psychological Review, 82*, 1–25.

Mace, J. H. (2005). Priming involuntary autobiographical memories. *Memory, 13*, 874–884.

Pham, L. B., & Taylor, S. E. (1999). From thought to action: Effects of process-versus outcome-based mental simulations on performance. *Personality and Social Psychology Bulletin, 25*, 250–260.

Rasmussen, A. S., & Berntsen, D. (2013). The reality of the past versus the ideality of the future: Emotional valence and functional differences between past and future mental time travel. *Memory & Cognition, 41*, 187–200. doi:10.3758/s13421-012-0260-y

Schacter, D. L. (2012). Adaptive constructive processes and the future of memory. *American Psychologist, 67*, 603–613. doi:10.1037/a0029869

Schacter, D. L., Addis, D. R., Hassabis, D., Martin, V. C., Spreng, R. N., & Szpunar, K. K. (2012). The future of memory: Remembering, imagining and the brain. *Neuron, 76*, 677–694. Retrieved from http://dx.doi.org/10.1016/j.neuron.2012.11.001

Schlagman, S., & Kvavilashvili, L. (2008). Involuntary autobiographical memories in and outside the laboratory: How different are they from voluntary autobiographical memories? *Memory & Cognition, 36*, 920–932.

Seligman, M. E. P., Railton, P., Baumeister, R. F., & Sripada, C. (2013). Navigating into the future or driven by the past. *Perspectives on Psychological Science, 8*, 119–141.

Sellen, J. L., McMurran, M., Cox, W. M., Theodosi, E., & Klinger, E. (2006). The personal concerns inventory (offender adaptation): Measuring and enhancing motivation to change. *International Journal of Offender Therapy and Comparative Criminology, 50*, 294–305.

Singer, J. A., & Salovey, P. (1993). *The remembered self: Emotion and memory in personality*. New York: The free press.

Strathman, A., Gleicher, F., Boninger, D. S., & Edwards, C. S. (1994). The consideration of future consequences: Weighing immediate and distant outcomes of behavior. *Journal of Personality and Social*

Psychology, 66(4), 742–752. doi:10.1037/0022-3514. 66.4.742

Suddendorf, T., & Corballis, M. C. (2007). The evolution of foresight: What is mental time travel and is it unique to humans? *Behavioral and Brain Sciences*, 30, 299–351. doi:10.1017/S0140525X07001975

Szpunar, K. K. (2010a). Episodic future thought: An emerging concept. *Perspectives on Psychological Science*, 5, 142–162.

Szpunar, K. K. (2010b). Evidence for an implicit influence of memory on future thinking. *Memory & Cognition*, 38, 531–540. doi:10.3758/MC.38.5. 531

Szpunar, K. K., & Jing, H. J. (2013). Memory-mediated simulations of the future: What are the advantages and pitfalls? *Journal of Applied Research in Memory and Cognition*, 2, 240–242. doi:10.1016/j.jarmac. 2013.10.004

Wang, Q. (2008). Being American, being Asian: The bicultural self and Autobiographical Memory in Asian Americans. *Cognition*, 107, 743–751. doi:10. 1016/j.cognition.2007.08.005

Wheeler, M. A., Stuss, D. T., & Tulving, E. (1997). Toward a theory of episodic memory: The frontal lobes and autonoetic consciousness. *Psychological Bulletin*, 121, 331–354.

Remembering the past and imagining the future: Selective effects of an episodic specificity induction on detail generation

Kevin P. Madore and Daniel L. Schacter

Department of Psychology, Harvard University, Cambridge, MA, USA

According to the constructive episodic simulation hypothesis, remembering past experiences and imagining future experiences both rely heavily on episodic memory. However, recent research indicates that nonepisodic processes such as descriptive ability also influence memory and imagination. We recently found that an episodic specificity induction—brief training in recollecting details of past experiences—enhanced detail generation on memory and imagination tasks but not a picture description task and thereby concluded that the induction can dissociate episodic processes involved in remembering the past and imagining the future from those nonepisodic processes involved in description. To evaluate the generality of our previous findings and to examine the role of generative search in producing those findings, we modified our paradigm so that word cues replaced picture cues, and a word comparison task that requires generation of sentences and word definitions replaced picture description. Young adult participants received either a specificity induction or one of two control inductions before completing the memory, imagination, and word comparison tasks. Replicating and extending our previous work, we found that the specificity induction increased detail generation in memory and imagination without having an effect on word comparison. The induction's selective effect on memory and imagination stemmed from an increase in internal (i.e., on-topic and episodic) details and had no effect on external (e.g., off-topic or semantic) details. The results point to the efficacy of the specificity induction for isolating episodic processes involved in remembering the past and imagining the future even when a nonepisodic task requires generative search.

Research examining the relationship between remembering past experiences and imagining or simulating future experiences has grown dramatically during recent years (for reviews, see Klein, 2013; Schacter, Addis, & Buckner, 2008; Schacter et al., 2012; Szpunar, 2010). According to the constructive episodic simulation hypothesis (Schacter & Addis, 2007), many of the striking similarities that have been documented between remembering the past and imagining the future reflect the influence of episodic memory (Tulving, 1983, 2002), which is thought to support the

We thank Sean Hardy, Erin McDonnell, Lucy Walsh, and Michael Zamora for their assistance with various aspects of the experiment, and Donna Addis for providing the word cue stimuli used in the experiment.

This research was supported by National Institute of Mental Health [grant number MH060941] and National Institute on Aging [grant number AG08441] to D.L.S.

retrieval of details about past experiences and the flexible recombination of those details into simulations of possible future scenarios (i.e., episodic future thinking; Atance & O'Neill, 2001).

However, recent research concerning age-related changes in memory and future thinking has suggested an alternative interpretation of observed similarities between remembering the past and imagining the future that highlights the role of nonepisodic influences. Several studies have reported that young adults recall more specific details from past experiences, and imagine more specific details about possible future experiences, than do older adults (e.g., Addis, Musicaro, Pan, & Schacter, 2010; Addis, Wong, & Schacter, 2008; Cole, Morrison, & Conway, 2013; Rendell et al., 2012; for review, see Schacter, Gaesser, & Addis, 2013). According to the constructive episodic simulation hypothesis, such findings are attributable mainly to age-related declines in episodic memory that result in comparable age-related declines during episodic simulation/future thinking (Addis et al., 2010; Addis et al., 2008). Contrary to this interpretation, however, Gaesser, Sacchetti, Addis, and Schacter (2011) reported a similar pattern of results using a picture description task that does not require and should not involve episodic memory: Older adults reported fewer specific details concerning the contents of presented pictures than did young adults. These findings suggest a role for nonepisodic factors in driving the observed age effects, such as age-related changes in narrative style, communicative goals, or inhibitory control (cf. Adams, Smith, Nyquist, & Perlmutter, 1997; Arbuckle & Gold, 1993; Labouvie-Vief & Blanchard-Fields, 1982) that could similarly impact performance on tasks that tap remembering past experiences, imagining future experiences, and describing pictures (for discussion, see Gaesser et al., 2011; Schacter et al., 2013). More generally, these observations raise the possibility that many of the similarities documented between remembering the past and imagining the future in both young and older individuals could reflect the influence of such nonepisodic factors rather than the influence of episodic memory (note, however, that this line of reasoning

primarily applies to tasks requiring verbal descriptions and is probably less relevant to studies measuring phenomenological similarities in remembering and imagining with Likert scales; e.g., D'Argembeau & Van der Linden, 2004).

In an attempt to distinguish episodic and nonepisodic influences on remembering the past and imagining the future, Madore, Gaesser, and Schacter (2014) recently developed an experimental approach involving an *episodic specificity induction*: brief training in recollecting specific details of past experiences. After viewing a brief video of an everyday scene, participants received an episodic specificity induction based on the well-established Cognitive Interview (CI; Fisher & Geiselman, 1992; for recent review, see Memon, Meissner, & Fraser, 2010). The specificity induction guided participants to generate a mental picture of the scenes they had viewed in the video and report everything they remembered about the scenes in as much detail as possible, including what people looked like and did, how objects were arranged, and related episodic information. After receiving the specificity induction or control inductions (one requiring participants to describe their general impressions of the video, another requiring completion of math problems), participants performed memory, imagination, and picture description tasks like those used by Gaesser et al. (2011) in which they had three minutes to remember a past experience related to a pictorial cue (a colour picture of an everyday scene), imagine a plausible future experience related to the pictorial cue, or simply describe the picture. As in Gaesser et al. (2011) and earlier related studies (Addis et al., 2010; Addis et al., 2008), protocols were scored using an adapted version of the Autobiographical Interview (AI; Levine, Svoboda, Hay, Winocur, & Moscovitch, 2002), which distinguishes between two types of detail that comprise memories of personal experiences: "internal" details that are on-topic and episodic in nature, concerning what happened during an experience, who was there, and when and where the event occurred; and "external" details that are mainly semantic in nature, such as related facts, reflections on the meaning of what happened, or off-topic commentary and references to other events.

Adapting this scoring procedure to memory, imagination, and picture description tasks using the methods of Gaesser et al. (2011), Madore et al. (2014) found that compared with control inductions, the episodic specificity induction selectively increased the number of internal details in remembered past experiences and imagined future experiences, while having no effect on the number of internal details reported during picture description and no effect on the number of external details provided during any of the tasks.

The observed pattern of results suggests that the specificity induction selectively targets and enhances episodic retrieval on the memory and imagination tasks. We therefore argued that, consistent with the constructive episodic simulation hypothesis, these data provide evidence that remembering the past and imagining the future both depend heavily on episodic memory, whereas picture description relies on nonepisodic processes that are unaffected by the specificity induction. However, the results of the Madore et al. (2014) study provide only limited support of these conclusions, for at least two reasons. First, it is unknown whether the key pattern of results that we reported (i.e., specificity induction selectively impacts memory and imagination tasks) is specific to the particular cues and tasks that we used, or whether the pattern applies more broadly. If the specificity induction indeed distinguishes episodic from nonepisodic processes, then the results reported by Madore et al. (2014) should generalize to cues and tasks other than those used in our initial study.

A second, related issue concerns our prior use of picture description as the "nonepisodic" task. We argued that in contrast to the memory and imagination tasks, picture description should not recruit episodic retrieval. However, the picture description task also differs from memory and imagination tasks in that the latter two tasks required participants to use pictures as a basis for engaging in a controlled, generative search (e.g., Addis, Knapp, Roberts, & Schacter, 2012; Conway & Pleydell-Pearce, 2000) to construct a remembered or imagined experience. By contrast, in the picture description task, responses are much more directly constrained by the properties of the picture itself, and generative search is not required. In other words, compared with the memory and imagination tasks, picture description can be said to provide more environmental support (Craik, 1983; Lindenberger & Mayr, 2014) for target responses. It is thus important to determine whether the specificity induction that we used in Madore et al. (2014) can dissociate episodic from nonepisodic processes under conditions in which the nonepisodic task also requires generative search and retrieval.

To accomplish the foregoing objectives, we compared the effects of the specificity induction on episodic memory and imagination/future thinking tasks and a nonepisodic task that uses word cues instead of the picture cues used by Madore et al. (2014). We call this latter task *word comparison,* and it is based on similar tasks developed in previous research (e.g., Addis, Pan, Vu, Laiser, & Schacter, 2009; Addis, Wong, & Schacter, 2007). The word comparison task requires participants to engage in generative search and retrieval by constructing a sentence containing words that refer to relatively larger and smaller objects than the object referred to by a cue word and then to generate definitions of each word (see Method for more details), but does not require remembering or imagining personal episodes. We analysed responses on the memory and imagination tasks using a version of the AI based on that used in our previous studies (Addis et al., 2010; Addis et al., 2008; Gaesser et al., 2011; Madore et al., 2014) and further adapted the AI scoring procedures to analyse the number of details provided in definitions that participants generated on the word comparison task (see Method). Internal and external details are comparable on the memory and imagination tasks, but because participants do not generate episodic details on the word comparison task, "internal" and "external" details on this task are not strictly comparable to internal and external details on the memory and imagination tasks (see Method). Thus, to compare performance across the three tasks, and to ensure that our conclusions regarding how the specificity induction impacts detail generation do not depend on the particular criteria used for distinguishing internal from external details, we collapsed across internal and external detail categories to generate a total detail score for each task. In addition, to assess whether results from the present paradigm replicate our previous findings concerning effects of the specificity induction on internal and external details in memory and imagination (Madore et al., 2014), we also analysed the effects of the specificity induction on these tasks with respect to internal versus external details.

We compared AI responses on the three critical tasks following viewing of a video and a subsequent specificity induction, or two control inductions: one that required participants to describe their general impressions of the video and the other that required participants to complete math problems (the two control conditions yielded similar results in our previous work, Madore et al., 2014, but were included here to determine the generality of this outcome). If the specificity induction selectively affects performance on episodic but not nonepisodic tasks, then performance should be enhanced on the memory and imagination tasks following the specificity induction compared with control inductions, but not on the word comparison task. By contrast, if the specificity induction affects generative retrieval regardless of whether the tasks tap episodic or nonepisodic processes, then performance should be enhanced on all three tasks following the specificity induction compared with control inductions.

EXPERIMENTAL STUDY

Method

Participants

Thirty-two young adults (age 18–27 years, $M = 20.69$, $SD = 1.91$, 21 female) participated in the study for pay or for course credit. They had attained an average of 14.06 years of education (12–18 years of education, $SD = 1.63$) at the time of study and were recruited via postings at Harvard University and Boston University. All participants had normal or corrected-to-normal vision and no history of neurological impairment. Participants provided written informed consent before the study commenced and were treated in accordance with guidelines supported by the Committee on the Use of Human Subjects Research at Harvard University.

Materials, design, and procedure

Overview. Participants completed the study in one session composed of two different segments separated by a 5-min break during which they performed an odd–even number judgement task. In each of the two segments, participants (a) watched different versions of a short video involving two people carrying out routine actions in a kitchen and then completed a filler task, (b) after completing these tasks, received the episodic specificity induction or one of two control inductions, and (c) following each of the inductions, completed memory, imagination, and word comparison tasks in response to a total of 48 word cues, where they recalled a past experience, imagined a future experience, or generated a size sentence and word definitions for each trial.

Inductions. The key manipulation occurred in the induction phase of each segment after participants had watched the video and done the filler task. All participants were randomly assigned to receive the episodic specificity induction in one of the two segments. They were also randomly assigned to receive the *impressions control induction* or the *math control induction* in the other segment. The induction manipulation itself (i.e., control vs. specificity) was a within-subjects factor while the control induction received was a between-subjects factor (i.e., impressions vs. math). The order of inductions presented was counterbalanced across participants, as was the induction–video pairing. The entire study took approximately 2.5 to 3 hours to complete.

Participants received the *episodic specificity induction* during either the first or the second segment of the study. As in our previous work (Madore et al., 2014; Madore & Schacter, 2014), the induction was a modified CI (Fisher & Geiselman, 1992). Participants were first told that they were the chief expert about the video they had seen and were then asked to verbally report about episodic details they remembered related to the setting, people, and actions in the video. For each category of information, participants were instructed to report everything they remembered in as much detail as possible using mental imagery probing:

Please close your eyes and get a picture in your mind about the people in the video you saw. . . . Once you have a really good picture I want you to tell me everything you remember about the people. Try to be as specific and detailed as you can.

The setting probe asked participants to focus on the environment, the objects in it, and how they were arranged; the people probe asked participants to focus on what they looked like and what they were wearing; and the actions probe asked participants to focus on what they were and how the people did them, starting with the first one and ending with the last one. Participants were also asked to expand on different aspects of the video they had mentioned with follow-up probes that were typically presented in an open-ended, "tell me more" manner (e.g., "You said the man first brought in flowers. Tell me more about the flowers."). For each category, participants were generally asked one follow-up probe. See Madore et al. (2014) for the specificity induction script.

Participants received one of two versions of the control induction during whichever segment of the study they did not receive the specificity induction, as in our previous work (e.g., Madore et al., 2014). Participants who received the *impressions control induction* were first asked to verbalize their general opinions, impressions, and thoughts about the video they had seen. They were then asked to provide their opinions of the setting, actions, and people in the video specifically and adjectives they would use to describe each. Participants also responded to a number of questions from a question bank. These included items such as when they thought the video was made and if it reminded them of anything from their own lives. After going through the question bank, participants were asked whether they wanted to say anything else about the video or about their impressions or opinions of it. This control induction required participants to reflect on and speak about the contents of the video they had seen, as in the specificity induction;

the main difference was that the control induction elicited general impressions from participants while the specificity induction elicited episodic details from participants. See Madore et al. (2014) for the impressions control induction script.

Participants who did not receive the impressions induction for the control instead received the *math control induction*. Here participants did not speak about the contents of the video they had seen; they simply filled out a packet of addition and subtraction math problems after watching the video and doing the filler task. We included two control inductions to ensure that any differences in performance from baseline on the main tasks could be attributed to a boost from the specificity induction and not to a reduction from the impressions control induction (for detailed discussion of this issue, see Madore et al., 2014). Participants did not significantly differ at the $p < .05$ level in terms of age, education level, or gender as a function of control induction.

On average, participants also spent similar amounts of time in the control induction ($M = 3$ minutes, 26 s, $SD = 1$ min, 22 s) and the specificity induction ($M = 3$ min, 48 s, $SD = 1$ min, 7 s), $F(1, 31) = 2.13$, $MSE = 3802.26$, $p = .16$, $\eta_P^2 = .06$.

Adapted autobiographical interview. After the induction phase, participants transitioned to an adapted AI task that involved *construction* and *elaboration* phases (e.g., Addis, Cheng, Roberts, & Schacter, 2011; Addis et al., 2009; Addis et al., 2007). In each of the two segments, participants saw 48 different word cues and were asked to remember an event from the past few years related to the cue, imagine an event that could occur in the next few years related to the cue, or complete a word comparison task. For the memory and imagination trials, participants were instructed to think of a single event lasting a few minutes to a few hours that spanned no longer than a day. They were instructed to think about the event through their own eyes and to think of everything they could in as much detail as possible, including the people involved in the event, their actions, and emotions. For the word comparison task, participants were instructed to use the cue in a size sentence with two related words and to then think of a definition for each word. They were also instructed to think of everything they could in as much detail as possible. For example, if the word cue were "Apple", then participants could come up with the size sentence "tree is larger than pie is larger than apple" and then define each of the three words. Participants were instructed to use the $X > Y > Z$ format for their size sentence specifically (i.e., larger than rather than smaller than) as in previous work (e.g., Addis et al., 2011). While all three tasks involve building on, integrating, and generating details that are related to the cue at hand, only for the memory and imagination tasks are these details primarily episodic in nature (see Addis et al., 2009; Addis et al., 2007).

After the presentation of each cue, participants had 26 s to mentally (i.e., silently) produce a memory, imagination, or word comparison before they verbalized the content generated during this time period. We used this procedure because the present study also served as a behavioural pilot for a functional magnetic resonance imaging (MRI) study that would require silent thought during scanning, and our 26-second trial length is based on the trial length used previously in relevant neuroimaging studies (e.g., Addis et al., 2011; Addis et al., 2009; Addis et al., 2007). During the 26-s silent period, participants hit the space bar when they had initially *constructed* their memory, imagination, or size sentence. They then mentally *elaborated* on the details of their memory or imagination, or the definitions of the words contained in their size sentence, for the remainder of the 26-s silent period. After the silent period ended, participants verbalized what they had thought about during the silent period. This latter phase was self-paced, and participants spoke without any input or probing from the experimenter. After participants had finished speaking, they hit the space bar to move on to the next trial.

The 96 word cues for the AI task were nouns used by Addis et al. (2011). The nouns were high in imageability, concreteness, and Thorndike–Lorge frequency according to Clark and Paivio's (2004) extended norms. The word cues were divided into different lists of 16, and the lists did not differ significantly in these characteristics (Fs $\leq$ 1.18, MSEs $\geq$ 0.02, ps $\geq$ .28, η_p^2s $\leq$ .07). There were six memory trials, six imagination trials, and four word comparison trials per list. In each segment, participants viewed three lists for a total of 18 memory trials, 18 imagination trials, and 12 word comparison trials per segment (i.e., 48 trials per segment). The order of the trials within each list was randomized across participants, and the word cues contained in each list were randomly paired with the different task types.

Coding

Participants' responses for the memory, imagination, and word comparison trials were audio-recorded during the study and later transcribed and scored. Scoring focused on segmenting the bits of information contained in participants' responses into meaningful units, as is typically done in studies using the AI (see Levine et al., 2002) and in our previous work with the induction (Madore et al., 2014; Madore & Schacter, 2014). Internal details for memory and imagination were episodic and on-topic in nature (e.g., time, setting, people, objects, actions, feelings, and/or thoughts about the one central event). External details for memory and imagination were semantic or off-topic in nature (e.g., facts, commentary, repetitive information, and/or disconnected from the one central event). Internal details for the word comparison task were details contained in the definitions of the three words that were on-topic and meaningful (e.g., for the "Apple" cue an internal detail could be describing an apple as a fruit, or an apple as red). External details for word comparison were bits that were off-topic, repetitive, disconnected from the definitions, or not meaningful (e.g., for the "Apple" cue an external detail could be describing the task as hard, or repeating that an apple is a fruit). As noted earlier, because the criteria for internal and external details were necessarily slightly different when applied to the memory and imagination tasks on the one hand and the word comparison task on the other, we calculated a total detail count on each of the three tasks based on scoring done for internal and external details. Scoring was completed by one of three raters blind to experimental hypotheses and to which induction had been received. Before viewing the main experimental responses, these raters completed training and scored a practice set of 20 separate responses from young adults. Interrater reliability was high, with a Cronbach's α of .95 for total details, .95 for internal details, and .91 for external details.

Results

Preliminary analyses

Reaction time. We first examined whether participants varied in how long it took them to mentally construct a memory, imagination, or word comparison during the AI as a function of the within-subjects factors of induction (control vs. specificity) and task (memory vs. imagination vs. word comparison) and the between-subjects factor of control condition (impressions vs. math). Previous work (e.g., Addis et al., 2011; Addis et al., 2009) has found that remembered and imagined events typically take similar amounts of time to construct, while word comparison slightly differs. Using a mixed-factorial analysis of variance (ANOVA) and follow-up tests in the form of repeated measures ANOVAs, we found a similar pattern. There was a significant main effect of task on reaction time during construction, $F(1.64, 49.27) = 34.08$, $MSE = 3.84$, $p < .001$, $\eta_p^2 = .53$ (the Huynh–Feldt correction was used to correct for violations of sphericity assumptions for the task variable). Collapsed across induction and control condition, participants took significantly longer to construct a word comparison than they did to construct a memory, $F(1, 31) = 41.42$, $MSE = 2.28$,

$p < .001$, $\eta_p^2 = .57$, or imagination, $F(1, 31) = 40.34$, $MSE = 1.58$, $p < .001$, $\eta_p^2 = .57$. The word comparison took approximately 2.5 s longer to construct than memory and approximately 2 s longer than imagination. Importantly, participants did not differ significantly in how long it took them to construct in memory or imagination, though there was a trend towards significance, $F(1, 31) = 41.42$, $MSE = 0.77$, $p = .06$, $\eta_p^2 = .11$. This pattern of results—with no significant differences in reaction times for construction in memory and imagination, and slightly different reaction times for construction in the semantic word task—indicates that participants were completing the study as has been done previously (e.g., Addis et al., 2009).[1]

The main effect of task on reaction time was the only significant finding in this model. The induction and control condition main effects were nonsignificant, and these variables did not interact significantly with each other or with the task variable ($Fs \leq 2.67$, $MSEs \geq 1.64$, $ps \geq .11$, $\eta_p^2 s \leq .08$). Of critical importance, this pattern indicates that participants spent similar amounts of time mentally constructing memories, imaginations, and word comparisons after they received the control and specificity inductions, regardless of the type of control induction used as the comparison. That is, participants spent similar amounts of time mentally elaborating on their memories, imaginations, and the definitions of the words in their comparisons with both the control induction and the specificity induction, and when both types of control inductions were used as the comparison (because each 26-s trial window was divided into self-paced construction and elaboration times). This finding is important because it suggests that any significant effects of the induction are attributable to factors other than the length of time spent mentally constructing and mentally elaborating after the control and the

---

[1]It should be noted that in Addis et al. (2009) participants took less time to construct a word comparison than to construct a remembered or imagined event, whereas here we found that participants took more time to construct a word comparison. We do not think this difference is important because participants overall did not significantly differ in times to construct for the episodic event tasks. Moreover, recent evidence from Hach, Tippett, and Addis (2014) with young adults shows the same pattern as that found in the current study, with time to construct longer for word comparison than for the episodic event tasks (which do not differ from each other).

Table 1. *Mean reaction times for construction in seconds*

Task	Control induction	Specificity induction	Collapsed
Memory	7.41 (0.63)	7.88 (0.67)	7.65 (0.63)
Imagination	8.06 (0.73)	8.10 (0.65)	8.08 (0.67)
Word comparison	9.90 (0.70)	10.26 (0.84)	10.08 (0.75)

Note: Standard errors in parentheses.

specificity inductions.[2] Table 1 displays the construction times collapsed across induction and split by induction.

Carryover. We also conducted a preliminary analysis to determine whether the total number of details that participants generated on the memory, imagination, or word comparison tasks differed as a function of whether they received the specificity induction in the first segment or in the second segment. Although we have not found carryover effects from either the specificity or control inductions in previous studies using counterbalanced designs (e.g., Madore et al., 2014; Madore & Schacter, 2014), this study differed from our prior studies in that participants received the second induction almost immediately after the first; in our previous studies, participants received the second induction approximately one week after the first induction. We conducted a series of mixed-factorial ANOVAs on details generated for the memory, imagination, and word comparison tasks similar to the analysis described above for reaction times that included induction order (i.e., carryover) as an independent variable. We found no significant main effects or interactions

associated with induction order on total details, internal details, or external details ($Fs \leq 2.31$, $MSEs \geq 2.33$, $ps \geq .11$, $\eta_p^2s \leq .08$). These results indicate that participants performed similarly in the study irrespective of whether they received the specificity induction in the first segment or second segment, and that any significant findings or lack thereof are not attributable to carryover effects.

Main analysis
Induction and details. To address our main hypotheses we first examined whether the episodic specificity induction impacted performance on the different components of the AI task in terms of total details generated, and whether any effects found depended on the control condition used as the comparison. To do this we performed a mixed-factorial ANOVA with the within-subjects factors of induction (control vs. specificity) and task (memory vs. imagination vs. word comparison) and the between-subjects factor of control condition (impressions vs. math). We computed the main effects and interactions for the different factors and focus here on the interactions found because they (a) trumped the main effects and (b) address our hypotheses most directly. Follow-up tests were repeated measures ANOVAs to ensure that the same effect size measure was used throughout the analyses. To anticipate our main findings, as shown in Figure 1 and Table 2, which display the results collapsed across control inductions and also split by control induction, these findings extend our previous work (Madore et al., 2014): Compared with the control inductions, the episodic specificity induction significantly increased total

[2]We also examined how long participants spent generating a description in response to each cue, as well as word count. We found that participants spent significantly longer describing, $F(1, 31) = 4.96$, $MSE = 6.99$, $p = .033$, $\eta_p^2 = .14$, and used more words on memory trials, $F(1, 31) = 6.65$, $MSE = 39.76$, $p = .015$, $\eta_p^2 = .18$, when they had received the specificity induction than when they had received the control induction. We did not find significant differences in timing or word count on the imagination trials or word comparison trials as a function of induction ($Fs \leq 2.03$, $MSEs \geq 5.17$, $ps \geq .17$, $\eta_p^2s \leq .06$). Thus, it seems unlikely that the differences in timing and word count on memory trials are critical to the specificity induction effect because we observed the same detail boost from the specificity induction on the imagination task, where timing and word count did not differ. Moreover, it is not unreasonable to expect that the specificity induction could help participants generate richer and more detailed events on a later task than they otherwise would, which could be reflected in timing and/or word count differences. We have also observed the same specificity induction benefit on memory and imagination tasks when time is held constant across conditions (i.e., Madore et al., 2014; Madore & Schacter, 2014) rather than self-paced.

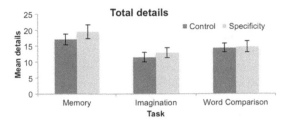

Figure 1. *Mean total details reported by participants as a function of induction and task collapsed across control condition used as the comparison. Error bars represent one standard error.*

details on memory and imagination tasks but had no effect on the word comparison task.

We found a significant two-way interaction of Induction × Task, $F(2, 60) = 3.56$, $MSE = 4.73$, $p = .035$, $\eta_P^2 = .11$; the three-way interaction of Induction × Task × Control Condition was non-significant, $F(2, 60) = 0.37$, $MSE = 4.73$, $p = .70$, $\eta_P^2 = .01$. The control condition variable did not interact significantly with induction or task separately ($Fs \leq 0.23$, $MSEs \geq 11.55$, $ps \geq .64$, $\eta_P^2s \leq .01$). Collapsed across control conditions, participants generated a significantly greater number of total details on the memory task, $F(1, 31) = 8.18$, $MSE = 11.69$, $p = .008$, $\eta_P^2 = .21$, and the imagination task, $F(1, 31) = 13.47$, $MSE = 2.43$, $p = .001$, $\eta_P^2 = .30$, when they received the specificity induction than when they received the control inductions; critically, total details on the word comparison task

did not differ significantly as a function of induction, $F(1, 31) = 0.39$, $MSE = 6.41$, $p = .54$, $\eta_P^2 = .01$.

Given that the specificity induction boosted total details generated in memory and imagination without affecting word comparison, we conducted a second analysis to examine whether there were differences in the type of detail the induction affected. We focused on memory and imagination because these were the two tasks that the induction impacted, and we divided the total details into internal and external details as in our previous work (e.g., Madore et al., 2014). We used another mixed-factorial ANOVA with the within-subjects factors of induction (control vs. specificity), task (memory vs. imagination), and detail (internal vs. external) and the between-subjects factor of control condition (impressions vs. math). We focused on the interactions found, and follow-up tests were repeated measures ANOVAs.

We found a significant two-way interaction of Induction × Detail, $F(1, 30) = 13.73$, $MSE = 5.58$, $p = .001$, $\eta_P^2 = .31$; the task and control condition variables were nonsignificant when added to this interaction separately or together ($Fs \leq 0.97$, $MSEs \geq 3.29$, $ps \geq .33$, $\eta_P^2s \leq .03$). The two-way interactions of Induction × Task and Induction × Control Condition were also nonsignificant ($Fs \leq 1.35$, $MSEs \geq 3.06$, $ps \geq .26$, $\eta_P^2s \leq .04$). Collapsed across task and control condition, the specificity induction significantly increased internal

Table 2. *Mean details generated split by control condition and task*

Detail type	Impressions control	Specificity[a]	Math control	Specificity[a]
Memory total details	17.36 (2.57)	20.11 (3.22)	16.72 (2.31)	18.86 (2.99)
Memory internal details	16.17 (2.50)	19.03 (3.13)	15.25 (2.10)	17.47 (2.81)
Memory external details	1.19 (0.31)	1.09 (0.26)	1.47 (0.30)	1.39 (0.32)
Imagination total details	11.66 (2.07)	12.96 (2.33)	11.30 (2.17)	12.86 (2.27)
Imagination internal details	10.53 (2.14)	11.88 (2.42)	9.79 (2.06)	11.61 (2.16)
Imagination external details	1.13 (0.31)	1.09 (0.33)	1.51 (0.29)	1.25 (0.26)
Word comparison total details	14.63 (2.15)	15.54 (2.71)	14.09 (1.88)	13.96 (2.16)
Word comparison internal details	14.33 (2.12)	15.19 (2.60)	13.80 (1.88)	13.48 (2.21)
Word comparison external details	0.29 (0.09)	0.35 (0.21)	0.29 (0.10)	0.48 (0.17)

Note: Standard errors in parentheses.

[a]Results for the specificity induction are reported separately for the 16 participants who received the impressions control and the 16 participants who received the math control.

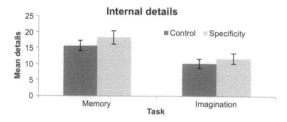

Figure 2. *Mean internal details reported by participants as a function of induction and task collapsed across control condition used as the comparison. Error bars represent one standard error.*

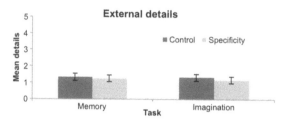

Figure 3. *Mean external details reported by participants as a function of induction and task collapsed across control condition used as the comparison. Error bars represent one standard error.*

details generated on memory and imagination compared with the control induction, $F(1, 31) = 14.74$, $MSE = 4.62$, $p = .001$, $\eta_P^2 = .32$, and had no effect on external details, $F(1, 31) = 2.14$, $MSE = 0.12$, $p = .15$, $\eta_P^2 = .06$. We also found the same pattern of results when the memory and imagination tasks were analysed separately. As displayed in Figures 2 and 3 and Table 2, the dissociable effect of the induction on memory and imagination internal details—but not external details—replicates and extends our previous work (e.g., Madore et al., 2014).

It should also be noted that while the main analysis indicated the specificity induction had no effect on total detail generation in word comparison, we also ran a secondary analysis including this task in the model and examined the follow-up tests for word comparison internal and external details, as defined in the Method section: As in total detail, the specificity induction had no effect on the internal, $F(1, 31) = 0.17$, $MSE = 6.63$, $p = .68$, $\eta_P^2 = .01$, or external details, $F(1, 31) = 1.99$, $MSE = 0.13$, $p = .17$, $\eta_P^2 = .06$, generated in word comparison (see Table 2).

Discussion

The results of the present study extend our previous findings concerning the effects of a specificity induction on episodic and nonepisodic processes (Madore et al., 2014) in two important ways. First, our finding that the specificity induction, compared with control inductions, produced a significant increase in internal but not external details when participants remembered the past and imagined the future in response to word cues indicates that our previous report of the same pattern with picture cues is not restricted to those cues and generalizes across cue types. Second, our finding that the specificity induction selectively affected details generated in the memory and imagination tasks, while having no effect on details generated in the word comparison task, indicates that the specificity induction can dissociate episodic from nonepisodic processes under conditions in which the nonepisodic task requires generative search and retrieval (Addis et al., 2012; Conway & Pleydell-Pearce, 2000), in contrast to the picture description task used in our previous study (Madore et al., 2014). As discussed in the introduction, the picture description task provides more environmental support (Craik, 1983; Lindenberger & Mayr, 2014) for responding than do the memory and imagination tasks: Responses on the picture description task are highly constrained by physical properties of the picture, such that generative search is not required, whereas memory and imagination tasks require generative search. The word comparison task used here also requires generative search to come up with details of word definitions, yet no effects of the specificity induction were observed on generation of definition details.

These results are consistent with the idea that the effects of the specificity induction are selective to tasks that draw on episodic retrieval, including tasks that involve imagining future personal experiences. More specifically, we hypothesized previously (Madore et al., 2014) that the induction affects a specific subcomponent of retrieval known as *episodic retrieval orientation* (Morcom & Rugg, 2012): retrieval cue processing that involves a focus on the specific episodic details (e.g., details of places, people, and actions) that comprise an episode. During the specificity induction, asking participants to generate a picture in their minds about the environment, people, and actions in the video should have led them to adopt a more specific retrieval orientation on subsequent tasks that benefit from focus on episodic details compared with the control inductions. We believe that this specific retrieval orientation impacts subsequent memory and imagination tasks because these tasks involve creating mental scenarios containing details similar to those focused on during the specificity induction, whereas the word comparison task emphasizes generating semantic details, which are not targeted by the specificity induction.

As noted in the Method section, to score detail generation on the word comparison task, we adapted the criteria used for "internal" details on the memory and imagination tasks, such that "internal" details on the word comparison task were details contained in the definitions of the three words that were on-topic and meaningful, in contrast to the episodic details that count as internal details on memory and imagination tasks. External details for word comparison were largely comparable to external details as scored on the memory and imagination tasks. Importantly, however, the observed pattern of results on the word comparison task did not depend critically on the criteria used for distinguishing internal from external details: There was no effect of the specificity induction on total detail generation in word comparison in contrast to the significant effects on total detail generation in memory and imagination. Further, when we split the AI responses into internal and external details we observed the same lack of a specificity induction effect on word comparison. This pattern contrasts with the reliable effects of the specificity induction on internal but not external details in the memory and imagination tasks. In light of the present results, we also examined total detail generation in Madore et al. (2014) for memory, imagination, and picture description as a function of induction, and obtained the same results as reported here: The specificity induction significantly and selectively enhanced total detail generation for memory and imagination without affecting picture description. As reported by Madore et al. (2014), dividing total details into internal and external ones revealed the same pattern, whereby the induction significantly and selectively boosted internal details in memory and imagination with no effect on external details, and no effect on either type of detail in picture description. Taken together, these findings indicate that the specificity induction targets episodic processes involved in memory and imagination that are distinct from those processes involved in word comparison or other tasks that measure nonepisodic processes (e.g., picture description) and thereby supports the constructive episodic simulation hypothesis (Schacter & Addis, 2007).

Another aspect of our findings that merits discussion concerns the amount of time and effort spent during the specificity induction. Although the data reported here replicate and extend our previous finding that the specificity induction reliably affects memory and imagination tasks, our observation that the induction enhanced performance whether it preceded the control induction or came directly after it suggests that the effect does not last very long. That is, there was no effect of induction order (i.e., carryover) on detail generation, indicating that whatever benefits are derived from the specificity induction are not evident only a few minutes later when participants perform memory and imagination tasks after receiving control inductions. One reason for this short-term impact could be the relatively brief amount of time spent in the specificity induction. Studies that have used specificity manipulations similar to the one reported here in attempts to increase the specificity of autobiographical memory retrieval in psychopathological populations often include multiple sessions spread out over several weeks, where participants complete homework assignments and meet individually and in groups to train on and discuss specificity (e.g., Neshat-Doost et al., 2013). This procedure culminates in hours of training compared with our approximately 4-min induction. Future work should examine more systematically the impact of extended specificity induction sessions on memory and imagination.

In a related vein, we have recently (Madore & Schacter, 2014) established that the specificity induction improves performance on a means–ends problem-solving task (Platt & Spivack, 1975) where participants are provided with beginning problems and ending solutions and are asked to fill in the steps they would take to solve each problem and reach the identified end state. Compared with the control induction, the specificity induction increased the number of relevant steps that participants generated in problem solving without increasing the number of irrelevant or off-topic steps. When the steps were scored for internal and external details, it was found that the specificity induction also increased the number of internal details generated in step solutions without increasing external details. This pattern of findings suggests that the specificity induction can improve performance on a cognitive task that taps everyday problem-solving skills, and where retrieving and recombining episodic details is important (Sheldon, McAndrews, & Moscovitch, 2011). It also indicates the potential usefulness of the specificity induction as a tool for isolating episodic contributions on a range of cognitive tasks that involve memory and imagination. Further research using a specificity induction to target a broader range of tasks that tap functionally useful cognitive processes and rely on episodic retrieval would be highly desirable. The present results are important for such research because they help to establish more securely the critical point that the specificity induction selectively boosts episodic retrieval across a range of situations.

REFERENCES

Adams, C., Smith, M. C., Nyquist, L., & Perlmutter, M. (1997). Adult age-group differences in recall for the literal and interpretive meanings of narrative text. *The Journals of Gerontology, Series 1, 52*, P187–P195. doi:10.1093/geronb/52B.4.P187

Addis, D. R., Cheng, T., Roberts, R. P., & Schacter, D. L. (2011). Hippocampal contributions to the episodic simulation of specific and general future events.

Hippocampus, 21, 1045–1052. doi:10.1002/hipo. 20870

Addis, D. R., Knapp, K., Roberts, R. P., & Schacter, D. L. (2012). Routes to the past: Neural substrates of direct and generative autobiographical memory retrieval. *NeuroImage, 59*, 2908–2922. doi:10.1016/j. neuroimage.2011.09.066

Addis, D. R., Musicaro, R., Pan, L., & Schacter, D. L. (2010). Episodic simulation of past and future events in older adults: Evidence from an experimental recombination task. *Psychology and Aging, 25*, 369–376. doi:10.1037/a0017280

Addis, D. R., Pan, L. P., Vu, M., Laiser, N., & Schacter, D. L. (2009). Constructive episodic simulation of the future and the past: Distinct subsystems of a core brain network mediate imagining and remembering. *Neuropsychologia, 47*, 2222–2238. doi:10.1016/j. neuropsychologia.2008.10.026

Addis, D. R., Wong, A. T., & Schacter, D. L. (2007). Remembering the past and imagining the future: Common and distinct neural substrates during event construction and elaboration. *Neuropsychologia, 45*, 1363–1377. doi:10.1016/j.neuropsychologia.2006. 10.016

Addis, D. R., Wong, A. T., & Schacter, D. L. (2008). Age-related changes in the episodic simulation of future events. *Psychological Science, 19*, 33–41. doi:10.1111/j.1467-9280.2008.02043.x

Arbuckle, T. Y., & Gold, D. P. (1993). Aging, inhibition, and verbosity. *The Journal of Gerontology, 48*, P225–P232. doi:10.1093/geronj/48.5.P225

Atance, C. M., & O'Neill, D. K. (2001). Episodic future thinking. *Trends in Cognitive Sciences, 5*, 533–539. doi:10.1016/S1364-6613(00)01804-0

Clark, J. M., & Paivio, A. (2004). Extensions of the Paivio, Yuille, and Madigan (1968) norms. *Behavior Research Methods, Instruments, & Computers, 36*, 371–383. doi:10.3758/BF03195584

Cole, S. N., Morrison, C. M., & Conway, M. A. (2013). Episodic future thinking: Linking neuropsychological performance with episodic detail in young and old adults. *Quarterly Journal of Experimental Psychology, 66*, 1687–1706. doi:10.1080/17470218. 2012.758157

Conway, M. A., & Pleydell-Pearce, C. W. (2000). The construction of autobiographical memories in the self-memory system. *Psychological Review, 107*, 261–288. doi:10.1037//0033-295X.107.2.261

Craik, F. I. M. (1983). On the transfer of information from temporary to permanent memory. *Philosophical*

Transactions B, 302, 341–359. doi:10.1098/rstb.1983. 0059

D'Argembeau, A., & Van der Linden, M. (2004). Phenomenal characteristics associated with projecting oneself back into the past and forward into the future: Influence of valence and temporal distance. *Consciousness and Cognition, 13,* 844–858. doi:10. 1016/j.concog.2004.07.007

Fisher, R. P., & Geiselman, R. E. (1992). *Memory-enhancing techniques for investigative interviewing: The cognitive interview.* Springfield, IL: Charles C. Thomas Books.

Gaesser, B., Sacchetti, D. C., Addis, D. R., & Schacter, D. L. (2011). Characterizing age-related changes in remembering the past and imagining the future. *Psychology and Aging, 26,* 80–84. doi:10.1037/a0021054

Hach, S., Tippett, L. J., & Addis, D. R. (2014). Neural changes associated with the generation of specific past and future events in depression. *Neuropsychologia, 65,* 41–55. doi:10.1016/j. neuropsychologia.2014.10.003

Klein, S. B. (2013). The complex act of projecting oneself into the future. *Wiley Interdisciplinary Reviews—Cognitive Science, 4,* 63–79. doi:10.1002/wcs.1210

Labouvie-Vief, G., & Blanchard-Fields, F. (1982). Cognitive ageing and psychological growth. *Ageing and Society, 2,* 183–209. doi:10.1017/S0144686X00009429

Levine, B., Svoboda, E., Hay, J. F., Winocur, G., & Moscovitch, M. (2002). Aging and autobiographical memory: Dissociating episodic from semantic retrieval. *Psychology and Aging, 17,* 677–689. doi:10.1037/0882-7974.17.4.677

Lindenberger, U., & Mayr, U. (2014). Cognitive aging: Is there a dark side to environmental support?. *Trends in Cognitive Sciences, 18,* 7–15. doi:10.1016/j.tics.2013.10.006

Madore, K. P., Gaesser, B., & Schacter, D. L. (2014). Constructive episodic simulation: Dissociable effects of a specificity induction on remembering, imagining, and describing in young and older adults. *Journal of Experimental Psychology: Learning, Memory, and Cognition, 40,* 609–622. doi:10.1037/a0034885

Madore, K. P., & Schacter, D. L. (2014). An episodic specificity induction enhances means-end problem solving in young and older adults.

Psychology and Aging, 29, 913–924. doi:10.1037/a0038209

Memon, A., Meissner, C. A., & Fraser, J. (2010). The cognitive interview: A meta-analytic review and study space analysis of the past 25 years. *Psychology, Public Policy, and Law, 16,* 340–372. doi:10.1037/a0020518

Morcom, A. M., & Rugg, M. D. (2012). Retrieval orientation and the control of recollection: An fMRI study. *Journal of Cognitive Neuroscience, 24,* 2372–2384. doi:10.1162/jocn_a_00299

Neshat-Doost, H. T., Dalgleish, T., Yule, W., Kalantari, M., Ahmadi, S. J., Dyregrov, A., & Jobson, L. (2012). Enhancing autobiographical memory specificity through cognitive training: An intervention for depression translated from basic science. *Clinical Psychological Science, 1,* 84–92. doi:10.1177/2167702612454613

Platt, J., & Spivack, G. (1975). *Manual for the means-end problem solving test (MEPS): A measure of interpersonal problem solving skill.* Philadelphia: Hahnemann Medical College and Hospital.

Rendell, P. G., Bailey, P. E., Henry, J. D., Phillips, L. H., Gaskin, S., & Kliegel, M. (2012). Older adults have greater difficulty imagining future rather than atemporal experiences. *Psychology and Aging, 27,* 1089–1098. doi:10.1037/a0029748

Schacter, D. L., & Addis, D. R. (2007). The cognitive neuroscience of constructive memory: Remembering the past and imagining the future. *Philosophical Transactions B, 362,* 773–786. doi:10.1098/rstb. 2007.2087

Schacter, D. L., Addis, D. R., & Buckner, R. L. (2008). Episodic simulation of future events: Concepts, data, and applications. *Annals of the New York Academy of Sciences, 1124,* 39–60. doi:10.1196/annals.1440.001

Schacter, D. L., Addis, D. R., Hassabis, D., Martin, V. C., Spreng, R. N., & Szpunar, K. K. (2012). The future of memory: Remembering, imagining, and the brain. *Neuron, 76,* 677–694. doi:10.1016/j. neuron.2012.11.001

Schacter, D. L., Gaesser, B., & Addis, D. R. (2013). Remembering the past and imagining the future in the elderly. *Gerontology, 59,* 143–151. doi:10.1159/000342198

Sheldon, S., McAndrews, M. P., & Moscovitch, M. (2011). Episodic memory processes mediated by the medial temporal lobes contribute to open-ended problem solving. *Neuropsychologia*, *49*, 2439–2447. doi:10.1016/j.neuropsychologia.2011.04.021

Szpunar, K. K. (2010). Episodic future thought: An emerging concept. *Perspectives on Psychological Science*, *5*, 142–162. doi:10.1177/1745691610362350

Tulving, E. (1983). *Elements of episodic memory*. Oxford: Clarendon Press.

Tulving, E. (2002). Episodic memory: From mind to brain. *Annual Review of Psychology*, *53*, 1–25. doi:10.1146/annurev.psych.53.100901.135114

You'll change more than I will: Adults' predictions about their own and others' future preferences

Louis Renoult[1], Leia Kopp[2], Patrick S. R. Davidson[2], Vanessa Taler[2], and Cristina M. Atance[2]

[1]School of Psychology, University of East Anglia, Norwich, UK
[2]School of Psychology, University of Ottawa, Ottawa, ON, Canada

It has been argued that adults underestimate the extent to which their preferences will change over time. We sought to determine whether such mispredictions are the result of a difficulty imagining that one's own current and future preferences may differ or whether it also characterizes our predictions about the future preferences of others. We used a perspective-taking task in which we asked young people how much they liked stereotypically young-person items (e.g., Top 40 music, adventure vacations) and stereotypically old-person items (e.g., jazz, playing bridge) now, and how much they would like them in the distant future (i.e., when they are 70 years old). Participants also made these same predictions for a generic same-age, same-sex peer. In a third condition, participants predicted how much a generic older (i.e., age 70) same-sex adult would like items from both categories today. Participants predicted less change between their own current and future preferences than between the current and future preferences of a peer. However, participants estimated that, compared to a current older adult today, their peer would like stereotypically young items *more* in the future and stereotypically old items *less*. The fact that peers' distant-future estimated preferences were different from the ones they made for "current" older adults suggests that even though underestimation of change of preferences over time is attenuated when thinking about others, a bias still exists.

It is always thus, impelled by a state of mind which is destined not to last, that we make our irrevocable decisions. (Marcel Proust, In Search of Lost Time, Vol. II: Within a Budding Grove)

Adults spend a considerable amount of time thinking about their futures (D'Argembeau, Renaud, & Van der Linden, 2011). Over the course of a typical day, we think about leisure activities, work, errands, and relationship issues that may occur both in our near and in our distant futures. Despite the prominence of these thoughts about the future in our daily lives, research consistently shows that adults mispredict their future preferences and values (for reviews, see Gilbert & Wilson, 2007; Loewenstein, O'Donoghue, & Rabin, 2003). For example, in a recent study, Quoidbach, Gilbert, and Wilson (2013) found that although adults reported that their personalities, values, and preferences had changed substantially in the past 10 years, they thought that they

Preparation of this manuscript was supported by grants from the Natural Sciences and Engineering Research Council (NSERC) of Canada to P.S.R.D., V.T., and C.M.A., a grant from the Alzheimer Society of Canada to V.T., and a fellowship from the Fonds de la Recherche en Santé du Québec (FRSQ) to L.R.

would change very little in the next 10 years. These authors proposed the term "end of history illusion" to capture adults' underestimation of the extent to which they would change in the future.

The mechanisms underlying this phenomenon are still debated. One possible mechanism is the "presentism bias", a tendency to interpret past and future selves in relation to present motives and knowledge (Cameron, Wilson, & Ross, 2004; Gilbert, Gill, & Wilson, 2002). A similar type of explanation is the "projection bias" (Loewenstein et al., 2003), which entails projecting our current preferences, values, or feelings into the future, even when these may no longer be relevant. Other explanations have focused on the characteristics of future simulations themselves, such as their unrepresentative, abbreviated, and decontextualized character (Gilbert & Wilson, 2007). Mispredictions about the future are not trivial because they can lead people to make decisions in the present that are based on preferences, emotions, and personality traits that may shift in the future. For example, the young person who gets a large neck tattoo today may, in 20 years, regret it.

An interesting question to ask is whether such biases are more prevalent—or appear exclusively—when considering one's own preferences, emotions, and personality traits, or whether these same biases also affect our judgements about others. This is of theoretical interest because it may explain the source of people's mispredictions and also qualify the particular explanation given to account for these mispredictions. For example, by the "end of history illusion" and "presentism bias" accounts, people have difficulty predicting change. However, is this difficulty situated solely in the context of self-predictions (i.e., predicting that one's *own* preferences will change), or is it situated in the context of predicting change more broadly (i.e., predicting that *everyone's* preferences will change)? To our knowledge, no previous study has directly compared how adults simultaneously predict their own and other people's changes in preferences in the future. Doing so will help to determine the parameters of "prediction" biases and may also shed light on their underlying mechanisms.

Research on perspective taking has shown that, when thinking about others, we typically use our own perspective as a starting point or "judgemental anchor" (Davis, Hoch, & Ragsdale, 1986; Epley, Keysar, Van Boven, & Gilovich, 2004; Nickerson, 1999) and represent what others would think or feel in a situation by imagining ourselves in this same situation (Decety & Grezes, 2006; Goldman, 2002; Gordon, 1986). Nonetheless, certain future-thinking biases like the optimism bias, by which we overestimate the likelihood of positive events in our future (Sharot, 2011), are attenuated when thinking about others (Baker & Emery, 1993; Grysman, Prabhakar, Anglin, & Hudson, 2013). Indeed, a vast body of research has shown that when comparing ourselves to others, we tend to think that we are better or less typical than average and have a brighter future (reviewed in Chambers, 2008). In the case of preferences, these "false uniqueness perceptions" may lead us to think that our own preferences are wise and therefore unlikely to change in the future. In contrast, the preferences of an average peer might be judged as less wise and therefore more likely to change as he or she gets older.

Reviewing relevant neuroimaging research, Buckner and Carroll (2007; see also Hassabis & Maguire, 2007) proposed that a core brain network, including frontal, medial temporal, and parietal cortices, support various forms of self-projection: remembering our past, thinking about our future (i.e., prospection), and taking the perspective of others (i.e., theory of mind). Accordingly, if similar neurocognitive mechanisms are involved in projecting ourselves in the future and in taking another person's perspective, one might speculate that mispredictions about the future will be similar when considering our own or someone else's perspective.

However, some differences in the neural correlates of self-versus-other judgements have been observed, with a number of brain regions responding preferentially to self-relevant information. Among these brain regions, the ventromedial prefrontal cortex is known to be important for future thinking, as patients with selective lesions in this area were reported to make decisions that illustrated "myopia for the future" (Bechara, Damasio, & Damasio, 2000) and an overestimation of self-monitoring abilities (Robinson, Calamia, Glascher, Bruss, & Tranel, 2014). Interestingly, a number of studies have reported evidence of a self-to-other gradient in this brain region: More ventral parts were shown to respond preferentially to information related to self, while more dorsal parts were more active when taking the perspective of others (D'Argembeau et al., 2007; Denny, Kober, Wager, & Ochsner, 2012; Murray, Schaer, & Debbane, 2012). Other brain regions like the insula and the caudate nucleus were also found to be more active for self- than for other-related judgements (Denny et al., 2012).

Taken together, the evidence shows that the neurocognitive processes that allow us to consider our own versus another person's perspective are largely—but not fully—overlapping. Considering our own perspective may have unique additional properties, as shown by the contribution of brain regions involved in reward (i.e., caudate nucleus), interoceptive awareness (i.e., insula), and personal value (i.e., ventromedial prefrontal cortex). Interestingly, this set of brain regions is also associated with the most emotional, visceral aspects of decision-making processes or the so-called "gut feeling" (Bechara & Damasio, 2005; Naqvi, Shiv, & Bechara, 2006). Accordingly, because of the specific properties of this self-relevance network, the projection bias that has repeatedly been described when projecting oneself in the future may manifest itself differently or be absent when thinking about others. For instance, the "present-ism bias" observed during future thinking could be partly due to the inherent qualities of this network of brain regions, coding for the visceral, somatosensory aspects of current self-relevance. Loewenstein (1996) has described how "immediately experienced visceral factors" may explain projection bias. However, specific characteristics of future simulations themselves, such as their unrepresentative, abbreviated, and decontextualized character (Gilbert & Wilson, 2007), may be equally operative when thinking about one's own and others' futures. Accordingly, the projection bias may be attenuated, but still present, when

thinking about another person's future as compared to thinking about one's own future. As noted above, such a pattern has been described for the optimism bias (Grysman et al., 2013).

The goal of the present study was to test this hypothesis. We investigated whether an underestimation of change of preferences over time is specific to self or is also present when making predictions about others. We used a perspective-taking task in which we asked young adults how much they liked stereotypically young-person things (e.g., Top 40 music, adventure vacations) and stereotypically old-person things (e.g., jazz, playing bridge) "now" and how much they will like them in the distant future (i.e., when they are 70). Participants also had to estimate how much a generic peer of their age and gender liked these same items now and in the future. Finally, participants had to rate how much an older adult (aged 70) liked these items. We hypothesized that participants would underestimate how much their preferences would change over time and that they would predict less change for themselves than for their peers. In addition, we predicted that this underestimation of change, although attenuated, would still be present when estimating their peer's preferences; that is, we expected that participants would judge that their peers would like stereotypically young items *more*, and stereotypically old items *less*, in the future than a generic older adult does now.

EXPERIMENTAL STUDY

Method

Participants

A total of 134 participants (28 males) took part in the perspective-taking task. They were recruited via the Integrated System of Participation in Research (ISPR) of the School of Psychology at the University of Ottawa. Participants obtained course credit for their participation. Their mean age was 19 years (± 2.54, range = 18–33), and they had completed 14 years (± 2.17, range = 11–23) of education on average. All participants signed an online informed consent form accepted by the Ethics Board of the University of Ottawa.

Perspective-taking task

Selection of stimuli. We asked 22 young adults who did not participate in the experiment (4 males, mean age = 24 ± 4.27 years, range = 19–34; mean level of education = 17 years ± 3.01, range = 12–22) how much they liked 21 stereotypically young-person things (e.g., Top 40 music, adventure vacations) and 21 stereotypically old-person things (e.g., jazz, playing bridge) now, and how much they would like these items when they are 70 years old. In each trial, participants responded using a 7-point Likert scale: 1–strongly dislike, 2–dislike, 3–dislike somewhat, 4–neutral, 5–like somewhat, 6–like, 7–strongly like. We then selected the items that differed most in ratings according to time (now versus future). Twelve stereotypically young-person things and 12 stereotypically old-person things were selected as differing significantly in their ratings (i.e., young items being preferred in the present relative to the future, and old items being preferred in the future relative to the present; see Appendix for the list of stimuli).

We subsequently verified that these items also showed this effect in the participants of our perspective-taking experiment ($N = 134$; see details above). All items differed significantly in their preference ratings according to time: Stereotypically young items were preferred in the present relative to the future, and stereotypically old items were preferred in the future relative to the present (all ps < .03).

Task design. In the perspective-taking task, we asked participants how much they liked stereotypically young-person things (e.g., Top 40 music, adventure vacations) versus stereotypically old-person things (e.g., jazz, playing bridge) now, and how much they would like them when they are 70 years old. We also asked them to rate these items for two generic same-sex adults: a same-age peer and an older adult. The same 7-point Likert scale as that in the norming study was used (from 1-strongly dislike to 7-strongly

like; see above). Participants completed the task online using an internet questionnaire (https://www.surveymonkey.com/). The following instructions were used:

In this experiment you are going to be asked about likes and dislikes. Sometimes you will be asked about your own likes and dislikes, and sometimes about somebody else's. Some of the questions will be about right now, and some will be about the future. For each trial you will be told whose likes and dislikes you are being asked about, and whether we are asking you about right now or the future. You will be asked to rate the extent to which you or someone else would like something or not. For the trials in which you will have to judge the likes and dislikes of someone else, you will have to think about: someone else your age or a 70-year-old.

Separate male and female versions of the task were created. In trials in which participants were asked to take a perspective other than their own, the photograph of an unknown face from the same gender as that of the participant was presented: It was either "Someone else your age" or "a 70-year-old". There were three perspectives (self, peer, older adult), and two times (now versus future) that corresponded to five separate blocks of trials: self–now, self–future, peer–now, peer–future, older adult–now. Blocks were presented in random order. Within each block, items (stereotypically young versus stereotypically old) were also presented randomly.

Statistical analyses

To test the extent to which predictions about one's own and a peer's future preferences were similar/different, we conducted a repeated measures analysis of variance (ANOVA) on preference ratings with perspective (self versus peer), time (now versus future), and item type (stereotypically young versus stereotypically old-person things) as within-subject factors.

To test the extent to which predictions of self and peer's preferences would differ from an older adult's perspective, we conducted two additional repeated measures ANOVAs. These ANOVAs included perspective (self versus peer versus older adult) and item type (stereotypically young versus stereotypically old-person things) as within-subject factors (time was not included because the older adult was only asked about in the now condition). One of these analyses was conducted with the now conditions, and the other with the future conditions of self and peer. Sex of participants was added as a between-subjects factor in all analyses.

Partial eta-squared (η_p^2) is indicated as a measure of effect size in all analyses.

Results

Main analysis

The repeated measures ANOVA on preference ratings revealed main effects of perspective, $F(1, 133) = 34.26$, $p < .001$, $\eta_p^2 = .21$, and time, $F(1, 133) = 151.35$, $p < .001$, $\eta_p^2 = .53$, and a three-way interaction between perspective, time, and item type, $F(1, 133) = 79.10$, $p < .001$, $\eta_p^2 = .37$. No interaction with sex of participants was found (all $ps < .2$).

Analyses by time period

Now condition. Analyses for the now conditions showed main effects of perspective, $F(1, 133) = 30.71$, $p < .001$, $\eta_p^2 = .19$, and item type, $F(1, 133) = 440.76$, $p < .001$, $\eta_p^2 = .77$, as well as an interaction between perspective and item type, $F(1, 133) = 75.12$, $p < .001$, $\eta_p^2 = .36$. No interaction with sex of participants was found (all $ps < .5$). Subsequent analyses for stereotypically young items revealed that participants attributed higher preference ratings to their peer (mean = 5.56) than to themselves (mean = 4.88), $F(1, 133) = 85.84$, $p < .001$, $\eta_p^2 = .39$. In contrast, for stereotypically old items, participants gave slightly higher preference ratings to themselves (mean = 3.81) than to their peer (mean = 3.58), $F(1, 133) = 15.06$, $p < .001$, $\eta_p^2 = .10$.

These analyses thus suggest that participants rated themselves as liking "young" items less now than their peers, whereas they rated themselves as liking "old" items more (see Figure 1).

Future condition. Analyses for the future conditions revealed main effects of perspective, $F(1, 133) = 15.59$, $p < .001$, $\eta_p^2 = .11$, and item type, $F(1, 133) = 211.33$, $p < .001$, $\eta_p^2 = .61$, as well as an interaction between perspective and item type, $F(1, 133) = 7.24$, $p = .008$, $\eta_p^2 = .05$. No

interaction with sex of participants was found (all $ps < .2$). Further analyses for stereotypically young items showed no significant difference in ratings between self (mean = 3.34) and peer (mean = 3.31). In contrast, for stereotypically old items, participants gave slightly higher preference ratings to peer (mean = 4.9) than to themselves (mean = 4.5), $F(1, 133) = 20.69$, $p < .001$, $\eta_p^2 = .14$.

Thus, although participants judged that both they and their peers would show similar preference levels for young items in the future, they judged that their peers would prefer old items more in the future than they would (Figure 1).

Analyses by item type

Stereotypically young items. Analyses of preference ratings for stereotypically young items revealed main effects of time, $F(1, 133) = 658.69$, $p < .001$, $\eta_p^2 = .83$, and perspective, $F(1, 133) = 27.71$, $p < .001$, $\eta_p^2 = .17$, and an interaction between these two factors, $F(1, 133) = 16.75$, $p < .001$, $\eta_p^2 = .31$. Subsequent analyses showed that, as expected, young items were preferred now relative to the future both for self, $F(1, 133) = 447.24$, $p < .001$, $\eta_p^2 = .77$, and for a peer, $F(1, 133) = 514.97$, $p < .001$, $\eta_p^2 = .79$. Importantly, however, the difference in ratings was greater for peer (5.56 vs. 3.31) than for self (4.88 vs. 3.34). Thus, as predicted, participants predicted less change in their own future preferences than they did for a same-age peer. No interaction with sex of participants was found for any of the variables (all $ps < .3$).

Stereotypically old items. Analyses of preference ratings for stereotypically old items revealed a main effect of time, $F(1, 133) = 261.21$, $p < .001$, $\eta_p^2 = .66$, and an interaction between time and perspective, $F(1, 133) = 52.57$, $p < .001$, $\eta_p^2 = .28$. Again, as expected, old items were rated as preferred in the future relative to now, both for self, $F(1, 133) = 138.05$, $p < .001$, $\eta_p^2 = .51$, and for a peer, $F(1, 133) = 236.78$, $p < .001$, $\eta_p^2 = .64$, but the difference in ratings was slightly greater for peer (4.9 vs. 3.6) than self (4.5 vs. 3.8), which again suggests that participants were predicting

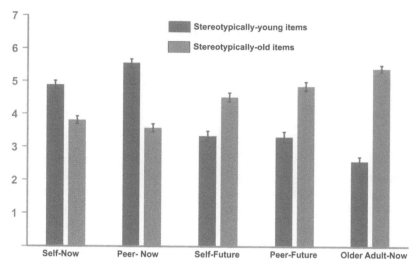

Figure 1. *Average preference ratings. Mean preference ratings (with 95% confidence interval bars) for self and a same-sex peer "now" and in the future are represented, as well as mean current preferences of an older adult. Participants indicated their preferences using a 7-point Likert scale (from 1–strongly dislike to 7–strongly like) separately for stereotypically young items (e.g., Top 40 music, adventure vacations) and stereotypically old items (e.g., jazz, playing bridge). To view this figure in colour, please visit the online version of this Journal.*

less change for self than for another generic peer (see Figure 1). No interaction with sex of participants was found for any of the variables (all $ps < .4$).

Comparisons of young and older adults

Comparison between self, peer, and older adult "now". To verify that participants did indeed judge that they and their peers would show higher preference ratings for young items now and, conversely, lower preference ratings for old items now than an older adult would, we conducted a repeated measures ANOVA comparing older adults, peer, and self in the "now" condition. This analysis revealed a main effect of perspective, $F(2, 266) = 80.98$, $p < .001$, $\eta_p^2 = .38$, and an interaction between perspective and item type, $F(2, 266) = 699.37$, $p < .001$, $\eta_p^2 = .84$.

Stereotypically young items. For stereotypically young items, there was a main effect of perspective, $F(2, 266) = 628.04$, $p < .001$, $\eta_p^2 = .83$. Follow-up analyses showed that, for these items, participants gave higher preference ratings to a peer (mean = 5.56), $F(1, 133) = 866.25$, $p < .001$, $\eta_p^2 = .87$, and to themselves (mean = 4.88),

$F(1, 133) = 687.90$, $p < .001$, $\eta_p^2 = .84$, than to an older adult (mean = 2.57).

Stereotypically old items. For stereotypically old items, there was also a main effect of perspective, $F(2, 266) = 335.02$, $p < .001$, $\eta_p^2 = .72$. Subsequent analyses showed that, for these items, participants attributed higher preference ratings to an older adult (mean = 5.40) than to a peer (mean = 3.58), $F(2, 266) = 418.29$, $p < .001$, $\eta_p^2 = .76$, and to themselves (mean = 3.81), $F(1, 133) = 406.42$, $p < .001$, $\eta_p^2 = .75$.

These analyses thus showed that participants judged that they and their peers currently liked young items better and old items less than an older adult did (see Figure 1).

Comparison between older adult "now" and self and peer in the future

To determine whether the extent of predicted change of preferences of self and peer in the future were comparable to those of an older adult, we conducted a repeated measures ANOVA comparing older adult "now" with peer and self in the future condition. This analysis revealed main effects of perspective, $F(2, 266) = 8.16$, $p < .001$, $\eta_p^2 = .06$, and item type, $F(1, 133) = 478.70$, $p < .001$, $\eta_p^2 = .78$, and an interaction between perspective and item type, $F(2, 266) = 99.21$, $p < .001$, $\eta_p^2 = .43$.

Stereotypically young items. For stereotypically young items, there was a main effect of perspective, $F(2, 266) = 65.54$, $p < .001$, $\eta_p^2 = .33$. Follow-up analyses showed that, for these items, participants gave higher preference ratings to self (mean = 3.34), $F(1, 133) = 111.212$, $p < .001$, $\eta_p^2 = .46$, and peer (mean = 3.31), $F(1, 133) = 99.77$, $p < .001$, $\eta_p^2 = .43$, than to an older adult (mean = 2.57).

Stereotypically old items. For stereotypically old items, there was also a main effect of perspective, $F(2, 266) = 82.91$, $p < .001$, $\eta_p^2 = .38$. Subsequent analyses revealed that, for these items, participants gave higher preference ratings to an older adult (mean = 5.40) than to a peer (mean = 4.86), $F(1, 133) = 73.66$, $p < .001$, $\eta_p^2 = .36$, or to themselves (mean = 4.52), $F(1, 133) = 161.19$, $p < .001$, $\eta_p^2 = .55$.

Together, these analyses suggest that even though participants estimated that their preferences and those of their peer would change in the future, these preferences would still differ from those of an older adult (see Figure 1).

Discussion

The goal of the present study was to test whether people's tendency to underestimate the extent to which their preferences will change in the future is specific to self or is also present when making predictions about others. Participants predicted less change between their own current and future preferences than between the current and future preferences of a generic same-sex peer. This was observed for preferences relating to stereotypically old (e.g., jazz, playing bridge) as well as stereotypically young (e.g., Top 40 music, adventure vacations) items.

However, differences in ratings between self and other were found both for current and future preferences. That is, participants thought that their peers currently liked stereotypically young items more and stereotypically old items less than they did themselves. They also reported that other young people would like more stereotypically old items than they would themselves in the future. These findings are reminiscent of the "false uniqueness" bias in reasoning that has been repeatedly demonstrated when comparing oneself to others (reviewed in Chambers, 2008). False uniqueness is our tendency to think that we are "better than average", unique, or, in the present case, less typical, with respect to our abilities and personalities. Here, we found that items that were most relevant in the present for our young participants (i.e., stereotypically young items) were rated as more preferred by other young peers than self. Conversely, items that were more relevant in the future (i.e., stereotypically old items) were rated as more preferred by others than by self in the future. Our findings therefore indicate that false uniqueness biases are similar for judgements relating to the present and to the future.

Interestingly, our participants did not have a strict "end of history illusion" (Quoidbach et al., 2013), as they still believed that their preferences would change over time. Indeed, judgements about the present as compared to the future were characterized by qualitatively similar inverted ratings for self and other: In both cases, participants gave higher preference ratings to stereotypically young items in the present and to stereotypically old items in the future. This supports the observation that people generally understand qualitatively the direction in which their preferences and tastes will change in the future (Loewenstein et al., 2003). If participants had a strict presentism bias (Cameron et al., 2004) and only interpreted future selves in relation to present motives, they would have predicted that their love of young items would be as strong in the future as in the present—though an important point, and one to which we return, is that we asked our participants to make predictions about a very distant future.

Participants predicted that their preferences would change less over time than would those of a generic peer. This difference in the magnitude of estimated change suggests that distinct mechanisms may underlie people's judgements about their own and others' current and future preferences. Moreover, such mechanisms may already differ early in development given that a recent study showed that preschoolers are better at predicting that a peer's preferences will change in the future than they are at making this same prediction for self (Bélanger, Atance, Varghese, Nguyen, & Vendetti, 2014). Nevertheless, our findings do not imply that an underestimation of change of preferences over time is specific to self. When participants were asked to judge the current preferences of an older adult, we found that the difference in preferences between stereotypically old and young items was rated as much more pronounced than for self *and* for a peer. More specifically, older adults were considered to currently like stereotypically old items more, and stereotypically young items less, than both self and another young adult would in the future. These results therefore indicate that even though underestimation of change of preferences over time is attenuated when thinking about others, the phenomenon is still present as the estimated change in preferences of others does not bring them to the level of "current" older adults' preferences. However, another possible interpretation of these findings is that our participants anticipated differences in preferences for different generations (i.e., current older adults versus future older adults). While our design does not allow us to rule out this hypothesis, a generational change in preferences would likely have been associated with a global decrease in ratings for all items, as compared to current older adults (as other types of items—yet unknown—would be preferred by the next generation). However, our participants reported that, compared to a current older adult, their peer would like stereotypically young items *more* in the future and stereotypically old items *less*. These findings therefore seem more compatible with an underestimation of change of preferences over time, such as a presentism bias, than with anticipated generational changes in preferences.

One important aspect of our experimental design is that we asked our young participants to

think about a very distant future (i.e., when they, or a peer, are 70). This allowed us to compare preferences of distant selves (as an older adult) with preferences of current older adults. Prior research on future thinking and projection bias has tended to focus on the immediate (i.e., next day; e.g., Gilbert et al., 2002) or relatively near future (i.e., about 10 years forward; e.g., Quoidbach et al., 2013). Our results thus illustrate that adults underestimate the extent to which their preferences will change over time even in the distant future. If anything, thinking about a distant future should have resulted in participants experiencing "discontinuity in their senses of self" (Lampinen, Odegard, & Leding, 2004) and envisaging larger scale changes as they grow old. The fact that our participants still predicted less change for themselves than for their peers indicates that this projection bias is robust to these discontinuities. It will be interesting in future studies to use intermediate future conditions to see whether participants predict a progressive change in preferences over time, even if the amount of change never matches that attributed to peers, as demonstrated here for projections into the distant future.

A few other studies have reported similar biases when thinking about the near or distant future. For instance, Remedios, Chasteen, and Packer (2010; see also Sharot, Korn, & Dolan, 2011) found an optimism bias when young adults (i.e., mean age: 19 years) had to describe themselves at the age of 70, consistent with past research on optimism bias about one's own future (Sharot, 2011; Weinstein, 1980). Broadly, as compared to temporally close future events, distant future events are more abstract (Liberman & Trope, 1998; Trope & Liberman, 2000), contain fewer sensory and contextual details (D'Argembeau & Van der Linden, 2004), and are often represented in a third-, rather than in a first-, person perspective (D'Argembeau & Van der Linden, 2004, 2012). This is compatible with the idea that thinking about distant future events may often involve semantic rather than episodic forms of future thinking (Atance & O'Neill, 2001; Martin-Ordas, Atance, & Louw, 2012; Szpunar, Spreng, & Schacter, 2014).

In the present case, thinking about our future preferences may depend on subtypes of personal semantics such as self-knowledge and knowledge of autobiographical facts (reviewed in Renoult, Davidson, Palombo, Moscovitch, & Levine, 2012), whereas thinking about the preferences of a generic other may rely on general semantics. The fact that the presentism bias and the optimism bias (Remedios et al., 2010) appear to be observed for the distant as well as for the near future suggests that the reliance on episodic (i.e., near future) versus personal semantics (i.e., distant future) may not be the crucial factor at work. Both of these types of memory handle self-relevant information and involve partly overlapping neural networks, such as medial prefrontal, restrosplenial, and temporo-parietal cortices (Renoult et al., 2012). It may thus be, both for near and for distant future simulations, that it is the use of our own personal perspective (accompanied by our sense of uniqueness) as a "judgemental anchor" (Davis et al., 1986; Epley et al., 2004; Nickerson, 1999) that contaminates our attempts at future simulations. Accordingly, using general semantics to simulate the future of others results in a reduction of future simulation biases.

As such, it will be interesting in future studies to include a familiar-other condition. For example, Grysman et al. (2013) have reported that the optimism bias was attenuated when thinking about a nonclose friend, but similar when considering ourselves or a close friend. Interestingly, a number of neuroimaging studies have reported that the magnitude of self–other differentiation in the ventromedial prefrontal cortex depends on perceived similarity in personality traits (Benoit, Gilbert, Volle, & Burgess, 2010) or in-group membership (Morrison, Decety, & Molenberghs, 2012). It is thus likely that the strength of underestimation of change of preferences over time would be more similar between self and close other than between self and a generic peer.

Taken together, our results indicate that people's underestimation of change in preferences over time is attenuated when thinking about others. However, bias still exists given that people's estimated change in others' preferences does not directly map onto their judgements about "current" older adults' preferences. These findings have practical implications for decision making because they suggest that simulating the future perspectives of a typical peer or, even better, the current perspectives of an older adult, prior to making simulations for self, may lead to improved decision making (e.g., saving more money for retirement). Importantly, reported similarities in judgements between self and close others suggest that more realistic future simulations may be attained when considering unknown/generic others, for whom we are more likely to escape the visceral aspects of self-relevance that are associated with presentism biases.

Mispredictions about the future are not trivial because they can lead people to make decisions in the present that are based on preferences, emotions, and personality traits that may shift in the future. Our results show that these mispredictions of change are attenuated, but still present, when thinking about others, in turn suggesting both overlapping and distinct mechanisms in how we think about our own and others' futures.

REFERENCES

Atance, C. M., & O'Neill, D. K. (2001). Episodic future thinking. *Trends in Cognitive Sciences*, *5*(12), 533–539. doi:10.1016/S1364-6613(00)01804-0

Baker, L. A., & Emery, R. E. (1993). When every relationship is above average: Perceptions and expectations of divorce at the time of marriage. *Law and Human Behavior*, *17*(4), 439–450.

Bechara, A., & Damasio, A. R. (2005). The somatic marker hypothesis: A neural theory of economic decision. *Games and Economic Behavior*, *52*(2), 336–372. doi:10.1016/j.geb.2004.06.010

Bechara, A., Damasio, H., & Damasio, A. R. (2000). Emotion, decision making and the orbitofrontal cortex. *Cerebral Cortex*, *10*(3), 295–307. doi:10.1093/cercor/10.3.295

Bélanger, M. J., Atance, C. M., Varghese, A. L., Nguyen, V., & Vendetti, C. (2014). What will i like best when i'm all grown up? Preschoolers' understanding of future preferences. *Child Development*, *85*(6), 2419–2431. doi:10.1111/cdev.12282

Benoit, R. G., Gilbert, S. J., Volle, E., & Burgess, P. W. (2010). When I think about me and simulate you: Medial rostral prefrontal cortex and self-referential processes. *NeuroImage*, *50*(3), 1340–1349. doi:10.1016/j.neuroimage.2009.12.091

Buckner, R. L., & Carroll, D. C. (2007). Self-projection and the brain. *Trends in Cognitive Sciences*, *11*(2), 49–57. doi:10.1016/j.tics.2006.11.004

Cameron, J. J., Wilson, A. E., & Ross, M. (2004). Autobiographical memory and self-assessment. In D. R. Beike, J. M. Lampinen, & D. A. Behrend (Eds.), *The self and memory* (pp. 207–226). New York: Psychology Press.

Chambers, J. R. (2008). Explaining false uniqueness: Why we are both better and worse than others. *Social and Personality Psychology Compass*, *2*(2), 878–894.

D'Argembeau, A., Renaud, O., & Van der Linden, M. (2011). Frequency, characteristics and functions of future-oriented thoughts in daily life. *Applied Cognitive Psychology*, *25*(1), 96–103. doi:10.1002/Acp.1647

D'Argembeau, A., Ruby, P., Collette, F., Degueldre, C., Balteau, E., Luxen, A., … Salmon, E. (2007). Distinct regions of the medial prefrontal cortex are associated with self-referential processing and perspective taking. *Journal of Cognitive Neuroscience*, *19*(6), 935–944. doi:10.1162/jocn.2007.19.6.935

D'Argembeau, A., & Van der Linden, M. (2004). Phenomenal characteristics associated with projecting oneself back into the past and forward into the future: Influence of valence and temporal distance. *Consciousness and Cognition*, *13*(4), 844–858. doi:10.1016/j.concog.2004.07.007

D'Argembeau, A., & Van der Linden, M. (2012). Predicting the phenomenology of episodic future thoughts. *Consciousness and Cognition*, *21*(3), 1198–1206. doi:10.1016/j.concog.2012.05.004

Davis, H. L., Hoch, S. J., & Ragsdale, E. K. E. (1986). An anchoring and adjustment model of spousal predictions. *Journal of Consumer Research*, *13*(1), 25–37. doi:10.1086/209045

Decety, J., & Grezes, J. (2006). The power of simulation: Imagining one's own and other's behavior. *Brain Research*, *1079*, 4–14. doi:10.1016/j.brainres.2005.12.115

Denny, B. T., Kober, H., Wager, T. D., & Ochsner, K. N. (2012). A meta-analysis of functional neuroimaging studies of self- and other judgments reveals a spatial gradient for mentalizing in medial prefrontal cortex. *Journal of Cognitive Neuroscience*, *24*(8), 1742–1752. doi:10.1162/jocn_a_00233

Epley, N., Keysar, B., Van Boven, L., & Gilovich, T. (2004). Perspective taking as egocentric anchoring and adjustment. *Journal of Personality and Social Psychology*, *87*(3), 327–339. doi:10.1037/0022-3514.87.3.327

Gilbert, D. T., Gill, M. J., & Wilson, T. D. (2002). The future is now: Temporal correction in affective forecasting. *Organizational Behavior and Human Decision Processes*, *88*(1), 430–444. doi:10.1006/obhd.2001.2982

Gilbert, D. T., & Wilson, T. D. (2007). Prospection: Experiencing the future. *Science*, *317*(5843), 1351–1354. doi:10.1126/science.1144161

Goldman, A. A. (2002). Simulation theory and mental concepts. In J. P. J. Dokic (Ed.), *Simulation and knowledge of action* (pp. 1–19). Amsterdam: John Benjamins Publishing Company.

Gordon, R. M. (1986). Folk psychology as simulation. *Mind & Language*, *1*(2), 158–171.

Grysman, A., Prabhakar, J., Anglin, S. M., & Hudson, J. A. (2013). The time travelling self: Comparing self and other in narratives of past and future events. *Consciousness and Cognition*, *22*(3), 742–755. doi:10.1016/j.concog.2013.04.010

Hassabis, D., & Maguire, E. A. (2007). Deconstructing episodic memory with construction. *Trends in Cognitive Sciences*, *11*(7), 299–306. doi:10.1016/j.tics.2007.05.001

Lampinen, J. M., Odegard, T. N., & Leding, J. K. (2004). Diachronic disunity. In D. R. Beike, J. M. Lampinen, & D. A. Behrend (Eds.), *The self and memory* (pp. 227–253). New York: Psychology Press.

Liberman, N., & Trope, Y. (1998). The role of feasibility and desirability considerations in near and distant future decisions: A test of temporal construal theory. *Journal of Personality and Social Psychology*, *75*(1), 5–18. doi:10.1037//0022-3514.75.1.5

Loewenstein, G. (1996). Out of control: Visceral influences on behavior. *Organizational Behavior and Human Decision Processes*, *65*(3), 272–292. doi:10.1006/obhd.1996.0028

Loewenstein, G., O'Donoghue, T., & Rabin, M. (2003). Projection bias in predicting future utility. *Quarterly Journal of Economics*, *118*(4), 1209–1248. doi:10.1162/003355303322552784

Martin-Ordas, G., Atance, C. M., & Louw, A. (2012). The role of episodic and semantic memory in episodic foresight. *Learning and Motivation*, *43*(4), 209–219. doi:10.1016/j.lmot.2012.05.011

Morrison, S., Decety, J., & Molenberghs, P. (2012). The neuroscience of group membership. *Neuropsychologia*,

50(8), 2114–2120. doi:10.1016/j.neuropsychologia. 2012.05.014

Murray, R. J., Schaer, M., & Debbane, M. (2012). Degrees of separation: A quantitative neuroimaging meta-analysis investigating self-specificity and shared neural activation between self- and other-reflection. *Neuroscience Biobehavioral Reviews, 36*(3), 1043–1059. doi:10.1016/j.neubiorev.2011.12.013

Naqvi, N., Shiv, B., & Bechara, A. (2006). The role of emotion in decision making: A cognitive neuroscience perspective. *Current Directions in Psychological Science, 15*(5), 260–264. doi:10.1111/j.1467-8721.2006.00448.x

Nickerson, R. S. (1999). How we know - and sometimes misjudge - What others know: Imputing one's own knowledge to others. *Psychological Bulletin, 125*(6), 737–759. doi:10.1037//0033-2909.125.6.737

Quoidbach, J., Gilbert, D. T., & Wilson, T. D. (2013). The end of history illusion. *Science, 339*(6115), 96–98. doi:10.1126/science.1229294

Remedios, J. D., Chasteen, A. L., & Packer, D. J. (2010). Sunny side up: The reliance on positive age stereotypes in descriptions of future older selves. *Self and Identity, 9*(3), 257–275. Pii 913840139. doi:10.1080/15298860903054175

Renoult, L., Davidson, P. S., Palombo, D. J., Moscovitch, M., & Levine, B. (2012). Personal semantics: At the crossroads of semantic and episodic memory. *Trends in Cognitive Sciences, 16*(11), 550–558. doi:10.1016/j.tics.2012.09.003

Robinson, H., Calamia, M., Glascher, J., Bruss, J., & Tranel, D. (2014). Neuroanatomical correlates of executive functions: A neuropsychological approach using the EXAMINER battery. *Journal of the International Neuropsychological Society, 20*(1), 52–63. doi:10.1017/S135561771300060X

Sharot, T. (2011). The optimism bias. *Current Biology, 21*(23), R941–R945.

Sharot, T., Korn, C. W., & Dolan, R. J. (2011). How unrealistic optimism is maintained in the face of reality. *Nature Neuroscience, 14*(11), 1475–1479. doi:10.1038/Nn.2949

Szpunar, K. K., Spreng, R. N., & Schacter, D. L. (2014). A taxonomy of prospection: Introducing an organizational framework for future-oriented cognition. *Proceedings of the National Academy of Science of the United States of America, 111*(52), 18414–18421. doi:10.1073/pnas.1417144111

Trope, Y., & Liberman, N. (2000). Temporal construal and time-dependent changes in preference. *Journal of Personality and Social Psychology, 79*(6), 876–889. doi:10.1037//0022-3514.79.6.876

Weinstein, N. D. (1980). Unrealistic optimism about future life events. *Journal of Personality and Social Psychology, 39*(5), 806–820. doi:10.1037/0022-3514.39.5.806

APPENDIX

List of stimuli used in the perspective-taking task

Stereotypically young-person things:

Energy drinks
Adventure vacations
Canoe camping
Rollerblading
Converse sneakers
Top 40 music
American Eagle Outfitters
Facebook
Texting
Living downtown
Bungee jumping
Whitewater rafting

Stereotypically old-person things:

Playing bridge
Dinner at 5
Jazz
Sears department store
Oldsmobile
Game shows
Suspenders
Scrapbooking
English breakfast tea
Living in the suburbs
Birdwatching
Gardening

The relationship between prospective memory and episodic future thinking in younger and older adulthood

Gill Terrett[1], Nathan. S. Rose[1], Julie D. Henry[2], Phoebe E. Bailey[3], Mareike Altgassen[4,5], Louise H. Phillips[6], Matthias Kliegel[7], and Peter G. Rendell[1]

[1]School of Psychology, Australian Catholic University, Melbourne, Australia
[2]School of Psychology, University of Queensland, Brisbane, Australia
[3]School of Social Sciences and Psychology, University of Western Sydney, Sydney, Australia
[4]Donders Institute for Brain, Cognition and Behaviour, Radboud University, Nijmegen, The Netherlands
[5]Department of Psychology, Technische Universitaet Dresden, Dresden, Germany
[6]School of Psychology, University of Aberdeen, Aberdeen, UK
[7]Department of Psychology, University of Geneva, Geneva, Switzerland

Episodic future thinking (EFT), the ability to project into the future to "preexperience" an event, and *prospective memory* (PM), remembering to perform an intended action, are both examples of future-oriented cognition. Recently it has been suggested that EFT might contribute to PM performance but to date few studies have examined the relationship between these two capacities. The aim of the present study was to investigate the nature and specificity of this relationship, as well as whether it varies with age. Participants were 125 younger and 125 older adults who completed measures of EFT and PM. Significant, positive correlations between EFT and PM were identified in both age groups. Furthermore, EFT ability accounted for significant unique variance in the young adults, suggesting that it may make a specific contribution to PM function. Within the older adult group, EFT did not uniquely contribute to PM, possibly indicating a reduced capacity to utilize EFT, or the use of compensatory strategies. This study is the first to provide systematic evidence for an association between variation in EFT and PM abilities in both younger and older adulthood and shows that the nature of this association varies as a function of age.

In everyday life, we spend a considerable amount of time thinking about and planning our future. This ability to disengage from the present to contemplate potential future scenarios is highly adaptive as it allows us to act in anticipation of future needs (Hanson & Atance, 2014). There are a number of different types of future-oriented cognitions, two of which will be the focus of the current study: *episodic future thinking* (EFT) and *prospective memory* (PM). While both have individually attracted attention in the literature in terms of, for example, their developmental trajectories (e.g.,

We would like to thank Clare Ryrie and Melissa Bugge for their assistance in recruiting and testing participants, Fiona Sparrow for transcribing, and Ashleigh Dever, Kimberly Mercuri, and Matthew Nangle for scoring, the Autobiographical Interview (AI) interviews.

This research was supported by a Discovery Project grant from the Australian Research Council [grant number DP110100652].

Busby & Suddendorf, 2005; Mahy, Moses, & Kliegel, 2014) and level of impairment in clinical groups (e.g., Henry et al., 2014; Terrett et al., 2013), little attention has been paid to the nature of the relationship between these constructs. This is a surprising omission, since clarifying this relationship would further extend our understanding of the cognitive mechanisms implicated in PM function and would also provide an important test of the claim that EFT is a key determinant of PM performance (Brewer & Marsh, 2010).

EFT refers to the ability to imagine experiencing future situations (Addis, Wong, & Schacter, 2008; Atance & O'Neill, 2001; Schacter, Addis, & Buckner, 2007). It is a complex process requiring the construction of mental scenes and as such is likely to rely on multiple cognitive processes including executive control (D'Argembeau, Ortoleva, Jumentier, & Van der Linden, 2010), semantic memory (Irish, Addis, Hodges, & Piguet, 2012), and self-projection (Buckner & Carroll, 2007).

Particular attention, however, has been paid to the role of episodic memory. According to the *constructive episodic simulation hypothesis*, episodic future thinking relies heavily on the retrieval of personally relevant memories of the past as these provide basic information that is then flexibly recombined to construct novel future events (Schacter et al., 2007). Evidence of a core neural network underpinning both episodic memory and EFT adds support to the claim that episodic memory and EFT are closely related (Weiler, Suchan, & Daum, 2010; see Schacter et al., 2012, for a review).

PM refers to the ability to remember to carry out a planned action. Successful completion of a PM task involves several sequential stages: (a) forming the intention; (b) retaining the intention in memory while engaged in other ongoing activities; (c) initiating the intended action at the appropriate time; and (d) evaluating the outcome of the action (Einstein & McDaniel, 1996; Ellis, 1996; Ellis & Freeman, 2008). A number of cognitive abilities have been implicated in PM, including executive functioning and retrospective memory (see Kliegel, Altgassen, Hering, & Rose, 2011, for a review). Retrospective memory in particular has

been shown to play a key role as PM tasks require retention of the content of the tasks as well as the circumstances in which it is to be executed (Kliegel, McDaniel, & Einstein, 2000; Zöllig et al., 2007).

Recently it has been proposed that EFT might be an important determinant of PM performance (e.g., Schacter, Addis, & Buckner, 2008; Szpunar, 2010). More specifically, it has been suggested that mentally simulating a future intended action strengthens encoding of the content of the PM intention during the intention formation stage (Brewer & Marsh, 2010; Schnitzspahn & Kliegel, 2009), which increases the likelihood that the PM task will be successfully completed. Currently two main lines of evidence support a possible link between EFT and PM. The first comes from research that has found future simulation to be an effective strategy for improving PM performance (e.g., Altgassen et al., 2015; Brewer & Marsh, 2010; Leitz, Morgan, Bisby, Rendell, & Curran, 2009; McFarland & Glisky, 2012; Neroni, Gamboz, & Brandimonte, 2014). For example, in a lab-based study, Paraskevaides et al. (2010) found that imagining performing PM tasks reduced the adverse effects of alcohol consumption. In addition, Brewer and Marsh (2010) found that young adults who were asked to form an implementation intention (i.e., "If X occurs, then I will do Y"; Gollwitzer, 1999) while imagining themselves undertaking a PM task performed better than those who were given standard instructions (i.e., "Please do Y whenever X occurs") with no additional future visualization component. In both studies it was argued that imagining successful completion of the PM task at the appropriate future time-point, a form of episodic foresight, allows participants to preexperience the retrieval cues. By increasing the salience of retrieval cues that indicate the point at which the PM task should be performed, it is suggested that they are detected more efficiently, which in turn increases the likelihood that PM tasks will be successfully completed.

The second source of support for a link between PM and EFT comes from neuroimaging studies. For example, it has been shown that imagining produces similar brain activation to that produced when experiencing a specific situation in real life (e.g., Stokes, Thompson, Cusack, & Duncan, 2009). Given that greater similarity in brain activation at encoding and retrieval is associated with better PM performance (e.g., West & Ross-Munro, 2002), it has therefore been argued that imagining the future cue when forming an intention increases the similarity in brain activity at encoding and retrieval (Gilbert, Armbuster, & Panagiotidi, 2012), thereby supporting successful PM performance. Furthermore, a high degree of overlap has been shown in brain areas, particularly BA10, that are activated during both PM (e.g., Burgess, Scott, & Frith, 2003) and EFT (Addis, Wong, & Schacter, 2007; Weiler et al., 2010), additionally supporting the claim that the two processes are related. However, it should be noted that this common neural network, as previously noted, is also activated during episodic memory (Addis et al., 2007; Weiler et al., 2010), raising the possibility that any association between EFT and PM may be a reflection of a common reliance, at least to some degree, on episodic memory.

Nevertheless, these two research areas have been used to support the argument that EFT plays a key role in PM. The first critical test of this hypothesis, however, is showing that there is a direct, positive association between these two constructs. Investigations using this approach are, however, surprisingly limited. To the best of our knowledge there have been only two such studies (Atance & Jackson, 2009; Nigro, Brandimonte, Cicogna, & Cosenza, 2013), and both involved young children. While these studies did not find a relationship between EFT and PM in children under the age of 5 years, Nigro et al. (2013) did report a significant, albeit moderate, correlation between EFT and PM amongst an additional group of older children (7-year-olds), suggesting that this relationship may not emerge until middle childhood. However, this association has not been directly investigated in adulthood. A further important test of this hypothesis is showing that EFT contributes to PM over and above the contribution of other variables previously shown to be associated with PM, in particular retrospective memory.

The aim of the current study was therefore to advance our understanding of the relationship between EFT and PM, as well as the contribution of EFT to PM, by conducting a large-scale assessment of these abilities in two distinct adult populations. Specifically, because of considerable prior evidence showing that the capacity for future-oriented cognition may be reduced for older relative to younger adults (Addis, Musicaro, Pan, & Schacter, 2010; Addis et al., 2008; Lyons, Henry, Rendell, Corballis, & Suddendorf, 2014; see Henry, MacLeod, Phillips, & Crawford, 2004, for a review), and because age is known to disrupt many of the cognitive abilities likely to be important to engage in both types of prospective thought (e.g., Prull, Gabrieli, & Bunge, 2000; Salthouse, 2000), we tested younger and older adulthood separately. First, we assessed the association between the two constructs and hypothesized that EFT and PM would be positively related in young adults. We also expected this to be the case for older adults, based on the assumption that the positive relationship between EFT and PM would be retained despite an anticipated reduction in both of these abilities in this group. Second, we investigated whether EFT ability would contribute to PM over and above individual differences in age, general cognitive ability, and retrospective memory for both young and older adults.

EXPERIMENTAL STUDY

Method

Participants

Participants were 125 young adults (18–30 years) and 125 older adults (65–85 years). Both groups were approximately 70% female. Participant characteristics for the two age groups are shown in Table 1. Both groups were recruited from the community and received $30 Australian, for participating in a two- to three-hour testing session. Older adults were screened for possible dementia using the Addenbrooke's Cognitive Examination Revised (ACE–R; Mathuranath, Nestor, Berrios,

Table 1. *Characteristics of the sample*

Characteristic	Young adults (age range 18–30 years) n = 125		Older adults (age range 65–85 years) n = 125		t test[a] (df = 248)	
	M	SD	M	SD	t	p
Number of women	89 (71.2)		87 (69.6)			
Age (in years)	22.90	3.45	73.80	5.57		
Education (in years)	13.94	2.16	13.30	3.61	1.72	.086
Self-rated health[b]	2.11	0.84	2.21	0.94	0.85	.395
Self-rated sleep	2.56	0.93	2.66	1.00	0.79	.432
Geriatric Depression Scale	7.21	4.75	4.04	3.84	5.80	<.001
NART Verbal IQ[c]	98.20	8.13	110.17	8.44	11.42	<.001
Verbal Fluency[d]	57.37	13.07	56.32	14.14	0.61	.543
ACE–R[e]			92.70	3.76		
MMSE[f]			29.37	0.86		

Note: Percentages in parentheses.

[a]Age groups were compared on each measure with separate independent-groups *t* tests. [b]Self-rated health on day of test and sleep over past month were rated on a 5-point scale: 1 = excellent; 2 = very good, 3 = good; 4 = not very good; 5 = poor. [c]Verbal IQ score as predicted from the number of errors made on the NART (National Adult Reading Test). [d]Verbal fluency was total number of words generated for P, R, W, and animals. [e]ACE–R refers to the Addenbrooke's Cognitive Examination Revised—range of scores was 84–100. [f]MMSE refers to the Mini-Mental State Examination—range of scores was 27–30.

Rakowics, & Hodges, 2000). All older adults had normal mental status as indexed by scores greater than 83 (out of a possible 100) on the ACE–R and scores greater than 26 (out of a possible 30) for the Mini-Mental State Examination (MMSE; Mathuranath et al., 2000). The age groups did not differ in years of education, verbal fluency, self-ratings of sleep, or self-ratings of health on day of testing. The older adults had higher mean predicted verbal IQ, based on the National Adult Reading Test (NART) vocabulary test, and had lower levels of depression than the young adults (see Table 1).

Materials and procedure
Episodic future thinking. The Autobiographical Interview (AI) procedure originally developed by Levine, Svoboda, Hay, and Winocur (2002) and adapted by Addis et al. (2008) was used to assess participants' ability to generate episodic content when describing past and future events. This is an interview format procedure during which participants are presented with a set of cue words. They are required to generate details

about either a past or a future event in response to each word. In the current study, each participant was asked to describe six events: three in each of two conditions (past three years and next three years). All three events relating to one temporal direction, past or future, were completed before commencing the further three events relating to the other temporal direction, and temporal direction was counterbalanced across participants. A total of six cue words were administered to participants. We followed a two-stage randomizing process to assign the words to the past and future conditions. First, for each participant, the six words were randomly selected from a list of 20 possible cue words (e.g., apple, letter, baby). These words were nouns that were all rated high on frequency, imageability, and concreteness (Clark & Paivio, 2004). Second, these six words were then randomly allocated to past and future conditions for each participant.

AI testing sessions. First, the interviewers were trained in the procedures outlined by Addis et al. (2008) and were given an interview administration manual.[1] Several pilot interviews conducted by the interviewers were then audiotaped and analysed for consistency with the procedures used by Addis et al. (2008). Feedback was provided. Following completion of this training, testing of participants began. Testing sessions commenced with participants receiving instructions about the task, including a demonstration of a cue word and sample response. In the interview that followed, for each of the three trials relating to a particular temporal direction (past or future), participants were given a cue word (e.g., baby) and had three minutes to generate as many details as possible about an event that they either experienced in the past or could imagine experiencing in the future. The event did not have to relate directly to the cue word, and participants were encouraged to freely associate. They were told that the event had to refer to a specific time and place, it had to be less than one day in duration, and future events had to be realistic and not previously experienced. Participants were also directed to describe the event from their own perspective rather than that of an observer. During the three minutes, the relevant cue word was displayed on a card along with the task instruction ("recall past event" or "imagine future event"). If necessary, the experimenter gave general probes in line with protocols set out by Addis et al. (2008) such as asking general questions to clarify instructions and facilitate further description of event details. Responses were recorded using a digital audio recorder and were sent to a professional transcriber who was independent of the study, and who was blind to participants' group membership and study hypotheses.

AI scoring. Standardized AI scoring procedures outlined by Addis et al. (2008) were applied. The following scoring procedure was applied. A central event was first identified in the transcription for each cue word trial, and this was then sectioned into key details, or unique chunks of information.

These details were categorized as either *internal* (episodic information related to central event) or *external* (nonepisodic details, including semantic information, information of other events not specific in time and place to the central event, and repetitions). The total number of internal details generated across all three future events was the primary measure of EFT.

Four scorers, independent from the study, blind to group membership, and blind to the experimental hypotheses, scored the transcripts. Prior to commencing, the scorers completed the training procedures set out by Addis et al. (2008) using the manual and training events provided by Donna Addis (see Footnote 1). Their scores during training were assessed for consistency with four experienced scorers from Donna Addis's lab (see Footnote 1). Inter-rater reliability for our four scorers was then assessed on the basis of a two-way mixed-design analysis of variance (ANOVA) intraclass correlation analysis of their scores on the first 12 participants' responses. The Cronbach alphas obtained for our scorers were .90 for internal details and .89 for external details.

Prospective memory. Virtual Week is a laboratory measure designed to represent PM in daily life (original version, Rendell & Craik, 2000; and for recent review, Rendell & Henry, 2009). It uses a computerized board game format in which participants move their token around the board by rolling a die and moving their token the number of squares indicated by the die. Each circuit of the board represents a day from 8 a.m. to 11 p.m., thus capturing the hours that participants are typically awake. As participants move their token around the board and pass each event "E" square, they must pick up an "event card". There are 10 event cards per virtual day, with each card presenting participants with a plausible event to pretend to be engaged in and three options for activities to participate in during the event. Based on the activity that the participant selects, a dice roll consequence is revealed,

---

[1] We thank Donna Addis for providing advice and detailed manuals for Autobiographical Interview administration and scoring procedures consistent with Addis et al. (2008), and for data and information allowing inter-rater reliabilities to be calculated between the scorer in the current study and those from her lab.

such as "roll an even number", or "roll a 6", which must be achieved before the participant can move their token and continue on with the game. In addition to selecting activities, rolling the die, and moving the token, participants are to remember to carry out several PM tasks. Thus, the PM tasks are embedded in the ongoing activity of simulating participation in the Virtual Week.

Each day of the Virtual Week includes 10 PM tasks (four regular, four irregular, and two time-check). The four regular PM tasks are the same each day and simulate the kinds of regular tasks that occur as one undertakes normal duties (such as taking medication every day at the same time); two of the regular PM tasks are event based (triggered by specific event cards: breakfast and dinner event cards), and two are time based (triggered by specific virtual times of the day: 11 a.m. and 9 p.m.). The four irregular PM tasks (also two event and two time based) are different each day and simulate the kinds of *occasional* tasks that can occur in daily life. These tasks thus place greater demands on retrospective memory than regular tasks that recur each day (Kliegel, Martin, McDaniel, & Einstein, 2002). In the original version of Virtual Week, which was used in prior studies (e.g., Rose et al., 2010), time of day was cued by having consecutive hours of the day marked on the squares on the board. A recent innovative feature of Virtual Week, which was used in the current study, is that both the regular and irregular time-based tasks require monitoring a virtual clock calibrated to the token position on the board (see Rendell et al., 2011). The two time-check tasks are the same each day but require the participant to disengage from the activities of the game to monitor real time on the stop-clock that is displayed on the board and indicate when a specified period of time had passed: 2 min 30 s and 4 min 15 s from the start of each virtual day. In this study, all participants completed the practice day and four virtual days. In order to maximize the reliability of PM measurement and variability in the samples, overall PM performance (i.e., a composite score of proportion correct across all types of PM tasks in Virtual Week) was the dependent measure under consideration in this study.[2]

As in recent computerized versions of Virtual Week, a standardized set of instructions with help messages and prompts were presented throughout the practice day. Participants were tested one at a time, and an experimenter sat with them throughout the practice day, highlighting the help messages and ensuring that the participants understood all procedures before starting their first virtual day. Just prior to starting the first virtual day, the experimenter informed the participant that there would be no more help messages on the screen or assistance provided by the experimenter.

Retrospective memory. Upon completion of each virtual day, participants were immediately asked to write down the PM tasks they remembered being required to carry out during the virtual days. Participants were instructed that they did not need to recall the target cue that prompted the task, only the content of the task. This recall task (scored as mean proportion of PM tasks correctly recalled) provides an index of the retrospective memory component of the PM tasks. As in Rendell et al. (2011), only recall of irregular PM tasks was used to measure retrospective memory because, as in previous studies using Virtual Week in adults (e.g., Rose, Rendell, McDaniel, Aberle, & Kliegel, 2010), postexperimental retrospective memory of regular PM tasks was at ceiling.

Verbal intelligence. Verbal intelligence was estimated using the National Adult Reading Test (NART; Nelson, 1982). Participants are required to read aloud 50 English words with phonetically atypical pronunciation (e.g., *cellist*). Standardized IQ scores

---

[2]As part of another study, participants were directed to use different encoding strategies when performing some of the PM tasks. Analyses of the overall PM score (the index of PM ability used in this study) revealed that there was no interaction between age and encoding condition, and, furthermore, a similar pattern of results to those reported below was obtained when encoding condition was controlled for in the multiple regression analyses. In the interests of parsimony, therefore, the analyses reported in this paper do not include effects of the encoding condition manipulation.

are calculated, with higher scores indicating better performance. The NART is a strong predictor of intellectual function in the normal population (Bright, Jaldow, & Kopelman, 2002). Reliability estimates for the NART range from .90 to .93, and it has been well documented as a valid and reliable estimate of verbal intelligence (Crawford, Henry, Crombie, & Taylor, 2001; Crawford, Parker, Stewart, Besson, & De Lacey, 1989).

Procedure. All participants were tested individually at the university. Testing was completed in one session of approximately two to three hours duration, with breaks provided as needed. Administration of cognitive assessments was counterbalanced.

Results

Initial analyses were undertaken to establish whether the older adults displayed the anticipated deficits in EFT and PM, as well as retrospective memory. In relation to EFT, an independent-samples t-test showed that older adults produced fewer internal details on the AI ($M = 32.98$, $SD = 19.39$) than young adults ($M = 52.59$, $SD = 19.71$), $t(244) = 7.87$, $p < .001$, $d = 1.00$, reflecting poorer EFT ability. In relation to overall PM score, an independent-samples t-test similarly showed that older adults ($M = .42$, $SD = .15$) scored significantly lower than young adults ($M = .72$, $SD = .17$), $t(248) = 14.74$, $p < .001$, $d = 1.87$. Older adults ($M = .25$, $SD = .16$) also scored significantly lower than young adults ($M = .52$, $SD = .18$), $t(248) = 12.41$, $p < .001$, $d = 1.58$, on retrospective memory.

To investigate the relationship between EFT and PM within each age band, we computed the Pearson product–moment correlations for the two groups separately (see Figure 1 for scatterplot of these relationships). These correlations were computed using overall PM scores in order to maximize the reliability of PM measurement and variability in the samples, and because we had no a priori

prediction regarding differences in the relationship between EFT and PM for different PM task types. The correlation between overall PM score and EFT was .274, $p = .002$, for the young adults, and .278, $p = .002$, for the older adults. Thus, there were significant, modest correlations between EFT ability and overall PM score, indicating that better episodic future thinking was associated with better PM performance for both young and older adults. (For the interested reader, correlations between EFT and each separate PM task type are reported in the Appendix, generally showing similar modest correlations for most task types for each group.)

Table 2 shows intercorrelations between all variables. To test whether the EFT–PM associations reflect a unique contribution of EFT to PM, hierarchical multiple regression analyses were conducted. The regression analyses were run for the young and older adult groups separately (see Table 3). Age and verbal intelligence were entered at Step 1 in the analyses, retrospective memory at Step 2, and EFT at Step 3.

The results of the regression analysis for the young adults showed that EFT accounted for a significant amount of variance (4%) in overall PM score, over and above that accounted for by the other predictors. Retrospective memory was the only other significant contributor, with better retrospective memory associated with better PM performance. It accounted for a larger proportion of the variance in PM (32%) than EFT. The predictors together accounted for a substantial amount of variance (37%). In the regression analysis for the older adults, age and retrospective memory were significant predictors, with increasing age and poorer retrospective memory associated with poorer PM performance. These variables uniquely accounted for 13% and 12% of the variance, respectively, in overall PM score. EFT, however, did not contribute any additional variance. For the older adults, the predictors together also accounted for substantial variance (28%) in PM.[3]

---

[3]To test for a nonlinear relation between EFT and PM, regression analyses were re-run to include both linear and quadratic EFT terms. The nonlinear effect did not account for a significant amount of unique variance in PM beyond the linear effect for either group: R^2 change young, $F(1, 119) = 3.05$, $p = .083$; old, $F(1, 119) = 3.05$, $p < .001$.

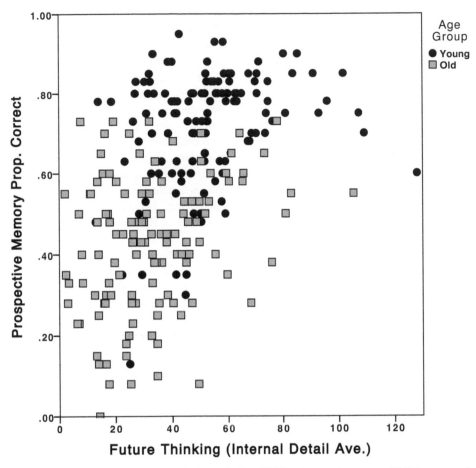

Figure 1. *Scatterplot of the relationship between episodic future thinking (EFT) and prospective memory (PM) for young and older adults.*

Table 2. *Pearson product–moment correlations for the two groups*

Measure	1	2	3	4	5	6
1. PM (Total)	—	.14	−.36***	.35***	.44***	.28**
2. IQ (Verbal)	.13	—	.05	.41***	.44***	.06
3. Age	.04	.33***	—	−.10	−.20*	−.28**
4. ACE–R (total)				—	.40***	.24**
5. Post VW recall	.57***	.10	−.03		—	.30**
6. EFT	.27**	−.02	−.05		.16	—

Note: EFT = episodic future thinking; PM = prospective memory; ACE–R = Addenbrooke's Cognitive Examination Revised; VW = Virtual Week. Intercorrelations for the older adult group (*n* = 125) are presented above the diagonal, and intercorrelations for younger adult group (*n* = 125) are presented below the diagonal. EFT: *n* = 121 for older adult group. ACE–R test was not administered to the younger adult group.

*p < .05. **p < .01. ***p < .001.

It should be noted that the retrospective memory variable in these analyses was the score on the recall task in Virtual Week (VW). Past internal details on the AI was another option as a measure of retrospective memory, but given that the regression analyses focused on the prediction of PM performance, memory for the content of the PM tasks provided a more methodologically rigorous test of the extent to which retrospective memory contributes to PM. Nevertheless, we did re-run the regression analyses substituting the past internal details score for the VW retrospective memory score. The pattern of results for the two age groups in relation to all other predictors in their respective models remained unchanged, including the continued significant contribution of EFT to PM for the young adults ($p = .010$) but not for the older adults ($p = .345$). Past internal details, however, did not account for significant variance in PM for either group. This is in contrast to the significant variance accounted for by the VW retrospective memory variable in the original regression analyses for both groups. We also re-ran the regression analyses adding past internal details to the original set of predictors (i.e., including VW retrospective memory). Again, the pattern of results did not change, and past internal details did not account for a significant amount of variance in PM for either young or older adults ($p = .640$, $p = .421$, respectively).

Discussion

The current study represents the first empirical assessment of the relationship between EFT and PM in an adult cohort. As predicted, a significant positive relationship was found between EFT and PM in both young and older adults. However, of particular importance here was the finding that EFT remained a significant contributor to young adults' PM even after accounting for the variance attributable to individual differences in age, general cognitive ability, and, of particular note, retrospective memory, which was the only other significant predictor in the regression model for the younger age group. The present results are consistent with the findings of Nigro et al. (2013), who reported a positive correlation between EFT and PM in middle childhood, and indeed extends them by showing that this relationship is not simply a reflection of shared variance with other variables, in particular retrospective memory. Although meaningful comparisons between these two studies are somewhat limited by the use of different developmental cohorts, measures, and methods, the findings of the current study are important in providing additional evidence that level of EFT ability predicts PM performance. While conclusions about the exact nature of the relationship between EFT and PM are beyond the scope of the current study, there are a number of possibilities that may be considered. For example, these findings support the claim that EFT might support encoding of the content of the PM intention during the intention formation stage (e.g., Brewer & Marsh, 2010; Schnitzspahn & Kliegel, 2009). In particular, EFT may contribute to PM performance by introducing a *preinstatement* of the context that will be encountered at the moment at which the intended action is to be retrieved and performed. These findings are also consistent with studies highlighting the effectiveness of encoding strategies such as implementation intentions, which involve episodic future simulation, in improving PM (e.g., Brewer & Marsh, 2010). The significant contributions of retrospective memory and EFT to PM also fit with evidence of shared neural substrates supporting all three of these cognitive processes (Addis et al., 2007).

The results for the older adults revealed the expected age-related reduction in both EFT and PM, as well as the hypothesized positive relationship between these two constructs. However, we also investigated whether EFT would make a unique contribution to PM in this age group. Interestingly, despite the significant bivariate correlation between them, the regression model showed that EFT did not explain unique variance in PM over and above that attributable to individual differences in age, general cognitive ability, and retrospective memory. More specifically, while for older adults PM was significantly associated with increasing age and poorer retrospective memory, individual differences in EFT did not additionally

Table 3. *Hierarchical multiple regression analyses predicting prospective memory from episodic future thinking, after controlling for age, verbal IQ, and retrospective memory, for young and older adults separately*

Predictor	Young (n = 125)			Old (n = 121[a])		
	ΔR^2	B	β	ΔR^2	B	β
Step 1	.02			.15***		
Age		0.00	.04		−0.01	−.26**
Verbal IQ		0.00	.07		0.00	−.02
Step 2	.32***			.12***		
Retrospective memory		0.21	.53***		0.20	.37***
Step 3	.04*			.01		
Episodic future thinking		0.00	.19*		0.00	.10
Total R^2	.37			.28		

Note: Adjusted R^2 for young = .35, old = .25.

[a]Four older adult participants did not perform the episodic future thinking (EFT) task.

*p < .05. **p < .01. ***p < .001.

explain PM performance. Thus it appears that the determinants of PM performance may change as we move into older adulthood. One possible explanation for this pattern of results may be that the well-established age-related reduction in attentional resources (see Craik & Rose, 2012, for a review) reduces the capacity of older adults to efficiently utilize both EFT and retrospective memory when performing PM tasks. Alternatively, it may be that the age-related deficits in both of these abilities lead older adults to use other strategies (possibly more externally based "bottom-up strategies" rather than "top-down", self-initiated cognitive strategies) to support their PM task performance. In both scenarios, the end result would be a weakening of the relationship between retrospective memory and PM and between EFT and PM. While these possibilities are of course speculative and await further empirical investigation, they are conceptually consistent with the literature addressing age-related changes in memory encoding, which not only highlights the deterioration with age in the neurological underpinnings of a range of cognitive processes resulting in a reduction in available processing resources, but also reports evidence of compensatory mechanisms occurring at both the neural and behavioural level in older adults when performing memory tasks (see Craik & Rose, 2012, for a review).

Strengths, limitations, and future directions

The current study used well-validated measures (AI and Virtual Week) to comprehensively assess two types of future-oriented cognition, EFT and PM, respectively. Furthermore, the sample size was substantial and allowed for analyses to address the relationship between individual differences in EFT and PM. However, the measure of EFT assessed only the phenomenological quality of future-directed thoughts. Given that PM involves remembering to carry out future intentions, one possible avenue for future research may be to consider behaviourally based measures of episodic future thinking such as the recently developed episodic foresight version of Virtual Week (Lyons et al., 2014), which taps the capacity to imagine the future and take steps in the present in anticipation of future needs. Conceptualizing EFT in this way in future studies may reveal even closer links to PM as a result of its contribution to the development of episodic plans (Atance & O'Neil, 2001), which impact how we initially choose, or develop, a mnemonic (a rehearsed, script-like plan) that will allow us to remember our intended action in the future. Furthermore, the inclusion of a measure of semantic future thinking would also be beneficial in future studies assessing the contribution of EFT to PM in order to tease out the extent to which an individual's ability to preexperience the future (EFT) contributes to PM over and above their ability to simply imagine some future state of the world. While the current study did have a measure of external details generated on the AI, which includes semantic details, it is severely limited as a measure of semantic future thinking. This is because participants are explicitly instructed and prompted to provide episodic rather than semantic details when completing the AI. Furthermore, in addition to semantic details, external details also include repetitions and information about events that are not specific in time and place (Addis et al., 2008). Finally, while the measure of PM used in the current study reflected a range of common PM tasks, assessing PM in everyday life where older adults can compensate for PM decline would also be valuable. Given the current findings for young adults, in addition to the initial positive correlation between EFT and PM among older adults, it is possible that EFT might account for unique variance in the PM performance of older adults in their everyday lives.

Conclusion

Overall, these data add to growing evidence for links between EFT and PM but, importantly, the results highlight that this relationship may vary with age across adulthood. As this is the first study assessing the contribution of EFT to PM in young and older adults, these findings require replication, but, nevertheless, they provide important information to help understand how the processes underlying PM may change as we age.

REFERENCES

Addis, D. R., Musicaro, R., Pan, L., & Schacter, D. L. (2010). Episodic simulation of past and future events in older adults: Evidence from an experimental recombination task. *Psychology and Aging, 25*, 369–376. doi:10.1037/a0017280

Addis, D. R., Wong, A. T., & Schacter, D. L. (2007). Remembering the past and imagining the future: Common and distinct neural substrates during event construction and elaboration. *Neuropsychologia, 45*, 1363–1377.

Addis, D. R., Wong, A. T., & Schacter, D. L. (2008). Age-related changes in the episodic simulation of future events. *Psychological Science, 19*, 33–41. doi:10.1111/j.1467-9280.2008.02043.x

Altgassen, M., Rendell, P. G., Bernhard, A., Henry, J. D., Bailey, P. E., Phillips, L. H., & Kliegel, M. (2015). Future thinking improves prospective memory performance and plan enactment in older adults. *Quarterly Journal of Experimental Psychology, 68*, 192–204. doi:10.1080/17470218.2014.956127

Atance, C. M., & Jackson, L. K. (2009). The development and coherence of future-oriented behaviors during the preschool years. *Journal of Experimental Child Psychology, 102*, 379–391.

Atance, C. M., & O'Neill, D. K. (2001). Episodic future thinking. *Trends in Cognitive Sciences, 5*, 533–539. doi:10.1016/S1364-6613(00)01804-0

Brewer, G. A., & Marsh, R. L. (2010). On the role of episodic future simulation in encoding of prospective memories. *Cognitive Neuroscience, 1*, 81–88. doi:10.1016/j.concog.2011.02.015

Bright, P., Jaldow, E., & Kopelman, M. D. (2002). The national adult reading test as a measure of premorbid intelligence: A comparison with estimates derived from demographic variables. *Journal of the International Neuropsychological Society, 8*, 847–854. doi:10.1017.S1355617702860131

Buckner, R. L., & Carroll, D. C. (2007). Self-projection and the brain. *Trends in Cognitive Sciences, 11*, 49–57. doi:10.1016/j.tics.2006.11.004

Burgess, P. W., Scott, S. K., & Frith, C. D. (2003). The role of the rostral frontal cortex (area 10) in prospective memory: A lateral versus medial dissociation. *Neuropsychologia, 41*, 906–918.

Busby, J., & Suddendorf, T. (2005). Recalling yesterday and predicting tomorrow. *Cognitive Development, 20*, 362–372. doi:10.1016/j.cogdev.2005.05.002

Clark, J. M., & Paivio, A. (2004). Extensions of the Paivio, Yuille, and Madigan (1968) norms. *Behaviour Research Methods, Instruments, & Computers, 36*, 371–383. doi:10.3758/BF03195584

Craik, F. I. M., & Rose, N. S. (2012). Memory encoding and aging: A neurocognitive perspective. *Neuroscience and Biobehavioral Reviews, 36*, 1729–1739.

Crawford, J. R., Henry, J. D., Crombie, C., & Taylor, E. P. (2001). Normative data for the HADS from a large non-clinical sample. *The British Journal of Clinical Psychology, 40*, 429–434. doi:10.1348/014466501163904

Crawford, J. R., Parker, D. M., Stewart, S. E., Besson, J. A., & De Lacey, G. (1989). Prediction of WAIS IQ with the national adult reading test: Cross-validation and extension. *British Journal of Clinical Psychology*, *28*, 267–273. doi:10.111/j.2044-8260.1989.tb01376.x

D'Argembeau, A., Ortoleva, C., Jumentier, S., & Van der Linden, M. (2010). Component processes underlying future thinking. *Memory and Cognition*, *38*, 809–819. doi:10.1073/pnas.0610561104

Einstein, G. O., & McDaniel, M. A. (1996). Retrieval processes in prospective memory: Theoretical approaches and some new empirical findings. In M. Brandimonte, G. Einstein, & M. McDaniel (Eds.), *Prospective memory: Theory and applications* (pp. 115–141). Hillsdale, NJ: Erlbaum.

Ellis, J. A. (1996). Prospective memory or the realization of delayed intentions: A conceptual framework for research. In M. Brandimonte, G. O. Einstein, & M. A. McDaniel (Eds.), *Prospective memory: Theory and applications* (pp. 1–22). Mahwah, NJ: Lawrence Erlbaum Associates Inc.

Ellis, J. A., & Freeman, J. (2008). Ten years on: Realizing delayed intentions. In M. Kliegel, M. A. McDaniel, & G. O. Einstein (Eds.), *Prospective memory: Cognitive, neuroscience, developmental, and applied perspectives* (pp. 1–28). Hove: Psychology Press.

Gilbert, S. J., Armbuster, D., & Panagiotidi, M. (2012). Similarity between brain activity at encoding and retrieval predicts successful realization of delayed intentions. *Journal of Cognitive Neuroscience*, *24*, 93–105. doi:10.1162/jocn_a_00094

Gollwitzer, P. M. (1999). Implementation intentions: Strong effects of simple plans. *American Psychologist*, *54*, 493–503. doi:10.1037/0003-066X.54.7.493

Hanson, L. K., & Atance, C. M. (2014). Brief Report: Episodic foresight in Autism spectrum disorders. *Journal of Autism and Development Disorders*, *44*, 674–684. doi:10.1007/s10803-013-1896-6

Henry, J. D., MacLeod, M. S., Phillips, L. H., & Crawford, J. R. (2004). A meta-analytic review of prospective memory and aging. *Psychology and Aging*, *19*, 27–39. doi:10.1037/0882-7974.19.1.27

Henry, J. D., Terrett, G., Altgassen, M., Raponi-Saunders, S., Ballhausen, N., Schnitzspahn, K. M., & Rendell, P. G. (2014). A Virtual Week study of prospective memory function in autism spectrum disorders. *Journal of Experimental Child Psychology*, *127*, 110–125. doi:10.1016/j.jecp.2014.01.001

Irish, M., Addis, D. R., Hodges, J. R., & Piguet, O. (2012). Considering the role of semantic memory in episodic future thinking: Evidence from semantic dementia. *Brain*, *135*(Pt 7), 2178–2191. doi:10.1093/brain/aws119

Kliegel, M., Altgassen, M., Hering, A., & Rose, N. S. (2011). A process-model based approach to prospective memory impairment in Parkinson's disease. *Neuropsychologica*, *49*, 2166–2177. doi:10.1016/j.neuropsychologia.2011.01.024

Kliegel, M., Martin, M., McDaniel, M. A., & Einstein, G. O. (2002). Complex prospective memory and executive control of working memory: A process model. *Psychologische Beiträge*, *44*, 303–318.

Kliegel, M., McDaniel, M. A., & Einstein, G. O. (2000). Plan formation, retention, and execution in prospective memory: A new approach and age-related effects. *Memory and Cognition*, *28*, 1041–1049. doi:10.3758/BF03209352

Leitz, J. R., Morgan, C. J. A., Bisby, J. A., Rendell, P. G., & Curran, H. V. (2009). Global impairments of prospective memory following acute alcohol. *Psychopharmacology*, *205*, 379–387. doi:10.1007/s00213-009-1546-z

Levine, B., Svoboda, E., Hay, J. F., & Winocur, G. (2002). Aging and autobiographical memory: Dissociating episodic from semantic retrieval. *Psychology and Aging*, *17*, 677–689. doi:10.1037/0882-7974.17.4.677

Lyons, A. D., Henry, J. D., Rendell, P. G., Corballis, M. C., & Suddendorf, T. (2014). Episodic foresight and aging. *Psychology and Aging*, *29*, 873–884. doi:10.1037/a0038130

Mahy, C. E. V., Moses, L. J., & Kliegel, M. (2014). The development of prospective memory in children: An executive framework. *Developmental Review*, *34*, 305–326. doi:10.1016/j.dr.2014.08.001

Mathuranath, P. S., Nestor, P. J., Berrios, G. E., Rakowics, W., & Hodges, J. R. (2000). A brief cognitive test battery to differentiate Alzheimer's disease and frontotemporal dementia. *American Academy of Neurology*, *55*, 1613–1620. doi:10.1212/01.wnl.0000434309.85312.19

McFarland, C. P., & Glisky, E. L. (2012). Implementation intentions and imagery: Individual and combined effects on prospective memory among young adults. *Memory and Cognition*, *40*, 62–69. doi:10.3758/s13421-011-0126-8

Nelson, H. E. (1982). *National adult reading test (NART): Test manual*. Windsor: NFER.

Neroni, M. A., Gamboz, N., & Brandimonte, M. A. (2014). Does episodic future thinking improve prospective remembering? *Consciousness and Cognition*, *23*, 53–62. doi:10.1016/j.concog.2013.12.001

Nigro, G., Brandimonte, M. A., Cicogna, P. C., & Cosenza, M. (2013). Episodic future thinking as a predictor of children's prospective memory. *Journal of Experimental Child Psychology*, *127*, 82–94. doi:10.1016/j.jecp.2013.10.013

Paraskevaides, T., Morgan, C. J. A., Leitz, J. R., Bisby, J. A., Rendell, P. G., & Curran, H. V. (2010). Drinking and future thinking: Acute effects of alcohol on prospective memory and future simulation. *Psychopharmacology*, *208*, 301–308. doi:10.1007/s00213-009-1731-0

Prull, M. W., Gabrieli, J. D. E., & Bunge, S. M. (2000). Age-related changes in memory: A cognitive neuroscience perspective. In F. I. M. Craik & T. A. Salthouse (Eds.), *Handbook of aging and cognition* (Vol. 2, pp. 91–153). Mahwah, NJ: Erlbaum.

Rendell, P. G., & Craik, F. I. M. (2000). Virtual and actual week: Age-related differences in prospective memory. *Applied Cognitive Psychology. Special Issue: New Perspectives in Prospective Memory*, *14*, S43–S62. doi:10.1002/acp.770

Rendell, P. G., & Henry, J. D. (2009). A review of Virtual Week for prospective memory assessment: Clinical implications. *Brain Impairment*, *10*, 14–22. doi:10.1375/brim.10.1.14

Rendell, P. G., Phillips, L. H., Henry, J. D., Brumby-Rendell, T., de la Piedad Garcia, X., Altgassen, M., & Kliegel, M. (2011). Prospective memory, emotional valence and aging. *Cognition and Emotion*, *25*, 916–925. doi:10.1080/02699931.2010.508610

Rose, N. S., Rendell, P. G., McDaniel, M. A., Aberle, I., & Kliegel, M. (2010). Age and individual differences in prospective memory during a "Virtual Week": The roles of working memory, vigilance, task regularity, and cue focality. *Psychology and Aging*, *25*, 595–605. doi:10.1037/a0019771

Salthouse, T. A. (2000). Aging and measures of processing speed. *Biological Psychology*, *54*, 35–54. doi:10.1016/S0301-0511(00)00052-1

Schacter, D. L., Addis, D. R., & Buckner, R. L. (2007). Remembering the past to imagine the future: The prospective brain. *Nature Reviews Neuroscience*, *8*, 657–661. doi:10.1038/nrn2213

Schacter, D. L., Addis, D. R., & Buckner, R. L. (2008). Episodic simulation of future events: Concepts, data, and applications. *Annals of the New York Academy of Sciences*, *1124*, 39–60. doi:10.1196/annals.1440.001

Schacter, D. L., Addis, D. R., Hassabis, D., Martin, V. C., Spreng, R. N., & Szpunar, K. (2012). The future of memory: Remembering, imagining, and the brain. *Neuron*, *76*, 677–694. doi:10.1016/j.neuron.2012.11.001

Schnitzspahn, K. M., & Kliegel, M. (2009). Age effects in prospective memory performance within older adults: The paradoxical impact of implementation intentions. *European Journal of Ageing*, *6*, 147–155. doi:10.1007/s10433-009-0116-x

Szpunar, K. K. (2010). Episodic future thought: An emerging concept. *Perspectives on Psychological Science*, *5*, 142–162.

Stokes, M., Thompson, R., Cusack, R., & Duncan, J. (2009). Top-down activation of shape-specific population codes in visual cortex during mental imagery. *Journal of Neuroscience*, *29*, 1565–1572. doi:10.1523/JNEUROSCI.4657-08.2009

Terrett, G., Rendell, P. G., Raponi-Saunders, S., Henry, J. D., Bailey, P. E., & Altgassen, M. (2013). Episodic future thinking in children with Autism spectrum disorder. *Journal of Autism and Developmental Disorders*, *43*, 2558–2568. doi:10.1007/s10803-013-1806-y

Weiler, J. A., Suchan, B., & Daum, I. (2010). When the future becomes the past: Differences in brain activation patterns for episodic memory and episodic future thinking. *Behavioural Brain Research*, *212*, 196–203. doi:10.1016/j.bbr.2010.04.013

West, R., & Ross-Munroe, K. (2002). Neural correlates of the formation and realization of delayed intentions. *Cognitive, Affective, and Behavioral Neuroscience*, *2*, 162–173. doi:10.3758/CABN.2.2.162

Zöllig, J., West, R., Martin, M., Altgassen, M., Lemke, U., & Kliegel, M. (2007). Neural correlates of prospective memory across the lifespan. *Neuropsychologia*, *45*, 3299–3314. doi:10.1016/j.neuropsychologia.2007.06.010

APPENDIX

Pearson product–moment correlations between EFT and PM task types for the two groups

| | PM task type | | | |
Group	Regular event	Irregular event	Regular time	Irregular time
Young adults				
EFT	.24**	.20*	.10	.31**
Older adults				
EFT	.29**	.07	.17†	.21*

Note: EFT = episodic future thinking; PM = prospective memory. Young adult group, $n = 125$; older adult group, $n = 121$.
*$p < .05$. **$p < .01$. †$p = .062$.

Scripts and information units in future planning: Interactions between a past and a future planning task

Aline Cordonnier[1,2], Amanda J. Barnier[1,2], and John Sutton[1,2]

[1]Department of Cognitive Science, Macquarie University, Sydney, NSW, Australia
[2]Australian Research Council Centre of Excellence in Cognition and Its Disorders, Macquarie University, Sydney, NSW, Australia

Research on future thinking has emphasized how episodic details from memories are combined to create future thoughts, but has not yet examined the role of semantic scripts. In this study, participants recalled how they planned a past camping trip in Australia (past planning task) and imagined how they would plan a future camping trip (future planning task), set either in a familiar (Australia) or an unfamiliar (Antarctica) context. Transcripts were segmented into information units that were coded according to semantic category (e.g., where, when, transport, material, actions). Results revealed a strong interaction between tasks and their presentation order. Starting with the past planning task constrained the future planning task when the context was familiar. Participants generated no new information when the future camping trip was set in Australia and completed second (after the past planning task). Conversely, starting with the future planning task facilitated the past planning task. Participants recalled more information units of their past plan when the past planning task was completed second (after the future planning task). These results shed new light on the role of scripts in past and future thinking and on how past and future thinking processes interact.

The future might be unknown, but it is predictable to some extent. Events tend to repeat themselves, people do not change drastically over time, and the laws of physics continue to operate in day-to-day life. This continuity of the self and of the world provides us with the framework needed to think about the future. With memories and knowledge as building blocks, humans have been shown to successfully simulate future events and to predict outcomes or plan actions in order to achieve specific goals (Atance & O'Neill, 2001). For example, a job seeker going to an interview might simulate the questions he will be asked, drawing both on similar past experiences and on specific knowledge of the company and the offered position; a teenager can predict how her parents will react if they find out she lied about her test results; or a couple may discuss a plan for the different steps needed to build their

We want to thank M. Irish for some helpful clarifications, as well as the anonymous reviewers for their constructive comments and suggestions, which have helped improve this manuscript.

This work was supported by a Macquarie University Research Excellence Scholarship to Aline Cordonnier; by an Australian Research Council (ARC) Discovery Project and ARC Future Fellowship to Amanda Barnier; by an ARC Discovery Project to John Sutton; and by the Memory Program of the ARC Centre of Excellence in Cognition and its Disorders and the Department of Cognitive Science, Macquarie University.

house. During the course of one day, people think about the future as often as they think about the past (Berntsen & Jacobsen, 2008; Finnbogadottir & Berntsen, 2012), with on average 60 future-oriented thoughts a day (D'Argembeau, Renaud, & Van der Linden, 2011).

Even though future thoughts are as common as memories in our daily life, our capacity to remember has received significantly more scientific attention than our capacity to think about the future. But in the last 20 years, researchers have shown an increased interest in future thinking, specifically in one particular aspect of it: *episodic future thinking* (also known as episodic foresight, episodic simulation, or prospection; for reviews, see Klein, 2013; Schacter, 2012; Szpunar, 2010). Episodic future thinking is usually defined as the capacity to project oneself into the future, and it is often studied in parallel with episodic memory for three principal reasons (Dudai & Carruthers, 2005; Schacter & Addis, 2007; Suddendorf & Corballis, 1997). First, the two phenomena partially rely on the same component processes and neural mechanisms (Addis, Musicaro, Pan, & Schacter, 2010; Addis, Pan, Vu, Laiser, & Schacter, 2009; D'Argembeau, Xue, Lu, Van der Linden, & Bechara, 2008; Tulving, 1985). Second, they draw on the same information stored in our episodic and semantic memory (Atance & O'Neill, 2001; Buckner & Carroll, 2007; Klein, Cosmides, Tooby, & Chance, 2002; Schacter, Addis, & Buckner, 2008; Suddendorf & Corballis, 2007). Third, researchers also have argued that thinking about the future is one of the major functions of memory (Klein, 2013; Schacter, 2012; Schacter et al., 2012; Suddendorf, Addis, & Corballis, 2009; Tulving, 2005).

Thinking about a future event can take many forms. A thought-sampling study showed that in everyday life, future thoughts served a wide array of functions such as dreaming about one's future, simulating an upcoming event, or making decisions (D'Argembeau et al., 2011). Notably, more than half of self-reported future thoughts were related to *planning* an event or an action (see also Baird, Smallwood, & Schooler, 2011). In their recent taxonomy, Szpunar, Spreng, and Schacter (2014)

defined four different modes of future thinking, namely simulation, prediction, intention and planning, and divided each mode into three forms (episodic, semantic, or hybrid). Similarly, in our own cognitive framework (Cordonnier & Sutton, 2016), we also identify four main processes of thinking about a potential upcoming event: (a) imagining, which does not entail any belief that the future event might happen; (b) simulating or predicting, which imply that the event will probably occur; (c) planning, a multicomponent goal-directed process that includes simulating the different steps and their consequences; and (d) forming and remembering intentions, also known as prospective memory. We distinguish these different forms of future thinking by the temporal distance between the moment the event is thought of and its possible realization, and by the subjective plausibility of the thought event. We also argue that the content of the future thoughts might rely more or less on episodic or on semantic memory depending on the accessibility of memories of similar past events, as well as on the sought plausibility of the thought future event.

However, the majority of studies to date have focused on how participants imagine future events without taking into account how plausible or probable they might be. Instead, research has emphasized the role of episodic details, influenced by the constructive episodic simulation hypothesis, which suggests that humans recombine episodic details from past events to simulate future ones (Schacter & Addis, 2007, 2009). This episodic focus can be observed at two levels: in experimental methods and in coding schemes. In terms of method, participants generally are instructed to imagine and describe specific episodic future events, cued with time periods such as "in five years" or "when you will retire" (e.g., MacLeod & Conway, 2007), with nouns such as "dog" or "birthday" (e.g., Addis, Wong, & Schacter, 2007; D'Argembeau & Demblon, 2012), or with a set of idiosyncratic cues, as in the episodic recombination paradigm (Addis et al., 2010; Addis, Pan, et al., 2009). Consequently, although the events produced by these different styles of cues are all imagined future events, their content, as well as their plausibility, may differ substantially contingent on the cue and the instructions received. Time period cues, especially when set in the distant future, would most likely trigger cultural life script events that might be highly plausible but lack episodic specificity (Berntsen & Bohn, 2010); noun cues might prompt mundane or repeated events; whereas idiosyncratic cues, as in the recombination paradigm, might potentially generate unlikely events that score high on episodic specificity but low on plausibility. In our research, we wanted to examine the use of episodic and semantic memory in plausible future thoughts. Following our framework, we thought that investigating future planning could help us achieve this goal.

In terms of coding schemes that specify and quantify details in transcripts of past and future thinking, episodic details have usually been regarded as more important than other types of details. One of the most widely used coding schemes concentrates on the quantity of internal (or episodic) details that participants generate while remembering past events or imagining future ones (e.g., Addis, Wong, & Schacter, 2008; Cole, Morrison, & Conway, 2013; De Brigard & Giovanello, 2012; Levine, Svoboda, Hay, Winocur, & Moscovitch, 2002). Any detail not considered internal is labelled external. External details include: repetitions, other episodic details not relevant to the specific episodic event, and semantic details. A reduction in internal details (often accompanied but not correlated with an increase of external details) has been interpreted either as an indication of age-related changes in older adults (Addis et al., 2008; Cole et al., 2013) or as symptomatic of memory deficits in patients with hippocampal damage (Addis, Sacchetti, Ally, Budson, & Schacter, 2009; Hassabis, Kumaran, & Maguire, 2007; Klein, Loftus, & Kihlstrom, 2002; Tulving, 1985). It is also important to note that to code for these internal and external details, the procedure requires identifying a clear central episodic event. This step can only be completed in episodic tasks; other types of future thinking, such as planning, cannot be analysed using this procedure.

If episodic details have been considered central in future thinking, a number of recent studies, although differing in aim and method, support the claim that semantic memory is also important for the simulation of future events and, in all likelihood, provides the scaffolding needed to give meaning and structure to the simulated event (Szpunar, 2010). D'Argembeau and Mathy (2011) explored the construction of mental representations of future events by asking healthy participants to report their thought flow while imagining specific future events. Participants usually first reported personal semantic information and/or general events before producing specific episodic details. Also, cueing them with personal goals facilitated the production of future events as well as access to episodic details. Cole, Gill, Conway, and Morrison (2012) examined the effect of trial duration on the production of episodic and semantic details. They showed that the amount of semantic details in past and future thinking was not related to the amount of episodic details, which indicates that semantic details are not generated at the expense of episodic details. Irish, Addis, Hodges, and Piguet (2012) investigated the role of semantic knowledge in past and future thinking by testing patients with semantic dementia. Irish et al. found that while these patients demonstrated intact retrieval of recent memories, they showed a compromised capacity to simulate novel events in the future. Neuroimaging results revealed that future thinking deficit in these semantic dementia patients was strongly correlated with atrophy in the anterior temporal lobes, which are critical for the representation of semantic knowledge (Irish et al., 2012). Interestingly, there also have been instances of patients with hippocampal amnesia who have shown preserved ability to imagine future events (Maguire, Vargha-Khadem, & Hassabis, 2010; Mullally, Hassabis, & Maguire, 2012). Together, these findings speak to the importance of semantic representations, such as general semantic knowledge but also personal semantics and semantic scripts, in the construction and simulation of future events. This has led Irish and Piguet (2013) to propose a new hypothesis— the semantic scaffolding hypothesis—which suggests that semantic memory helps scaffold both past and future thinking.

While the constructive episodic simulation and the semantic scaffolding hypothesis agree that episodic and semantic memory contribute to future thinking, the former emphasizes the role of episodic details whereas the latter emphasize the importance of semantic information to provide the framework of the event, which is then populated by episodic details and semantic knowledge depending on familiarity. Familiar events would more likely draw upon episodic details, whereas unfamiliar or novel events would more likely draw upon semantic knowledge and scripts. However, distinguishing the separate contribution of episodic and semantic memory can be challenging, mainly because the distinction between the two is not always well defined. For example, scripts, defined by Schank and Abelson (1977, p. 210) as "a structure that describes appropriate sequences of events in a particular context", are categorized as semantic memory but can derive from repeated episodic experiences (Abelson, 1981; Bower, Black, & Turner, 1979; Hudson, Fivush, & Kuebli, 1992; Nelson & Gruendel, 1981). As it scaffolds and cues episodic remembering, script knowledge becomes intertwined with episodic details in an almost indistinguishable way. Although scripts have been shown to support the remembering process, even more so when recalling goal-directed events (Lichtenstein & Brewer, 1980), we suggest their particular structures make them essential for any type of future thinking requiring the event to have a good level of plausibility. They provide knowledge of how events tend to unfold, and can be derived both from semantic knowledge and from episodic memories, especially when they are about recurring events. However, research on future thinking has yet to integrate scripts into analysis, as they cannot be coded with the traditional internal/external coding scheme. For example, when analysing a restaurant script, what matters is not specific details such as what food was selected on the menu or how the bill was paid; what is important is that the person considered these topics. Therefore, one way to investigate scripts is to compare higher order categories of details regardless of specific content.

In this introduction, we have discussed the need to expand research on future thinking by considering the role of scripts in plausible future events, depending on familiarity of the context. Therefore, in our study, we created a novel experimental paradigm that focused on the mental simulation of planning, which means we used planning as the content of both the remembering and imagining tasks (participants remembered planning a past event or imagined planning a future one). This was done for three reasons. First, planning is an important part of our daily life and seems to be a major aspect of future thinking (Baird et al., 2011; D'Argembeau et al., 2011). At the same time, planning has evolutionary benefits (Klein, 2013; McCormack, Hoerl, & Butterfill, 2011) and is regarded as an important developmental achievement, with many studies investigating planning in children (McCormack & Atance, 2011). Indeed, being able to simulate and anticipate what could occur as well as the consequences for our actions offers a unique advantage in our day-to-day lives. This makes it ideal for expanding research on future thinking, as our design might capture aspects of real-life future thinking not yet tapped by other paradigms.

Second, future planning is a multicomponent goal-directed process (Hayes-Roth & Hayes-Roth, 1979). However, we wanted to look at one particular component of the process that relates more to episodic future thinking: simulation of the planning (Szpunar et al., 2014). Unlike other aspects of future thinking, such as daydreaming, planning requires anticipation of the future by inferring how things might plausibly unfold in a given situation. To do so, one needs general knowledge of the context of the event, of the causal relationships between actions and their consequences, and even knowledge about one's own self and others. An easy way to obtain this knowledge is to bring to mind a similar situation that has been encountered previously and compare it to the current one. Therefore, using planning as the content of the simulation allowed us to constrain and control the type of content in each telling to make it comparable across tasks, and to examine how a past planning experience influences the planning of a similar future event.

Finally, scripts can be of great value in successful planning, especially when no comparable event has been planned before, and the context is unfamiliar. They provide general knowledge about the sequences of events that can be expected in a given situation, regardless of personal experience. Consequently, it gave us the opportunity to explore the use of scripts in past thinking and future thinking in familiar but also unfamiliar contexts.

To investigate how participants rely on past memories and on semantic scripts to simulate planning a future event, we divided our experiment into two tasks. We asked participants to remember how they planned a past camping trip and how they imagined they would plan a future one. The past camping trip was always set in Australia (as all our participants had been on a camping trip in Australia) whereas the context of the future trip was either identical to the past planning task—in Australia—or totally new to the participant—in Antarctica. This last condition (imagining how to plan a camping trip in Antarctica) was created to explore the construction of future thoughts when participants could not rely on an episodic recollection of having planned a similar event in the past. To the best of our knowledge, the only study investigating the quantity of details in familiar and unfamiliar future events found that familiar events contained more internal details than unfamiliar events (de Vito, Gamboz, & Brandimonte, 2012). However, scripts and semantic knowledge might have a bigger role to play in future planning than in future simulation. If so, we would expect participants to provide as many details in familiar and unfamiliar context by relying on scripts, and that these scripts would be similar in past and future thinking tasks. Furthermore, as we counterbalanced the

order of presentation of the tasks, we hypothesized that past and future planning scripts in a familiar context would be more alike when the past planning task was completed first.

EXPERIMENTAL STUDY

Method

Participants
We recruited 40 undergraduate university students (28 female and 12 male, mean age = 20.05 years, $SD = 2.42$; range = 18–30 years) enrolled in an introductory psychology course at Macquarie University (Sydney, Australia) as participants for this experiment. We selected them from a participant pool if English was their first language, and they had been on a camping trip in Australia in the past 5 years. They gave informed consent prior to testing, including agreement to be audio recorded, and received course credit as compensation for their time, in accordance with the Macquarie University Ethics Committee.

We tested participants in a 2 (task order: past planning first, then future planning vs. future planning first, then past planning) × 2 (familiarity of future planning context: familiar, Australia vs. unfamiliar, Antarctica) mixed design.

Materials and procedure
Upon arrival, participants gave their consent and received the following instructions:

You will carry out two main tasks followed by some questions. One of the tasks consists of remembering how you planned a past event; the other consists of imagining how you would plan a future event. I will ask you to tell me your answers out loud, so I can record them, and then to write a summary of them.

For their first task, we randomly assigned participants to one of three scenarios, adapted from Klein, Robertson, and Delton (2010):[1] (a) *a past planning scenario*, where we told participants to remember the different steps they had to undertake

---

[1]We designed the paradigm in a way that would let us include a subtest at the end of the session replicating Klein et al. (2010)'s experiment. However, the condition of the testing and the sample size were relatively different from those in the original. As we did not replicate the results, we decided not to include them in this article. The data for the replication were collected after the data presented in this paper; therefore it cannot be considered as a confounding factor.

to successfully plan and prepare for their past camping trip in Australia; (b) *a future planning in a familiar context scenario*, where we told participants to imagine the different steps they would undertake to successfully plan and prepare for their future camping trip in Australia; (c) *a future planning in a unfamiliar context scenario*, where we told participants to imagine the different steps they would undertake to successfully plan and prepare for their future camping trip in Antarctica. Participants had 1 min to remember or imagine the planning before they described it to the experimenter for up to 5 min. We audiorecorded their answers using the freeware computer-recording programme Audacity. Subsequently, participants summarized their answer on a sheet of paper. Once they completed their summary, a distractor task was presented to them in the form of a set of mazes.

At the end of the distractor task, we gave participants a second scenario. If they received the past planning scenario (a) as their first task, they were given one of the two future planning scenarios (b or c); and if they received one of the two planning scenarios (b or c) as their first task, they were given the past planning scenario (a). In summary, each participant completed a past planning task and a future planning task that was set either in a familiar context or in an unfamiliar context. We counterbalanced the order of the tasks across participants. Therefore we had four conditions in total: "past then future—familiar future context (Australia)", "past then future—unfamiliar future context (Antarctica)", "future then past—familiar future context (Australia)", "future then past—unfamiliar future context (Antarctica)".

To conclude the experiment, participants completed a short questionnaire about their demographic details, camping habits, and knowledge of Antarctica. They also provided ratings on a 10-point Likert scale on how difficult they found both the past planning and future planning tasks. We then fully debriefed participants and thanked them for their time.

Data analysis

With help from the written summaries, we transcribed each audio recording. Because of the nature of our tasks but also because we wanted to investigate script similarities between past and future planning, we created a new coding scheme. We divided sentences into small segments, each containing one new piece of information or "information unit". To avoid inflating results, we scored adjectives and nouns in the same noun phrase as one single information unit (e.g., windproof jacket was scored 1). Then, we scored these information units according to a higher order semantic category (to record what the information unit was about), such as information about *who* would be coming, *where* they were planning to go, or the type of *material* they were going to bring. We also coded actions that needed to be undertaken and conditions (such as health conditions, time constraints, etc.) to take into account. Overall there were 19 semantic categories, which represented the type of information that could be found in a general script of how to plan a camping trip. These semantic categories were not chosen a priori but were derived from the data. We used a dynamic process of creating the coding scheme by adding new categories when needed and reanalysing the transcripts with the modified coding scheme until every information unit could be placed in a category. In the final analyses, we did not include repetitions (information units previously mentioned by participants), and we also excluded event details that were not related to the planning of the camping trips (such as details about how the trip itself went) as these details were only found in the past planning task and were not relevant to the planning itself.

The categories were: (a) who, (b) where, (c) when, (d) duration, (e) transport, (f) why, (g) weather, (h) money, (i) food, (j) accommodation, (k) personal items, (l) general material, (m) security concerns, (n) leisure activities, (o) chores, (p) seeking information, (q) general knowledge, (r) actions, (s) conditions. For example, the sentences "The *four of us* and *my little brother* planned to go camping *around Umina beach*, which is north of Sydney to learn how to surf. We decided to go by car and we would take Jack's tent." contain seven information units from six different semantic categories (in order): *who (×2), where, general knowledge, why, transport,* and *accommodation.*

To check interrater reliability, the first author scored all transcripts, and one extra independent judge, blind to the aims of the study and trained on the coding technique, scored 53.75% of all transcripts (at least 50% in each condition). The initial agreement percentages were adequate (75.4%). Any discrepancy was resolved by discussion between coders until agreement.

These categories could be taken to constitute a complete script of what needs to be considered and done when planning a camping trip. A good planner would not necessarily provide more information units; however, they would provide information units coded under many different categories in order to cover the different steps of the plan. We therefore analysed both the total number of information units and the number of semantic categories mentioned in a transcript.

Furthermore, we wanted to analyse script similarity for each participant across the two tasks they completed. For example, if one participant considered place, time, food, weather, security, and leisure activities when planning his past trip, would he consider the same categories when planning his future trip? In other words, would participants retain a similar script of their plan, regardless of the specific content (e.g., food that you can cook on a fire vs. dry food that does not need to be cooked) and the quantity of information units (full list of items vs. mentioning planning for food), or would they provide information units from other categories depending on context? Therefore, we calculated for each participant the number of categories mentioned in both of their tasks, in their past planning task only, in their future planning task only, or in neither their past nor their future planning task.

Results

Camping experiences, knowledge of Antarctica, and difficulty ratings of the tasks

To ensure there was no discrepancy in prior knowledge and experience across conditions ("past planning then future planning in Australia, "past planning then future planning in Antarctica", "future planning in Australia then past planning", "future planning in Antarctica then past planning"), participants filled in a questionnaire about their camping experiences and general knowledge of Antarctica at the end of the study. There were no significant differences in the frequency of camping trips taken in Australia, how long ago was the camping they described in the past planning task, and their general knowledge of Antarctica (see Table 1). Therefore, subsequent differences between our conditions cannot be explained by differences in camping experiences or general knowledge of Antarctica.

Participants also rated how difficult they found both past and future planning tasks on a scale from 1 to 10. A 2 (task order) × 2 (familiarity of future planning context) univariate analysis of variance (ANOVA) revealed no significant differences for the self-rated difficulty of the past planning task ($M = 4.55$, $SD = 1.84$). However, there was a significant main effect for the self-rated difficulty of the future planning task when the context was unfamiliar to participants (Antarctica), $F(1, 36)$ = 4.88, $MSE = 2.95$, $p = .034$, $\eta_p^2 = .119$. The future planning task was rated as more difficult when the scenario was set in Antarctica ($M = 5$, $SD = 1.95$) than when it was set in Australia ($M = 3.8$, $SD = 1.36$), regardless of the order the tasks were presented in. Finally, it is worth noting that participants did not find the past planning task easier or more difficult than the future planning task, $t(1, 39) = 0.51$, $p = .613$.

Quantity of information units produced in past and future planning tasks

First, we analysed the quantity of information units produced in past and future planning tasks. We removed two outliers (with a z score of at least +2.5) from the initial sample for this set of analyses.[2]

As participants rated the future planning task with the unfamiliar context (Antarctica) harder than the future planning task with the familiar

---

[2]One participant was in the "past planning then future planning in Antarctica" condition; the other was in the "future planning in Australia then past planning" condition. These two participants had a significant number of information units in a single category that inflated their total

Table 1. *Camping experiences, knowledge of Antarctica, and difficulty ratings of the tasks, as a function of the task order and the future scenario familiarity*

Measure		*Task order: Past–future. Future scenario familiarity: Familiar (n = 10)*	*Past–future: Unfamiliar (n = 10)*	*Future–past: Familiar (n = 10)*	*Future–past: Unfamiliar (n = 10)*	*Significant differences*
No. of participants for each frequency of camping trips taken in Australia	Rarely (1–3)	4	3	3	3	*ns*
	Often (4–6)	3	3	3	3	*ns*
	Regularly (7+)	3	4	4	4	*ns*
How long ago was the camping they described in the experiment (months)		27.2 (24.5)	19.3 (17.0)	28.1 (14.9)	34.3 (23.9)	*ns*
Knowledge of Antarctica (10-point scale)		2.5 (1.4)	4.3 (1.8)	4 (1.7)	3.5 (1.7)	*ns*
Difficulty of the past planning task (10-point scale)		4.6 (1.3)	5.0 (2.0)	3.8 (1.8)	4.8 (2.1)	*ns*
Difficulty of the future planning task (10-point scale)		3.7 (1.5)	4.8 (2.0)	3.9 (1.3)	5.2 (2.0)	*

$^{*}p < .05$.

context (Australia), we started by comparing the average number of information units produced in the two future planning scenarios only. A 2 (task order) × 2 (familiarity of future planning context) univariate ANOVA revealed no main effect of task order, but more importantly, no main effect of the familiarity of the future scenario, $F(1, 34) = 1.02$, $MSE = 153.10$, $p = .320$. However, there was a significant two-way interaction between task order and familiarity of the context, $F(1, 34) = 6.06$, $MSE = 153.10$, $p = 0.019$, $\eta_p^2 = .15$.While there was no difference in quantity of information units produced between the two contexts (Australia and Antarctica) when the future planning task was completed first, participants provided fewer information units when the future planning task was completed second and when the context was familiar (Australia), $F(1, 34) = 6.03$, $MSE = 153.10$, $p = .019$, $\eta_p^2 = .19$.

Subsequently, we investigated the difference in quantity of information units produced in each task. We ran a mixed-design ANOVA with type of task (past planning vs. future planning) as a within-subject factor and task order (past–future vs. future–past) and familiarity of future planning context (familiar vs. unfamiliar) as between-subject factors. Means and standard deviations are summarised in Table 2.

Although there were no significant main effects, the analysis yielded a two-way interaction between the tasks and the order they were presented in, $F(1, 34) = 21.70$, $MSE = 36.22$, $p < .001$, $\eta_G^2 = .07$.[3] For the past planning task, participants recalled more information units about the way they planned their past camping trip if they imagined planning a future camping trip first, $F(1, 34) = 8.74$, $MSE = 36.22$, $p = .006$, $\eta_p^2 = .20$. For the future planning task, participants recalled a similar number of information units irrespective of when they completed the task: first or second.

The three-way interaction between the type of task, task order, and the familiarity of the future planning scenario also was significant, $F(1, 34) = 6.09$, $MSE = 36.22$, $p = .019$, $\eta_G^2 = .02$. Bonferroni

number of information units. For example, one participant simply listed every item of food he would pack. Including them in our analyses would not alter their outcome; on the contrary, it would increase our effect size and therefore gives an erroneous inflated view of our results.

[3]We used generalized eta squared to report effect sizes as this design has a within-subject variable (Bakeman, 2005; Lakens, 2013; Olejnik & Algina, 2003).

Table 2. *Total number of information units in each task as a function of the order of tasks presentation and the familiarity of the future planning scenario*

		Task			
		Past planning		Future planning	
Task order	Future scenario familiarity	M (SD)	95% CI	M (SD)	95% CI
Past–future	Familiar	31.20 (14.00)	[23.19, 39.01]	29.60 (12.84)	[21.65, 37.55]
	Unfamiliar	33.56 (9.49)	[25.22, 41.89]	43.56 (8.52)	[35.17, 51.94]
	Total	32.33 (11.81)	[26.58, 38.08]	36.58 (12.88)	[30.80, 42.36]
Future–past	Familiar	46.00 (13.34)	[37.66, 54.34]	38.44 (14.21)	[30.06, 46.83]
	Unfamiliar	42.30 (11.77)	[34.39, 50.21]	32.60 (13.01)	[24.65, 40.55]
	Total	44.15 (12.33)	[38.40, 49.90]	35.52 (13.54)	[29.74, 41.30]

CI = confidence interval.

contrasts found a global increase of information units generated between the first task and the second task in all conditions but one. That is, participants invariably gave more details on the second task than on the first, except when they started by remembering how they planned a past camping trip and then imagined planning a camping trip also in Australia.

To summarize, this set of analyses shows that participants recalled more information units if they imagined planning a future event first, regardless of the familiarity of the context of the future event. However, remembering how they planned the past event first did not help participants plan the future event, especially when the context of the future event was familiar.

General use of semantic categories in past and future planning tasks

Second, we investigated the general use of semantic categories in both past and future planning tasks by analysing the presence or absence of at least one information unit in each semantic category of our coding system. For each category, participants received either a score of 1 if they provided at least one information unit coded in this category or 0 if they provided no information unit related to the category, giving them a total maximum of 19 and a total minimum of 0. On average, participants mentioned information units belonging to 12.7 categories for the past planning task ($SD =$

2.57) and 13.1 categories for the future planning task ($SD = 2.81$). There were no significant outliers so we used the whole sample for this analysis. Means and standard deviations are summarized in Table 3.

As in the previous set of analysis, we first wanted to investigate the effect of the familiarity of the context on the number of categories used in the future planning task only. Similarly to the analysis run on the quantity of information units, the 2 (task order) × 2 (familiarity of future planning context) univariate ANOVA showed no main effect of the familiarity of the future scenario, $F(1, 36) = 0.01$, $MSE = 6.53$, $p = .947$, and no order effect. Furthermore, we found a similar two-way interaction between task order and familiarity of the context, $F(1, 36) = 8.82$, $MSE = 6.53$, $p = .005$, $\eta_p^2 = .20$. While there was no difference in the number of semantic categories used between the two contexts (Australia and Antarctica) when the future planning task was completed first, participants provided details coded in fewer semantic categories when the future planning tasks was completed second and when the context was familiar (Australia), $F(1, 36) = 9.92$, $MSE = 6.53$, $p = .003$, $\eta_p^2 = .22$.

We also ran another mixed-design ANOVA with type of task (past planning vs. future planning) as a within-subject factor and task order (past–future vs. future–past) and familiarity of future planning context (familiar vs. unfamiliar) as between-subject

Table 3. *Total number of semantic categories used in each task as a function of the order of task presentation and the familiarity of the future planning scenario*

		Task			
		Past planning		Future planning	
Task order	Future scenario familiarity	M (*SD*)	95% CI	M (SD)	95% CI
Past–future	Familiar	11.20 (2.78)	[9.61, 12.79]	11.20 (3.15)	[9.56, 12.84]
	Unfamiliar	12.60 (1.58)	[11.01, 14.19]	14.80 (2.04)	[13.16, 16.44]
	Total	11.90 (2.31)	[10.78, 13.02]	13.10 (3.18)	[11.84, 14.16]
Future–past	Familiar	13.80 (2.86)	[12.21, 15.39]	13.80 (2.20)	[12.16, 15.44]
	Unfamiliar	13.20 (2.49)	[11.61, 14.76]	12.60 (2.67)	[10.96, 14.24]
	Total	13.50 (2.63)	[12.38, 14.62]	13.20 (2.46)	[12.04, 14.36]

Note: CI = confidence interval.

factors, on the total number of semantic categories used. There were no significant main effects or interactions, which indicates that participants used the same average number of semantic categories in each scenario, regardless of the condition they were in or the task they were completing.

This set of analyses shows that participants used most of the semantic categories in both past and future planning tasks, as predicted by the literature on scripts. Yet, the number of categories used in the future planning task was significantly lower when the context of the event was familiar and when future planning was completed after the past planning task.

Analysis of script similarity across past and future planning tasks

To investigate script similarities across tasks, for each participant we counted how many of our 19 semantic categories were present *in both tasks* (past and future planning), *in the past planning task only*, *in the future planning task only*, and *in neither task* (categories that were never mentioned). For example, if a participant mentioned planning for food in his/her past and future plans (regardless of him/her stating 1 food item or 10 food items), the food category would be placed in the *both tasks* variable. We therefore compiled four values for each participant that represent script similarity (or lack of) between both completed tasks. These

values were then averaged across participants to create these new dependent variables. As some assumptions were violated, we ran 2 (task order) × 2 (familiarity of future planning context) factorial ANOVAs with bootstrapping procedures[4] on each of these four dependent variables. Results are shown in Table 4.

There were no significant main effects or interactions when the dependent variables were the number of semantic categories present in both tasks, in the past planning task only, or in the future planning task only. There was, however, a strong two-way interaction for the analysis with *in neither task* as the dependent variable, $F(1, 36)$ = 9.64, MSE = 4.15, p = .004, η_p^2 = .21. When the scenario was familiar (Australia) for both tasks, not mentioning certain semantic categories during the past planning task completed first made them less likely to be mentioned during the future planning task completed second. In the other three conditions ("past planning then future planning in Antarctica", "future planning in Australia then past planning", "future planning in Antarctica then past planning"), most categories were at least discussed in one of the two tasks, if not in both.

Together, these results show that when participants provided details coded in one category in the first task, they usually provided details coded in that same category in the second task. However, semantic categories not mentioned

[4]Bootstrapping procedures are robust methods that can be used when some assumptions are violated (Field, 2009).

Table 4. *Average presence of semantic categories for each participant in neither task, in both tasks, in the past planning task only, or in the future planning task only, as a function of the order of task presentation and the familiarity of the future planning scenario*

	Task order			
	Past–future		Future–past	
	Future scenario familiarity			
Measure	Familiar M (SD)	Unfamiliar M (SD)	Familiar M (SD)	Unfamiliar M (SD)
Categories present in *neither* task	5.3 (2.41)**	2 (1.05)	2.4 (1.84)	3.1 (2.51)
Categories present in *both* tasks	8.7 (2.95)	10.4 (1.90)	11 (3.50)	9.9 (2.51)
Categories present only in the *past planning* task	2.5 (1.72)	2.2 (1.40)	2.8 (1.69)	3.3 (1.16)
Categories present only in the *future planning* task	2.5 (1.96)	4.4 (1.84)	2.8 (1.93)	2.7 (1.34)

**$p < .01$.

when recalling past planning were less likely to be discussed when planning a future camping trip in Australia.

Discussion

In the present study, we investigated the quantity of information units as well as the use of semantic scripts in a past and a future planning task, with the future planning task set in either a familiar (Australia) or an unfamiliar (Antarctica) context. First, we investigated potential differences between the two contexts of the future planning task. Similarly to previous studies (Arnold, McDermott, & Szpunar, 2011), participants rated the unfamiliar scenario as harder than the familiar scenario. Yet our results showed no differences between the two conditions in terms of the quantity of information units produced. In spite of the fact that participants had experiences of camping in Australia and had never been to Antarctica, they produced as many information units when imagining their plans for camping in Antarctica as in Australia when they completed the future planning task first. We also found no difference between the two contexts when investigating the number of semantic categories. The inconsistency of our results with de Vito et al. (2012)'s study—who found that familiar events contained more internal details than unfamiliar events—might be because they focused on the simulation of events, whereas

we focused on the role of simulation in planning. Therefore, our results could indicate that future planning is not so constrained by familiarity, and that scripts and semantic knowledge could suffice to plan a future event. Our results also support the semantic scaffolding hypothesis, which suggests that depending on the familiarity of the event, one would be more likely to draw upon either on episodic details or on semantic memory (Irish & Piguet, 2013).

Second, we analysed the similarities and interactions across both past and future planning tasks. Our results found interactions between the tasks and the order they were presented in. This order effect can be divided into two separate findings. The first major finding from task interaction can be found in the higher number of information units in the past planning task when completed second than when completed first. It is important to note that familiarity of the future context did not influence our results. This finding was relatively surprising as until now research had only investigated the influence of memory on future thinking and not the opposite. We can therefore propose a tentative account of this effect through the concept of scripts, but further research is needed to investigate the underlying processes at play. Similar to Bartlett's (1932) concept of schemas, scripts represent a general sequence of events and can originate from repeated events, as well as from general semantic knowledge. Because of this, even if they are conventionally considered semantic memory, scripts might have more in common with concepts such as Neisser's (1981) "repisodic memory", Barsalou's (1988) "extended events", or Conway and Pleydell-Pearce's (2000) "general events", which are neither truly semantic nor episodic (Greenberg & Verfaellie, 2010; Martin-Ordas, Atance, & Louw, 2012). They are representative versions of similar events and can support both remembering and future thinking processes. Thus, when asked, as a first task, to imagine how they would plan for a future camping trip, participants could have relied on scripts to help them decide what they needed to plan, as well as on memories of how they planned past camping trips. Subsequently, when asked to recall how they planned a camping trip in the past, these scripts were accessible and facilitated the remembering process. Therefore they recalled more details than participants who did not already have these extensive scripts activated. We also find a similar influence of scripts when participants imagined planning a future camping trip in Antarctica as their second task. Script as well as episodic details could also have facilitated their future plans. However, as they had never camped or even been in Antarctica, on top of the activated scripts, they had to actively think of new details to consider and rely also on semantic knowledge they had about Antarctica.

Yet, we did not find the same influence of scripts when the future planning task was set in Australia and completed second. Compared to the other conditions where there was an increase in the amount of information units produced from the first to the second task, participants imagining how they would plan a future camping trip in Australia as their second task generated no more information units than they did when they remembered how they planned a similar past trip during the first task. The numbers even showed a small but non-significant reduction in the quantity of information units generated between the two tasks. A possible interpretation would be that in this case, participants relied more on episodic memory as a complete relevant plan had just been produced. Consequently when participants imagined how

they would plan for a camping trip also in Australia, they might have simply produced a very similar version to what they had just told us but in a future tense, without trying to think of new details or possible changes in the context. If their plans were successful in the past, repeating them in the future could be an efficient approach.

However, we could also find another potential interpretation of this result in the retrieval-induced forgetting phenomenon, where the act of remembering an item can inhibit the retrieval of related items later on (Anderson, Bjork, & Bjork, 1994). It is possible that in this case, remembering the past plan inhibited the search for additional similar details. Nevertheless, we found converging evidence for the first interpretation in our analysis of script similarities across past and future planning tasks. Our results indicated that, in general, if a category was mentioned in one of the tasks, participants also generated information units from that category in the second task. This result shows that similar scripts were used in both tasks. Moreover, the analysis further revealed that if a category was not mentioned in the first task, participants might still generate information units from that category in the second task. This was true of all conditions except when participants had to imagine planning a future event in a familiar context as the second task. In this case, categories not mentioned during the past planning task were also unlikely to be mentioned during the future planning task set in the familiar context. This indicates that participants simply followed the same script as the one they had just mentioned without adding categorically new details to their plan.

In summary, if the simulating subcomponent of the planning process is a type of episodic future thinking, then our study shows that the interaction between past and future thinking goes both ways. On the one hand, our findings suggest that starting by remembering the planning of a past event can influence the capacity to plan a future one in a similar context. On the other hand, starting by planning a future event might activate semantic categories that could later support the remembering process. Hence, future thinking seems to rely on episodic memory—especially when the information has been recently recalled—but is not constrained by it. However, episodic details are not the only components of future thinking; our results also indicate that semantic knowledge and script-knowledge play important roles when imagining and planning future events.

Importantly, our findings also highlight that the order of presentation of past and future thinking tasks matters, as they can influence one another. Future studies should keep in mind this order effect, as we know now that past memories can affect the way we think about similar future events and vice versa. Randomizing the task order might not be enough to control for the effect and might even confound results. Depending on the goal of the study, this order effect could potentially be reduced by running conditions on different days, by avoiding similar events in past and future tasks, or by accounting for it when running statistics. However, future studies should investigate the extent to which past and future interact and the underlying processes, depending on the familiarity of the events, their occurrence in everyday life, and the prevalence of cultural semantic scripts. Finally, research should continue to explore the role of semantic knowledge in the formation of future thoughts and planning as a function of the quantity of related memories available and the need to make the future thought or the future plan as plausible as possible.

REFERENCES

Abelson, R. P. (1981). Psychological status of the script concept. *American Psychologist*, *36*(7), 715–729. doi:10.1037/0003-066x.36.7.715

Addis, D. R., Musicaro, R., Pan, L., & Schacter, D. L. (2010). Episodic simulation of past and future events in older adults: Evidence from an experimental recombination task. *Psychology and Aging*, *25*(2), 369–376. doi:10.1037/a0017280

Addis, D. R., Pan, L., Vu, M. A., Laiser, N., & Schacter, D. L. (2009). Constructive episodic simulation of the future and the past: Distinct subsystems of a core brain network mediate imagining and remembering. *Neuropsychologia*, *47*(11), 2222–2238. doi:10.1016/j.neuropsychologia.2008.10.026

Addis, D. R., Sacchetti, D. C., Ally, B. A., Budson, A. E., & Schacter, D. L. (2009). Episodic simulation of future events is impaired in mild Alzheimer's disease. *Neuropsychologia*, *47*(12), 2660–2671. doi:10.1016/j.neuropsychologia.2009.05.018

Addis, D. R., Wong, A. T., & Schacter, D. L. (2007). Remembering the past and imagining the future: common and distinct neural substrates during event construction and elaboration. *Neuropsychologia*, *45*(7), 1363–1377. doi:10.1016/j.neuropsychologia.2006.10.016

Addis, D. R., Wong, A. T., & Schacter, D. L. (2008). Age-related changes in the episodic simulation of future events. *Psychological Science*, *19*(1), 33–41. doi:10.1111/j.1467-9280.2008.02043.x

Anderson, M. C., Bjork, R. A., & Bjork, E. L. (1994). Remembering can cause forgetting: retrieval dynamics in long-term memory. *Journal of Experimental Psychology: Learning, Memory, and Cognition*, *20*(5), 1063–1087. doi:10.1037/0278-7393.20.5.1063

Arnold, K. M., McDermott, K. B., & Szpunar, K. K. (2011). Imagining the near and far future: The role of location familiarity. *Memory & Cognition*, *39*(6), 954–967. doi:10.3758/s13421-011-0076-1

Atance, C. M., & O'Neill, D. K. (2001). Episodic future thinking. *Trends in Cognitive Sciences*, *5*(12), 533–539. doi:10.1016/s1364-6613(00)01804-0

Baird, B., Smallwood, J., & Schooler, J. W. (2011). Back to the future: Autobiographical planning and the functionality of mind-wandering. *Consciousness and Cognition*, *20*(4), 1604–1611. doi:10.1016/j.concog.2011.08.007

Bakeman, R. (2005). Recommended effect size statistics for repeated measures designs. *Behavior Research Methods*, *37*(3), 379–384. doi:10.3758/bf03192707

Barsalou, L. W. (1988). The content and organization of autobiographical memories. In U. Neisser & E. Winograd (Eds.), *Remembering reconsidered: Ecological and traditional approaches to the study of memory* (pp. 193–243). New York: Cambridge University Press.

Bartlett, F. C. (1932). *Remembering: A study in eperimental and social psychology*. Cambridge, UK: Cambridge University Press.

Berntsen, D., & Bohn, A. (2010). Remembering and forecasting: The relation between autobiographical memory and episodic future thinking. *Memory & Cognition*, *38*(3), 265–278. doi:10.3758/MC.38.3.265

Berntsen, D., & Jacobsen, A. S. (2008). Involuntary (spontaneous) mental time travel into the past and future. *Consciousness and Cognition, 17*(4), 1093–1104. doi:10.1016/j.concog.2008.03.001

Bower, G. H., Black, J. B., & Turner, T. J. (1979). Scripts in memory for text. *Cognitive Psychology, 11* (2), 177–220. doi:10.1016/0010-0285(79)90009-4

Buckner, R. L., & Carroll, D. C. (2007). Self-projection and the brain. Trends in Cognitive Sciences, *11*(2), 49–57. doi:10.1016/j.tics.2006.11.004

Cole, S. N., Gill, N. C., Conway, M. A., & Morrison, C. M. (2012). Mental time travel: Effects of trial duration on episodic and semantic content. *Quarterly Journal of Experimental Psychology, 65*(12), 2288–2296. doi:10.1080/17470218.2012.740053

Cole, S. N., Morrison, C. M., & Conway, M. A. (2013). Episodic future thinking: Linking neuropsychological performance with episodic detail in young and old adults. *Quarterly Journal of Experimental Psychology, 66*(9), 1687–1706. doi:10.1080/17470218.2012.758157

Conway, M. A., & Pleydell-Pearce, C. W. (2000). The construction of autobiographical memories in the self-memory system. *Psychological Review, 107*(2), 261. doi:10.1007/978-94-015-7967-4_16

Cordonnier, A., & Sutton, J. (2016). Autobiographical Thinking.

D'Argembeau, A., & Demblon, J. (2012). On the representational systems underlying prospection: Evidence from the event-cueing paradigm. *Cognition, 125*(2), 160–167. doi:10.1016/j.cognition.2012.07.008

D'Argembeau, A., & Mathy, A. (2011). Tracking the construction of episodic future thoughts. *Journal of Experimental Psychology: General, 140*(2), 258–271. doi:10.1037/a0022581

D'Argembeau, A., Renaud, O., & Van der Linden, M. (2011). Frequency, characteristics and functions of future-oriented thoughts in daily life. *Applied Cognitive Psychology, 25*(1), 96–103. doi:10.1002/acp.1647

D'Argembeau, A., Xue, G., Lu, Z.-L., Van der Linden, M., & Bechara, A. (2008). Neural correlates of envisioning emotional events in the near and far future. *Neuroimage, 40*(1), 398–407. doi:10.1016/j.neuroimage.2007.11.025

De Brigard, F., & Giovanello, K. S. (2012). Influence of outcome valence in the subjective experience of episodic past, future, and counterfactual thinking. *Consciousness and Cognition, 21*(3), 1085–1096. doi:10.1016/j.concog.2012.06.007

de Vito, S., Gamboz, N., & Brandimonte, M. A. (2012). What differentiates episodic future thinking from complex scene imagery? *Consciousness and Cognition, 21*(2), 813–823. doi:10.1016/j.concog.2012.01.013

Dudai, Y., & Carruthers, M. (2005). The Janus face of mnemosyne. Nature, *434*(7033), 567–567. doi:10.1038/434567a

Field, A. (2009). *Discovering statistics using SPSS.* London: Sage publications.

Finnbogadottir, H., & Berntsen, D. (2012). Involuntary future projections are as frequent as involuntary memories, but more positive. *Consciousness and Cognition, 22*(1), 272–280. doi:10.1016/j.concog.2012.06.014

Greenberg, D. L., & Verfaellie, M. (2010). Interdependence of episodic and semantic memory: Evidence from neuropsychology. *Journal of the International Neuropsychological Society, 16*(5), 748–753. doi:10.1017/S1355617710000676

Hassabis, D., Kumaran, D., & Maguire, E. A. (2007). Using imagination to understand the neural basis of episodic memory. *The Journal of Neuroscience, 27*(52), 14365–14374. doi:10.1523/JNEUROSCI.4549-07.2007

Hayes-Roth, B., & Hayes-Roth, F. (1979). A cognitive model of planning. *Cognitive Science, 3*(4), 275–310. doi:10.1016/s0364-0213(79)80010-5

Hudson, J. A., Fivush, R., & Kuebli, J. (1992). Scripts and episodes: The development of event memory. *Applied Cognitive Psychology, 6*(6), 483–505. doi:10.1002/acp.2350060604

Irish, M., Addis, D. R., Hodges, J. R., & Piguet, O. (2012). Considering the role of semantic memory in episodic future thinking: Evidence from semantic dementia. Brain, *135*(Pt 7), 2178–2191. doi:10.1093/brain/aws119

Irish, M., & Piguet, O. (2013). The pivotal role of semantic memory in remembering the past and imagining the future. *Frontiers in Behavioral Neuroscience, 7*, 1–11. doi:10.3389/fnbeh.2013.00027

Klein, S. B. (2013). The complex act of projecting oneself into the future. *Wiley Interdisciplinary Reviews: Cognitive Science, 4*(1), 63–79. doi:10.1002/wcs.1210

Klein, S. B., Cosmides, L., Tooby, J., & Chance, S. (2002). Decisions and the evolution of memory: Multiple systems, multiple functions. *Psychological Review, 109* (2), 306–329. doi:10.1037//0033-295x.109.2.306

Klein, S. B., Loftus, J., & Kihlstrom, J. F. (2002). Memory and temporal experience: The effects of episodic memory loss on an amnesic patient's ability to remember the past and imagine the future. *Social Cognition, 20*(5), 353–379. doi:10.1521/soco.20.5.353.21125

Klein, S. B., Robertson, T. E., & Delton, A. W. (2010). Facing the future: Memory as an evolved system for planning future acts. *Memory & Cognition, 38*(1), 13–22. doi:10.3758/MC.38.1.13

Lakens, D. (2013). Calculating and reporting effect sizes to facilitate cumulative science: A practical primer for t-tests and ANOVAs. *Frontiers in Psychology, 4*, 1–12. doi:10.3389/fpsyg.2013.00863

Levine, B., Svoboda, E., Hay, J. F., Winocur, G., & Moscovitch, M. (2002). Aging and autobiographical memory: Dissociating episodic from semantic retrieval. *Psychology and Aging, 17*(4), 677–689. doi:10.1037//0882-7974.17.4.677

Lichtenstein, E. H., & Brewer, W. F. (1980). Memory for goal-directed events. *Cognitive Psychology, 12*(3), 412–445. doi:10.1016/0010-0285(80)90015-8

MacLeod, A. K., & Conway, C. (2007). Well-being and positive future thinking for the self versus others. *Cognition & Emotion, 21*(5), 1114–1124. doi:10.1080/02699930601109507

Maguire, E. A., Vargha-Khadem, F., & Hassabis, D. (2010). Imagining fictitious and future experiences: Evidence from developmental amnesia. *Neuropsychologia, 48*(11), 3187–3192. doi:10.1016/j.neuropsychologia.2010.06.037

Martin-Ordas, G., Atance, C. M., & Louw, A. (2012). The role of episodic and semantic memory in episodic foresight. *Learning and Motivation, 43*(4), 209–219. doi:10.1016/j.lmot.2012.05.011

McCormack, T., & Atance, C. M. (2011). Planning in young children: A review and synthesis. *Developmental Review, 31*(1), 1–31. doi:10.1016/j.dr.2011.02.002

McCormack, T., Hoerl, C., & Butterfill, S. (2011). *Tool use and causal cognition.* New York: Oxford University Press.

Mullally, S. L., Hassabis, D., & Maguire, E. A. (2012). Scene construction in amnesia: An fMRI study. *The Journal of Neuroscience, 32*(16), 5646–5653. doi:10.1523/JNEUROSCI.5522-11.2012

Neisser, U. (1981). John Dean's memory: A case study. *Cognition, 9*(1), 1–22.

Nelson, K., & Gruendel, J. (1981). Generalized event representations: Basic building blocks of cognitive development. In A. L. Brown & M. E. Lamb (Eds.), Advances in *developmental psychology* (Vol. *1*, pp. 131–158). Hillsdale, NJ: Erlbaum.

Olejnik, S., & Algina, J. (2003). Generalized eta and omega squared statistics: Measures of effect size for some common research designs. *Psychological Methods, 8*(4), 434–447. doi:10.1037/1082-989x.8.4.434

Schacter, D. L. (2012). Adaptive constructive processes and the future of memory. *American Psychologist, 67*(8), 603–613. doi:10.1037/e502412013-188

Schacter, D. L., & Addis, D. R. (2007). The cognitive neuroscience of constructive memory: Remembering the past and imagining the future. Philosophical transactions of the Royal Society of London. *Series B, Biological Sciences, 362*(1481), 773–786. doi:10.1098/rstb.2007.2087

Schacter, D. L., & Addis, D. R. (2009). Remembering the past to imagine the future: A cognitive neuroscience perspective. *Military Psychology, 21*(Suppl 1), S108–S112. doi:10.1080/08995600802554748

Schacter, D. L., Addis, D. R., & Buckner, R. L. (2008). Episodic simulation of future events: Concepts, data, and applications. *Annals of the New York Academy of Sciences, 1124*, 39–60. doi:10.1196/annals.1440.001

Schacter, D. L., Addis, D. R., Hassabis, D., Martin, V. C., Spreng, R. N., & Szpunar, K. K. (2012). The future of memory: Remembering, imagining, and the brain. *Neuron, 76*(4), 677–694. doi:10.1016/j.neuron.2012.11.001

Schank, R. C., & Abelson, R. P. (1977). *Scripts, plans, goals, and understanding: An inquiry into human knowledge structures.* Oxford, England: Erlbaum.

Suddendorf, T., Addis, D. R., & Corballis, M. C. (2009). Mental time travel and the shaping of the human mind. *Philosophical Transactions of the Royal Society of London. Series B, Biological sciences, 364*(1521), 1317–1324. doi:10.1098/rstb.2008.0301

Suddendorf, T., & Corballis, M. C. (1997). Mental time travel and the evolution of the human mind. *Genetic, Social, and General Psychology Monographs, 123*(2). Retrieved from http://cogprints.org/725/

Suddendorf, T., & Corballis, M. C. (2007). The evolution of foresight: What is mental time travel, and is it unique to humans? *Behavioral and Brain Sciences, 30*(03), 299–313. doi:10.1017/s0140525×07001975

Szpunar, K. K. (2010). Episodic future thought an emerging concept. *Perspectives on Psychological Science, 5*(2), 142–162. doi:10.1177/1745691610362350

Szpunar, K. K., Spreng, R. N., & Schacter, D. L. (2014). A taxonomy of prospection: Introducing an organizational framework for future-oriented cognition. *Proceedings of the National Academy of Sciences, 111*(52), 18414–18421. doi:10.1073/pnas.1417144111

Tulving, E. (1985). Memory and consciousness. *Canadian Psychology, 26*(1), 1–12. doi:10.1037/h0080017

Tulving, E. (2005). Episodic memory and autonoesis: Uniquely human. In H. Terrace & J. Metcalfe (Eds.), The missing link in cognition (pp. 4–56). New York: Oxford University Press.

Thinking about the future can cause forgetting of the past

Annie S. Ditta and Benjamin C. Storm

Department of Psychology, University of California, Santa Cruz, CA, USA

People are able to imagine events in the future that have not yet happened, an ability referred to as episodic future thinking. There is now compelling evidence that episodic future thinking is accomplished via processes similar to those that underlie episodic retrieval. Drawing upon work on retrieval-induced forgetting, which has shown that retrieving some items in memory can cause the forgetting of other items in memory, we show that engaging in episodic future thinking can cause related autobiographical memories (Experiments 1–3) and episodic event descriptions (Experiments 3–4) to become less recallable in the future than they would have been otherwise. This finding suggests that episodic future thinking can serve as a memory modifier by changing the extent to which memories from our past can be subsequently retrieved.

Many people plan for the next day as they fall asleep or imagine how the day will unfold as they lie in bed in the morning. A person might see themselves taking a shower, eating breakfast, and driving to work to make the early quarterly meeting. This future simulation process—called episodic future thinking (EFT)—is seemingly easy to accomplish, yet none of the imagined events have actually happened. How is it that people are able to so rapidly and flexibly construct visions of the future?

According to the constructive episodic simulation hypothesis (e.g., Schacter & Addis, 2007; Schacter, Addis, & Buckner, 2007; Szpunar, 2010), elements from memory are combined in new ways to simulate possible future events in a manner similar to how autobiographical memories are reconstructed. Indeed, research has shown that the process of imagining the future relies on much of the same cognitive and neural mechanisms as does recalling the past (e.g., Addis, Pan, Vu, Laiser, & Schacter, 2009; Addis & Schacter, 2008; Addis, Wong, & Schacter, 2007; Botzung, Denkova, & Manning, 2008; Hach, Tippett, & Addis, 2014; Szpunar, Watson, & McDermott, 2007). Despite the increased focus on delineating the cognitive and neural underpinnings of EFT, we do not yet fully understand the way in which imagining the future affects other information in memory.

Storm and Jobe (2012) examined this issue in the context of retrieval-induced forgetting, a phenomenon in which selectively retrieving some items in memory causes the forgetting of other items in memory (Anderson, Bjork, & Bjork, 1994). The inhibitory account of retrieval-induced forgetting assumes that nontarget

We thank J. Brown, M. Freeman, A. Klassen, A. Luo, A. Meltzer, A. Mogensen, K. Ramakrishnan, and S. Shrikanth for their assistance with data collection.

information competes with the retrieval of target information, and that inhibition is elicited to help resolve that competition, rendering the nontarget items less recallable in the future than they would have been otherwise. Similar dynamics may be at play in EFT. Specifically, when one is attempting to imagine a future event, related past events that are not quite appropriate for use in constructing the future event may cause competition, and it is possible that such competition is resolved through inhibition. If so, inhibition—and the resulting effect of forgetting—may act to potentiate the construction of novel future events that are context appropriate and that deviate from specific past events. Such a process would seem particularly important because although the past may provide building blocks to construct the future, the past is rarely, if ever, perfectly repeated. Hence, it seems likely that imagining the future would often require the inhibition of certain past events and, thus, that thinking about the future should be capable, if not likely, to cause the forgetting of those past events.

Storm and Jobe (2012) tested this hypothesis by having participants study lists of events concerning individuals in specific contexts (e.g., Mario feeding hot dogs to pigeons in the park) and then having them either retrieve or generate new events associated with half of those contexts. Participants in the retrieval condition attempted to recall autobiographical instances in which they did something in a park, whereas participants in the generation condition imagined doing something in the park either in the future (Experiments 1 and 2) or in the past (Experiment 3). Although attempting to recall past events caused participants to forget the studied events associated with the same contexts (relative to studied events associated with other contexts), generating new events did not. Thus, the generation of novel episodic events—whether directed towards the future or past—failed to cause the forgetting of related information in memory.

The fact that EFT failed to cause forgetting is quite surprising, as it stands in direct contrast to a growing body of research showing that there are many types of generative processes that are capable of causing forgetting (thinking of new uses for objects, Storm & Patel, 2014; solving problems, Storm & Angello, 2010; solving word fragments, Healey, Campbell, Hasher, & Ossher, 2010; semantic generation, Bäuml, 2002; Storm, Bjork, Bjork, & Nestojko, 2006; for a review, see Storm et al., 2015), and it is unclear why EFT would behave differently. Moreover, the finding seems to suggest an important distinction in the processes underlying EFT and retrieval, a suggestion that does not fit well with the evidence cited above that there is substantial overlap in such processes.

Given the surprising nature of the results reported by Storm and Jobe (2012), and the fact that they do not fit well with other findings in related research, we felt it was important to investigate the issue further. To increase the likelihood of observing forgetting, participants were prompted to retrieve their own autobiographical memories before engaging in EFT. According to the inhibitory account of retrieval-induced forgetting, retrieval—or any other type of generative process—should only cause the forgetting of nontarget information when that information causes competition (for reviews, see Anderson, 2003; Murayama, Miyatsu, Buchli, & Storm, 2014; Storm & Levy, 2012), and it is possible that the events describing fictional individuals failed to cause such competition. By having participants retrieve their own personal autobiographical memories before attempting to think of their own personal autobiographical futures, we hoped that there would be greater competition and thus greater need for inhibition. Thus, we predicted that imagining the future would cause participants to forget related autobiographical memories from their past.

EXPERIMENT 1

Participants were prompted to recall autobiographical events associated with specific contexts before being asked to imagine future events associated with half of those contexts. Participants were then given a final test in which they were asked to recall each of the autobiographical memories they had initially retrieved. If EFT does cause forgetting, then the memories associated with contexts used to imagine the future should be less recallable on the final test than memories associated with contexts that had not been used to imagine the future. This finding would indicate that imagining the future can cause the forgetting of autobiographical past events.

In addition to causing autobiographical memories to be forgotten, or made less accessible, EFT might also distort or modify the way in which memories are subsequently reconstructed (when they can be retrieved). That is, it is possible that constructing episodic future events causes people to remember past events differently than they would have otherwise. Thus, Experiment 1 had two goals: first, to explore whether EFT can cause forgetting; and second, to examine the consequences of EFT on how participants subsequently reconstruct and reexperience their memories. Specifically, after the conclusion of the final test, participants were reexposed to strong, identifying cues for every autobiographical event and were asked to remember that event once again before rating it with regard to detail, emotional intensity, and perspective. If imagining the future modifies the way in which past memories are retrieved then events that are associated with prior EFT should be experienced differently than events that are not associated with prior EFT.

Method

Participants
Thirty-six University of California, Santa Cruz (UCSC) undergraduates ($M_{age} = 20.5$ years, age range = 18–26 years) participated for partial course credit.

Materials
The materials were adapted from Storm and Jobe (2012). Specifically, context words were used as cues to prompt memory recall and EFT. Most of the context words were taken directly from Storm and Jobe, and all contexts were places or activities that participants could be expected to have

experienced as undergraduates at a large university (i.e., college housing, mall, car, park, work, coffee shop, sleeping, friend's house, lecture hall, movie theatre, beach, and gym).

Procedure
We used a memory probe technique similar to that used by Storm and Jobe (2012), but modified it such that participants were asked to generate their own memories before thinking about the future (instead of studying facts about a fictional person). Participants were shown each context word and were asked to think of four associated memories. Specifically, they were given 15 s to bring each memory to mind that was both autobiographical (defined to the participants as: "about you, and not someone else") and episodic (defined to the participants as: "a specific event that you can mentally time travel back to and reexperience"), and the instructions were further clarified with an example. After retrieving the memory, participants were given an additional 5 s to provide a short title that represented that memory (e.g., for a memory retrieved in the context of "college housing", a title might be "Friendly Roommates"). This procedure was repeated four times for each context word, thus resulting in a total of 48 memories.

Next, participants entered the future thinking phase. They were re-presented with one of the context words and were told to imagine an event in the future that had not yet happened, but that could potentially happen. Again, they were told that the future event should be autobiographical and episodic. They were also presented with two additional cue words to incorporate into their imagined future event (e.g., for "college housing", participants might also see "party" and "soda"). This procedure, which was also employed by Storm and Jobe (2012), was used to prevent participants from simply reimagining an event from their past reoccurring in the future, thus requiring participants to construct a future event that was new and different. Participants were given 20 s to imagine the event quietly in their head and an additional 5 s to title the event. Participants imagined three future scenarios for half of the original 12 contexts (with the particular subset of contexts receiving EFT counterbalanced across participants), resulting in a total of 18 EFT trials. The order of the trials was block-randomized such that participants generated one event for each context, then a second event for each context, and then a third, with a different pair of additional cue words being provided on each trial. In this experiment, as well as in each of the experiments that followed, participants reported successfully generating future events on 99% of the trials.

Participants then entered the final recall phase, where they were re-presented with all 12 contexts again, one at a time, and in a new randomized order. They were instructed to recall, in any order, the four titles of the memories they had brought to mind during the initial portion of the study, but not the titles of any of the future events. They were given 16 s per context.

After the final test, participants were verbally re-presented with their titles (and visually re-presented with the associated context words) for each of the autobiographical memories and were asked to rate the memories in relation to three phenomenological characteristics. Specifically, they were instructed to spend 10 s bringing each memory to mind before rating it with regard to detail (1–5 scale; 1 is vague, 5 is vivid), emotional intensity (1–5 scale; 1 is detached, 5 is highly emotional), and perspective (first person was coded as 0; observer was coded as 1). Participants were given 5 s to complete each rating before being allowed to move on to the next item. This procedure was repeated for each of the 48 memories regardless of whether a given memory had been successfully recalled during the final recall phase.

Results

Memories recalled

A cut-off was established such that participants needed to successfully retrieve autobiographical memories on 75% of the initial trials. Four participants failed to reach this criterion, and so their data were removed. It should be noted that in this experiment, as well as in all subsequent experiments, we analysed the data using different cut-off criteria, and the resulting pattern of results was always the

same. Of the remaining 32 participants, an autobiographical event was retrieved 98% of the time.

Final recall performance

The mean proportion of memories successfully recalled on the final recall test was analysed as a function of item type (EFT vs. baseline). Participants recalled significantly fewer memories in the EFT condition ($M = .62$, $SE = .03$) than in the baseline condition ($M = .72$, $SE = .02$), $F(1, 31) = 13.59$, $MSE = .01$, $p < .001$, $\eta_p^2 = .31$. That is, participants were less able to recall autobiographical events associated with contexts that had been used to prompt EFT than they were for contexts that had not been used to prompt EFT. This finding demonstrates that imagining future events can cause the forgetting of related past events.

Rating analyses

We next examined whether participants reexperienced their autobiographical memories differently after engaging in EFT for related events. Multiple one-way analyses of variance (ANOVAs) were run as a function of item type (EFT vs. baseline), and no significant differences were observed—not with regard to detail, $F(1, 31) = 0.01$, $MSE = .07$, $p = .91$, $\eta_p^2 = .00$, emotionality, $F(1, 31) = 0.00$, $MSE = .09$, $p = .95$, $\eta_p^2 = .00$, or perspective, $F(1, 31) = .02$, $MSE = .01$, $p = .89$, $\eta_p^2 = .00$. Indeed, as can been seen in the top half of Table 1, the similarity between the two conditions was remarkable. The same analysis was conducted separately for memories that were recalled on the final test and for those that were not recalled on the final test. Once again, no significant differences were observed (all $p > .10$). These results suggest that although EFT can cause past events to be forgotten, it may not alter the way in which those events are subsequently experienced when they can be effectively reconstructed.

EXPERIMENT 2

The primary goal of Experiment 2 was to replicate the forgetting effect observed in Experiment 1. A secondary goal, however, was to explore whether

Table 1. *The mean detail, emotion, and perspective ratings as a function of condition in Experiments 1 and 2*

Experiment	Type of Rating Condition	Condition		Effect p-value
		EFT	Baseline	
Experiment 1	Detail	3.99 (.53)	3.99 (.43)	.91
	Emotion	3.24 (.63)	3.24 (.61)	.95
	Perspective	.16 (.18)	.16 (.15)	.89
Experiment 2	Detail			
	+Mem/+EFT	3.77 (0.70)	3.79 (0.54)	.79
	−Mem/+EFT	3.69 (0.64)	3.66 (0.58)	.69
	+Mem/−EFT	3.74 (0.59)	3.79 (0.54)	.48
	−Mem/−EFT	3.73 (0.67)	3.66 (0.58)	.34
	Emotion			
	+Mem/+EFT	3.31 (0.74)	3.28 (0.65)	.73
	−Mem/+EFT	3.19 (0.70)	3.28 (0.63)	.34
	+Mem/−EFT	3.20 (0.71)	3.28 (0.65)	.30
	−Mem/−EFT	3.36 (0.71)	3.28 (0.63)	.37
	Perspective			
	+Mem/+EFT	.28 (.28)	.27 (.26)	.89
	−Mem/+EFT	.29 (.26)	.28 (.22)	.66
	+Mem/−EFT	.25 (.28)	.27 (.26)	.52
	−Mem/−EFT	.28 (.23)	.28 (.22)	.87

Note: EFT = episodic future thinking. Detail and emotion ratings were rated on a scale from 1–5 (1 is low, 5 is high), and perspective was coded as either a 0 (first person) or 1 (observer). Experiment 2 shows the ratings additionally as a function of memory valence (positive, + vs. negative, −) and EFT valence (positive, + vs. negative, −). Standard deviations are displayed in the parentheses.

certain types of memories are more susceptible to forgetting than others. Specifically, participants were asked to retrieve both positive and negative autobiographical memories before being asked to generate future events that were either positive or negative. This procedure allowed us to explore whether negative memories are differentially susceptible to forgetting than positive memories and, moreover, to explore whether such a difference might vary as a function of the valence of EFT.

Method

Sixty-nine UCSC undergraduates ($M_{age} = 20.0$ years, age range = 18–28 years) participated for partial course credit. The materials and procedure were the same as those of Experiment 1 but with a few important exceptions. For the memory retrieval phase, participants were presented with a context word on the screen, as well as an emotional valence word ("positive" or "negative"). Participants

were informed that the memory they brought to mind should have the valence presented on the screen, where a positive memory was defined as an event that made the person feel "happy, proud, pleased, or gratified", and a negative memory was defined as an event that made the person feel "sad, embarrassed, or hurt". The participants were reminded that the memories should be recalled in their head silently. Two of the four memories per context word were prompted to be positive, and the other two were prompted to be negative, with the order of the promptings randomly determined and counterbalanced across trials and participants.

Each trial in the future thinking phase involved one of the context words, two additional cue words, and an emotional valence word. For half of the context words associated with EFT, participants were told to generate positive future events on all three trials; for the other half, participants were told to generate negative future events on all three trials. The particular set of context words associated with positive versus negative future thinking was counterbalanced across participants. The final cued recall phase proceeded as described in the previous experiment, with participants being cued with each context word and being asked to recall the four associated memories they had initially retrieved. The same post-final-test rating task as that administered in Experiment 1 was also administered.

Results

Memories recalled
The 75% criterion cut-off was again enforced, resulting in six participants being removed. Of the remaining 63 participants, an autobiographical event was retrieved on 96% of the trials.

Final recall performance
Final recall performance was assessed via a 2 (item type: EFT vs. baseline) × 2 (memory valence: positive vs. negative) × 2 (EFT valence: positive vs. negative) repeated measures ANOVA. There was a main effect of item type, $F(1, 62) = 41.00$, $MSE = .05$, $p < .001$, $\eta_p^2 = .40$, which replicated the forgetting effect found in Experiment 1— specifically, participants recalled fewer memories associated with EFT ($M = .55$, $SE = .02$) than they did for memories that were not associated with EFT ($M = .67$, $SE = .02$). Additionally, a main effect of memory valence was observed such that positive memories were recalled at a higher rate ($M = .64$, $SE = .02$) than negative memories ($M = .58$, $SE = .02$), $F(1, 62) = 16.25$, $MSE = .03$, $p < .001$, $\eta_p^2 = .21$, a finding consistent with prior research showing a positivity bias in autobiographical memory (for a review, see Walker, Skowronski, & Thompson, 2003). However, there were no significant two-way or three-way interactions (all $p > .10$), thus failing to provide any evidence of differential forgetting as a function of the emotional valence of the memories or EFT.

To examine specific forgetting effects, four 1-way ANOVAs were performed as a function of item type, memory valence, and EFT valence. As can be seen in Figure 1, the forgetting effects were all very similar. In all comparisons, memories associated with EFT were recalled at a significantly lower rate than were memories not associated with EFT. It should be noted that there was numerically more forgetting for negative memories ($M_{\text{forgetting}} = .13$, $SE = .03$) than for positive memories ($M_{\text{forgetting}} = .11$, $SE = .03$), though this difference was not statistically significant, $F(1, 62) = 0.61$, $MSE = .03$, $p = .44$, $\eta_p^2 = .01$.

Ratings analyses
We conducted three separate 2 (item type: EFT vs. baseline) × 2 (memory valence: positive vs. negative) × 2 (EFT valence: positive vs. negative) repeated measures ANOVAs on each of the ratings (detail, emotional intensity, and perspective). The relevant data are shown in the bottom section of Table 1. When controlling for multiple comparisons, no significant differences or interactions were observed. Additionally, none of the appropriate one-way ANOVAs for each of the ratings in each of the conditions were significant (all $p > .10$). As in Experiment 1, this pattern of results did not vary as a function of whether a given memory was recalled at final test. Neither memories that were recalled nor memories that were not recalled exhibited significant differences in phenomenology as a result of EFT (all $p > .10$).

EXPERIMENT 3

The results of Experiments 1 and 2 stand in significant contrast to those reported by Storm and Jobe (2012). Whereas we found that thinking of the future can cause forgetting, they did not. The most obvious difference between our study and their study is that we examined the consequences of EFT on the accessibility of one's own autobiographical memories, whereas Storm and Jobe examined the consequences of EFT on the accessibility of events associated with fictional individuals. Without a direct comparison within the same experiment, however, it is impossible to know whether it is this difference that is responsible for the conflicting pattern of results, or whether the discrepancy can be attributed to some other,

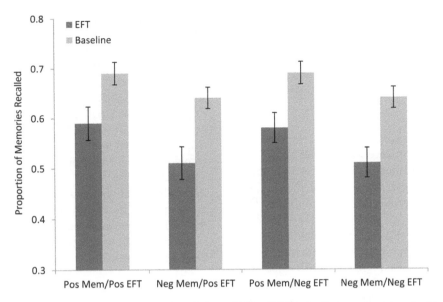

Figure 1. *Recall performance as a function of item type (episodic future thinking, EFT vs. baseline), memory valence (positive vs. negative), and EFT valence (positive vs. negative) in Experiment 2. "Pos" = "positive"; "neg" = "negative"; mem = memory. The two sets of outer paired columns represent matches in memory and EFT valence, whereas the two sets of inner paired columns represent mismatches in memory and EFT valence. Error bars represent standard errors of the mean.*

perhaps more subtle, difference in the methodologies. The goal of Experiment 3 was to provide such a comparison. Specifically, half of the participants retrieved autobiographical memories before simulating future events, and the other half studied other-person relevant events before simulating future events. Everything else was kept constant, thus allowing us to test whether autobiographical memories are susceptible to forgetting as a consequence of EFT whereas studied events are not.

Method

Sixty-eight UCSC undergraduates ($M_{age} = 19.5$ years, age range = 18–23 years) participated for partial course credit. We used the same contexts and cue words as those that had been used in the previous experiments. In the initial phase of the experiment, participants were asked either to recall four autobiographical events associated with each context (as in Experiment 1) or to study four events associated with a fictional character in each context (e.g., for the context word "park",

one of the four events was "Mario fed hot dogs to the pigeons"). We used many of the same events as those employed by Storm and Jobe (2012), as well as a few new events created using the same set of criteria. Participants in the recall condition were given a total of 20 s to recall and title each memory. Similarly, to control for time, participants in the study condition were given 20 s to study each event. As in Experiment 1, the future thinking phase involved presenting participants with half of the context words (with the particular subset counterbalanced) and participants being asked to generate three distinct future episodic simulations. As in the previous experiments, three sets of two cue words were presented with each context cue to ensure that participants generated simulations that were different from each other and that were different from the events they had either recalled or studied earlier. The final recall phase was the same as that of Experiment 1, with participants attempting to recall either their own memories or the facts that they had studied. There was no rating task in this experiment.

Results

Memories recalled

One participant was removed for failing to meet the 75% cut-off criterion. The remaining 67 participants retrieved an autobiographical event on 99% of the trials.

Final recall performance

Final recall performance was assessed via a 2 (item type: EFT vs. baseline) × 2 (memory type: studied vs. autobiographical) mixed-design ANOVA, with memory type manipulated between subjects. There was a main effect of trial type, $F(1, 65) = 26.86$, $MSE = .01$, $p < .001$, $\eta_p^2 = .29$, such that participants recalled fewer items associated with contexts used to prompt EFT ($M = .55$, $SE = .03$) than with contexts that had not been used to prompt EFT ($M = .66$, $SE = .02$). Two 1-way ANOVAs examined the specific forgetting effects. In the studied condition, events associated with EFT were recalled at a lower rate ($M = .43$, $SE = .04$) than were events not associated with EFT ($M = .52$, $SE = .04$), $F(1, 33) = 11.51$, $MSE = .01$, $p = .002$, $\eta_p^2 = .26$. Similarly, in the autobiographical condition, memories associated

with EFT were recalled at a lower rate ($M = .67$, $SE = .02$) than were memories not associated with EFT ($M = .75$, $SE = .02$), $F(1, 32) = 18.57$, $MSE = .01$, $p < .001$, $\eta_p^2 = .37$. The interaction was not significant, $F(1, 65) = 0.18$, $MSE = .01$, $p = .67$, $\eta_p^2 = .003$. Thus, as can be seen in Figure 2, we failed to find any evidence that the forgetting effect differs as a function of whether participants studied events or retrieved their own autobiographical events prior to engaging in EFT. The fact that we found significant forgetting in the study condition stands in direct contrast with the results of Storm and Jobe (2012), suggesting that EFT can cause people to forget studied events even if those events are not one's own autobiographical memories.

EXPERIMENT 4

In our final experiment, we sought to replicate the results of Experiment 3's studied condition while making the experimental design more similar to that used by Storm and Jobe (2012). One potentially important methodological difference

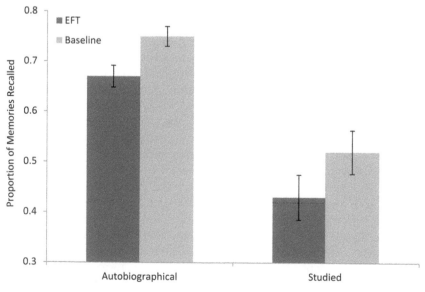

Figure 2. *Final recall performance as a function of item type (episodic future thinking, EFT vs. baseline) and memory type (autobiographical vs. studied) in Experiment 3. Error bars represent standard errors of the mean.*

between the two studies is that we had participants study each event for 20 s, whereas Storm and Jobe had participants study each event for 10 s. It is unclear why increasing study time would have increased the forgetting effect, but one possibility is that it led participants to imagine the studied events in their minds, creating episodic simulations similar to those created by participants in the autobiographical retrieval conditions of Experiments 1 and 2. Thus, in Experiment 4, participants were only given 10 s to study each event. If the amount of study time is responsible for the difference in forgetting across the two studies then the forgetting effect should be reduced or even eliminated.

Method

Twenty-seven UCSC undergraduates (M_{age} = 20.5 years, age range = 18–31 years) participated for partial course credit. The materials and procedure were identical to those employed in the studied condition of Experiment 3 except that participants were shown all four events simultaneously for 40 s—giving them 10 s to study each event—thus making the study conditions much more comparable to those employed by Storm and Jobe (2012).

Results

Final recall performance was assessed via a one-way ANOVA. Significant forgetting was observed such that events associated with EFT were recalled at a lower rate ($M = .24$, $SE = .03$) than were events not associated with EFT ($M = .42$, $SE = .04$), $F(1, 26) = 34.35$, $MSE = .01$, $p < .001$, $\eta_p^2 = .57$, a finding that demonstrates once again that studied information can be susceptible to forgetting as a consequence of thinking about the future.

GENERAL DISCUSSION

The results of the present research show that thinking about the future can cause the forgetting of the past, an effect that was observed in all four experiments and under a variety of methodological conditions. Both positive and negative autobiographical memories were susceptible to forgetting, as were episodic events related to other individuals. In all cases, information or experiences associated with cues used to construct future simulations became less recallable in the future than they would have been otherwise. This finding suggests that EFT has the potential to act as a memory modifier in much the same way that episodic retrieval acts as a memory modifier, affecting the extent to which memories from the past can be subsequently retrieved (e.g., Anderson et al., 1994; Bjork, 1975).

To date, the majority of research exploring episodic future thinking has suggested that imagining the future and remembering the past rely on highly similar or overlapping cognitive and neurobiological mechanisms (see, e.g., Schacter et al., 2007; Szpunar, 2010, for a review). The present results provide further support for this claim. Just as remembering some items in memory causes other items to be forgotten, a phenomenon referred to as retrieval-induced forgetting (Anderson et al., 1994; for reviews, see e.g., Murayama et al., 2014; Storm et al., 2015; Storm & Levy, 2012), imagining new events in the future causes other events to be forgotten. This finding is largely consistent with the growing body of work on retrieval-induced forgetting, which has found forgetting in just about every context in which it has been studied (for a review, see Storm et al., 2015). In fact, there are numerous demonstrations of effects similar to retrieval-induced forgetting when participants are prompted to generate or think of something new (e.g., Bäuml, 2002; Healey et al., 2010; Storm & Angello, 2010; Storm et al., 2006; Storm & Patel, 2014), and so the finding that thinking about the future causes forgetting is consistent with this previous work.

Given the abundance of evidence supporting the inhibitory account of retrieval-induced forgetting (Anderson, 2003; Murayama et al., 2014; Storm & Levy, 2012), and given the similarity between EFT and episodic retrieval, it stands to reason that inhibition plays at least some role in the process by which EFT causes related information to be forgotten. Future research should explore this possibility more directly, however, as there is still some disagreement regarding the mechanism or mechanisms that underlie retrieval-induced forgetting (for various viewpoints, see e.g., Anderson, 2003; Jonker, Seli, & MacLeod, 2013; Murayama et al., 2014; Raaijmakers & Jakab, 2013; Storm & Levy, 2012; Verde, 2012). However, the logic for why inhibition might be expected to play a role in EFT is relatively straightforward. The goal of EFT is to flexibly combine elements of past experiences to construct future events that serve some goal-directed purpose. In such efforts, information is likely to become activated that is not particularly useful, and just as inhibition is presumed to act to facilitate the retrieval of target items by making nontarget items less accessible, so too may it act to facilitate the construction of new and unexperienced episodic events by making nonuseful or contextually inappropriate information less accessible. In the current experiments, owing to the restrictive cues dictating the nature of the future events that participants needed to construct, the autobiographical memories and studied events would have been likely to impede the constructive process, thus making it more difficult to think of events that were new and different during EFT. Inhibition may have facilitated EFT by suppressing the accessibility of the memories and events.

Regardless of the exact underlying mechanism, the fact that thinking about the future causes forgetting holds important theoretical implications for understanding the way in which autobiographical memory is updated over time. Autobiographical memory serves a number of important functions, and its workings are guided by an individual's goals, personal knowledge, and experience (e.g., Conway, 2005; Conway & Pleydell-Pearce, 2000). When we engage in EFT, we do so in a way that suits certain goals, which may have the consequence of rendering information that is inconsistent with such goals less accessible in the future than it would have been otherwise. In this way, past events and information inconsistent with our current goals may over time become

increasingly difficult to remember. As such, EFT may play a very general role in cognition, acting not only to provide visions of the future, but updating the extent to which information from our past remains accessible.

Several aspects of the current results deserve further discussion. First, despite observing substantial effects of forgetting, we did not observe any evidence that the phenomenological qualities of the memories being forgotten differed as a function of thinking about the future. That is, although imagining potential future events decreased access to related memories, once the memories became recallable again, the way in which they were reconstructed was similar to the way in which memories not associated with EFT were reconstructed. This result is intriguing, but not without limitations. For example, we measured possible differences in the level of detail, the level of emotion, and the experiential perspective of how participants remembered the memories, but it remains possible that differences in other phenomenological aspects of the memories would have been observed had we attempted to measure them.

Another aspect of the current results that deserves discussion is the fact that forgetting did not differ as a function of valence. Specifically, negative memories were just as susceptible to forgetting as were positive memories, and thinking about the future caused forgetting regardless of whether participants thought of positive or negative future events, results that fit nicely with previous work demonstrating that negative and emotional materials are just as susceptible to forgetting as are positive and nonemotional materials (Barnier, Hung, & Conway, 2004; Kuhbandner, Bäuml, & Stiedl, 2009; Wessel & Hauer, 2006; but for further discussion, see Storm et al., 2015). We hesitate to make too much of this finding, however, as it is a null result observed in a single experiment. If replicated, however, it would suggest that most memories, if not all memories, are susceptible to forgetting as a function of EFT.

Finally, it is important to note that our results differ so strikingly from those of Storm and Jobe (2012). Whereas Storm and Jobe failed to find evidence of EFT causing forgetting in each of three experiments, we found evidence of EFT causing forgetting in each of four experiments. Although we strove to employ materials and procedures that matched those of Storm and Jobe, it remains possible that some methodological difference (or combination of differences) prevented Storm and Jobe from observing forgetting.[1] Although speculative, one possibility is that compared to our participants, Storm and Jobe's participants were more likely to use the studied events as constructive elements during EFT. We were very careful in our instructions to ensure that participants attempted to imagine future events that were different from the events they had either studied or retrieved during the previous phase of the experiment. Moreover, we had participants study/retrieve events for all context cues before moving on to a separate EFT phase, a methodological decision that may have made participants more likely to refrain from using the events as elements in the constructive process. In comparison, Storm and Jobe had participants engage in EFT for a given context cue immediately after studying events related to that context cue. Research in retrieval-induced forgetting has shown that when information is highly similar or useful in the retrieval or generation of something new, forgetting can be eliminated (e.g., Anderson, Green, & McCulloch, 2000; Chan, McDermott, & Roediger, 2006; Goodmon & Anderson, 2011; Storm & Patel, 2014), and similar dynamics may have occurred in the study by Storm and Jobe. Indeed, there are likely to be many instances in which past events and autobiographical memories are helpful in the constructive process of EFT, and presumably, in such instances, those events and memories should not be susceptible to forgetting.

REFERENCES

Addis, D. R., Pan, L., Vu, M. A., Laiser, N., & Schacter, D. L. (2009). Constructive episodic simulation of the future and the past: Distinct subsystems of a core brain network mediate imagining and remembering. *Neuropsychologia, 47,* 2222–2238.

Addis, D. R., & Schacter, D. L. (2008). Constructive episodic simulation: Temporal distance and detail of past and future events modulate hippocampal engagement. *Hippocampus, 18,* 227–237.

Addis, D. R., Wong, A. T., & Schacter, D. L. (2007). Remembering the past and imagining the future: Common and distinct neural substrates during event construction and elaboration. *Neuropsychologia, 45,* 1363–1377.

Anderson, M. C. (2003). Rethinking interference theory: Executive control and the mechanisms of forgetting. *Journal of Memory and Language, 49,* 415–445.

Anderson, M. C., Bjork, R. A., & Bjork, E. L. (1994). Remembering can cause forgetting: Retrieval dynamics in long-term memory. *Journal of Experimental Psychology: Learning, Memory, and Cognition, 20,* 1063–1087.

Anderson, M. C., Green, C., & Mculloch, K. C. (2000). Similarity and inhibition in long-term memory: Evidence for a two-factor theory. *Journal of Experimental Psychology: Learning, Memory, and Cognition, 26,* 1141–1159.

Barnier, A. J., Hung, L., & Conway, M. A. (2004). Retrieval-induced forgetting of emotional and unemotional autobiographical memories. *Cognition and Emotion, 18,* 457–477.

Bäuml, K-H. (2002). Semantic generation can cause episodic forgetting. *Psychological Science, 13,* 356–360.

Bjork, R. A. (1975). Retrieval as a memory modifier. In R. Solso (Ed.), *Information processing and cognition: The Loyola Symposium* (pp. 123–144). Hillsdale, NJ: Lawrence Erlbaum Associates.

Botzung, A., Denkova, E., & Manning, L. (2008). Experiencing past and future personal events:

---

[1]Methodological differences between Storm and Jobe (2012) and Ditta and Storm: total number of items studied per context cue (SJ: three events; DS: four events); study time per context cue (SJ: 30 seconds; DS: 40 seconds); total number of future thought practice trials (SJ: two; DS: none); filler task between context cues (SJ: an out-loud number reading task; DS: none); total number of future thoughts generated per context cue (SJ: four; DS: three); future thought labeling (SJ: no titles, multiple phenomenological ratings; DS: short title; no ratings); structure of procedure (SJ: study phase and EFT phase interleaved such that participants generated future thoughts for some contexts immediately after studying associated events, but not others; DS: study phase and EFT phase blocked such that participants studied events for all contexts before generating future thoughts associated with some of those contexts); participants (SJ: undergraduate students at the University of Illinois at Chicago; DS: undergraduate students at the University of California, Santa Cruz)

Functional neuroimaging evidence on the neural bases of mental time travel. *Brain and Cognition, 66,* 202–212.

Chan, J. C. K., McDermott, K. B., & Roediger, H. L. III. (2006). Retrieval-induced facilitation: Initially nontested material can benefit from prior testing of related material. *Journal of Experimental Psychology: General, 135,* 553–571.

Conway, M. A. (2005). Memory and the self. *Journal of Memory and Language, 53,* 594–628.

Conway, M. A., & Pleydell-Pearce, C. W. (2000). The construction of autobiographical memories in the self-memory system. *Psychological Review, 107,* 261–288.

Goodmon, L. B., & Anderson, M. C. (2011). Semantic integration as a boundary condition on inhibitory processes in episodic retrieval. *Journal of Experimental Psychology: Learning, Memory, and Cognition, 37,* 416–436.

Hach, S., Tippett, L. J., & Addis, D. R. (2014). Neural changes associated with the generation of specific past and future events in depression. *Neuropsychologia, 65,* 41–55.

Healey, M. K., Campbell, K. L., Hasher, L., & Ossher, L. (2010). Direct evidence for the role of inhibition in resolving interference in memory. *Psychological Science, 21,* 1464–1470.

Jonker, T. R., Seli, P., & MacLeod, C. M. (2013). Putting retrieval-induced forgetting in context: An inhibition-free, context-based account. *Psychological Review, 120,* 852–872.

Kuhbandner, C., Bäuml, K-H., & Stiedl, F. C. (2009). Retrieval-induced forgetting of negative stimuli: The role of emotional intensity. *Cognition and Emotion, 23,* 817–830.

Murayama, K., Miyatsu, T., Buchli, D., & Storm, B. C. (2014). Forgetting as a consequence of retrieval: A meta-analytic review of retrieval-induced forgetting. *Psychological Bulletin, 140,* 1383–1409.

Raaijmakers, J. G., & Jakab, E. (2013). Rethinking inhibition theory: On the problematic status of the inhibition theory for forgetting. *Journal of Memory and Language, 68,* 98–122.

Schacter, D. L., & Addis, D. R. (2007). The cognitive neuroscience of constructive memory: Remembering the past and imagining the future. *Philosophical Transactions of the Royal Society B: Biological Sciences, 362,* 773–786.

Schacter, D. L., Addis, D. R., & Buckner, R. L. (2007). Remembering the past to imagine the future: The prospective brain. *Nature Reviews Neuroscience, 8,* 657–661.

Storm, B. C., & Angello, G. (2010). Overcoming fixation creative problem solving and retrieval-induced forgetting. *Psychological Science, 21,* 1263–1265.

Storm, B. C., Angello, G., Buchli, D. R., Koppel, R. H., Little, J. L., & Nestojko, J. F. (2015). A review of retrieval-induced forgetting in the contexts of learning, eyewitness memory, social cognition, autobiographical memory, and creative cognition. In B. Ross (Ed.), *The Psychology of Learning and Motivation* (pp. 141–194). Academic Press: Elsevier.

Storm, B. C., Bjork, E. L., Bjork, R. A., & Nestojko, J. F. (2006). Is retrieval success a necessary condition for retrieval-induced forgetting? *Psychonomic Bulletin & Review, 13,* 1023–1027.

Storm, B. C., & Jobe, T. A. (2012). Remembering the past and imagining the future: Examining the consequences of mental time travel on memory. *Memory, 20,* 224–235.

Storm, B. C., & Levy, B. J. (2012). A progress report on the inhibitory account of retrieval-induced forgetting. *Memory & Cognition, 40,* 827–843.

Storm, B. C., & Patel, T. N. (2014). Forgetting as a consequence and enabler of creative thinking. *Journal of Experimental Psychology: Learning, Memory, and Cognition, 40,* 1594–1609.

Szpunar, K. K. (2010). Episodic future thought: An emerging concept. *Perspectives on Psychological Science, 5,* 142–162.

Szpunar, K. K., Watson, J. M., & McDermott, K. B. (2007). Neural substrates of envisioning the future. *Proceedings of the National Academy of Sciences, 104,* 642–647.

Verde, M. F. (2012). 2 Retrieval-induced forgetting and inhibition: A critical review. *Psychology of Learning and Motivation-Advances in Research and Theory, 56,* 47–80.

Walker, W. R., Skowronski, J. J., & Thompson, C. P. (2003). Life is pleasant—And memory helps to keep it that way! *Review of General Psychology, 7,* 203–210.

Wessel, I., & Hauer, B. J. A. (2006). Retrieval-induced forgetting of autobiographical memory details. *Cognition and Emotion, 3,* 430–447.

Retrieval-induced forgetting is associated with increased positivity when imagining the future

Saskia Giebl[1], Benjamin C. Storm[2], Dorothy R. Buchli[3], Elizabeth Ligon Bjork[1], and Robert A. Bjork[1]

[1]Department of Psychology, University of California, Los Angeles, CA, USA
[2]Department of Psychology, University of California, Santa Cruz, CA, USA
[3]Department of Psychology, Mercer University, Macon, GA, USA

People often think of themselves and their experiences in a more positive light than is objectively justified. Inhibitory control processes may promote this positivity bias by modulating the accessibility of negative thoughts and episodes from the past, which then limits their influence in the construction of imagined future events. We tested this hypothesis by investigating the correlation between retrieval-induced forgetting and the extent to which individuals imagine positive and negative episodic future events. First, we measured performance on a task requiring participants to imagine personal episodic events (either positive or negative), and then we correlated that measure with retrieval-induced forgetting. As predicted, individuals who exhibited higher levels of retrieval-induced forgetting imagined fewer negative episodic future events than did individuals who exhibited lower levels of retrieval-induced forgetting. This finding provides new insight into the possible role of retrieval-induced forgetting in autobiographical memory.

... the man who remembers or hopes must always be haunted by a certain image of that which he remembers or hopes ... so it follows that all pleasures consist either in perceiving things present, or in remembering things past, or in hoping things future. Now remembered things are pleasant, not only in those cases in which they were pleasant at the time, but sometimes, though they were unpleasant; provided that their sequel be noble and good. (Aristotle, trans. Jebb, 1909)

Of the many cognitive skills that humans possess, one of the most intriguing is the ability to engage in mental time travel: delving into the past, through the present, and on to the future. We can consciously reexperience, through *autonoetic* awareness and subjective time (see Tulving, 1985, 2001, 2002), past happenings and engage in episodic future thinking (Atance & O'Neill, 2001, 2005; Buckner & Carroll, 2007; Schacter & Addis, 2007a; Suddendorf & Corballis, 2007; Wheeler, Stuss, & Tulving, 1997). This ability to direct our

We thank Narcis Marshall, an Honors student at University of California, Los Angeles (UCLA), for helping with data collection and providing helpful insights. We also thank Toshiya Miyatsu, Carole Yue, Michael Cohen, and the members of the Bjork Lab for contributions made to the preparation of the manuscript.

This research was supported, in part, by the James S. McDonnell Foundation.

attention toward a specific event in the past and to construct a hypothetical episode in the future, using the mechanisms and resources of episodic and semantic memory, allows us to regulate our future behaviour in ways that would otherwise be impossible (e.g., Davies & Stone, 1995; Kahneman & Miller, 1986; Pham & Taylor, 1999; Taylor, Pham, Rivkin, & Armor, 1998; Taylor & Schneider, 1989). The ability to conceptualize alternative past and future episodes, for instance, may be used to achieve what Nietzsche (see Ramadanovic, 2001) called the greatest happiness of humans: the ability to forget and discern what was advantageous in the past and what is disadvantageous for the present and future. That is, if the information retrieved is important or the events and experiences recalled are positive and self-affirming, then it is advantageous to keep such knowledge accessible.

Several researchers have suggested that memory has evolved to retain information when it is likely to be useful or important in the future (J. R. Anderson & Schooler, 1991; Bjork & Bjork, 1988; Schacter, 2001). Indeed, a growing literature has built on this argument by highlighting the adaptive nature of a constructive memory system that allows for flexible extraction, recombination, and reassembly of past and present elements to aid our preparation for the future (Schacter & Addis, 2007b; Suddendorf & Corballis, 2007). This process, sometimes referred to as prospection, or mental time travel, is believed to rely on much of the same cognitive and neural resources whether we are attempting to remember the past or imagine the future (e.g., Addis, Wong, & Schacter, 2007, 2008; D'Argembeau & Van der Linden, 2004, 2006; Schacter, Addis, & Buckner, 2007, 2008; Szpunar, 2010; Szpunar, Watson, & McDermott, 2007).

The flexible nature of such a constructive memory system, however, also provides a means —through incomplete data gathering, imagination-induced memory distortions, biases, selective forgetting, and so forth—by which to remember the past (and imagine the future) in ways that better suit our personal or subjective needs even if it is at the expense of objective accuracy (Conway,

2005; Conway & Pleydell-Pearce, 2000; Fiske & Taylor, 1984; Schacter, 2012; Schacter, Guerin, & St. Jacques, 2011; Taylor, 1989; Taylor & Brown, 1988). Individuals, for example, often seek out positive experiences and avoid negative ones, leading them to view past (and future) life events in an overly positive light (Sharot, 2011; Sharot, Riccardi, Raio, & Phelps, 2007; Walker, Skowronski, & Thompson, 2003; Weinstein, 1980). Indeed, information or feelings associated with negative experiences tend to fade more quickly over time than information or feelings associated with positive experiences (Holmes, 1970; Szpunar, Addis, & Schacter, 2012; Taylor, 1991; Thompson, 1930; Walker & Skowronski, 2009).

Recent research has suggested that this positivity bias could arise, in part, through the inhibitory processes that underlie retrieval-induced forgetting (Storm & Jobe, 2012). Retrieval-induced forgetting is observed when the selective retrieval of some information causes the forgetting of other information (M. C. Anderson, Bjork, & Bjork, 1994; for a review of some of the many instantiations of retrieval-induced forgetting, see Storm et al., 2015). According to the inhibitory account of retrieval-induced forgetting, nontarget items can become activated in response to a retrieval cue, causing competition, and inhibition acts to resolve this competition by rendering the nontarget items less accessible—both in the moment and after a delay (M. C. Anderson, 2003; Murayama, Miyatsu, Buchli, & Storm, 2014; Storm & Levy, 2012).

In the study by Storm and Jobe (2012), a common version of the retrieval-practice paradigm was administered to measure individual differences in retrieval-induced forgetting. Then, to examine differences in autobiographical memory, participants were presented with 20 neutral keywords (e.g., "pool", "medicine") and were asked to generate either positive or negative memories associated with those keywords. Participants who exhibited high levels of retrieval-induced forgetting recalled significantly fewer negative memories than did participants who exhibited low levels of retrieval-induced forgetting, whereas a nonsignificant correlation was observed in the opposite direction for positive memories. In fact, participants who exhibited high levels of retrieval-induced forgetting were found to show a strong positivity bias, whereas participants who exhibited low levels of retrieval-induced forgetting were found to show a nonsignificant negativity bias.

Storm and Jobe (2012) speculated that the inhibitory process underlying retrieval-induced forgetting may play a direct role in facilitating a positivity bias in autobiographical memory. More specifically, the argument is that when people encounter a given retrieval cue, inhibitory processes may prevent negative or otherwise undesirable memories from coming to mind—in much the same way that inhibition is presumed to prevent nontarget items from coming to mind in the retrieval-practice paradigm. A cue such as *birthday party*, for example, is likely to be associated to an array of memories (some of which are positive and others that are negative)—and to the extent that autobiographical memory is biased toward remembering positive events over negative events, inhibition may act to reduce the accessibility of the negative memories in order to better facilitate access to the positive memories (see also Bjork, Bjork, & Anderson, 1998, for similar speculations regarding possible inhibitory mechanisms underlying varieties of goal-directed forgetting). Of course, it is also possible that some other factor—such as executive control or working memory capacity—is related to both retrieval-induced forgetting and positivity in autobiographical memory, and it may be this third factor that is responsible for the observed correlation. Regardless of the exact underlying mechanism, the results by Storm and Jobe suggest that individuals who are more susceptible to retrieval-induced forgetting remember the past more positively than do individuals who are less susceptible to retrieval-induced forgetting.

Our goal was to extend the work of Storm and Jobe (2012) by investigating whether individual differences in retrieval-induced forgetting would also predict differences in how people imagine the future. If similar processes underlie remembering and episodic future thinking, and if retrieval-induced forgetting (or some related factor) facilitates the remembering of more positive than negative past experiences, then it stands to reason that retrieval-induced forgetting (or some related factor) may also facilitate the construction of more positive than negative future experiences. Indeed, to the extent that memories of the past serve as building blocks in the construction of future thoughts, any mechanism that influences the type of memories that can come to mind should, in so doing, influence the type of episodic simulations that can be constructed. To test this hypothesis, we measured each participant's ability to imagine either positive or negative events taking place in the future and examined the correlation between that measure and retrieval-induced forgetting. We expected our results to mirror those reported by Storm and Jobe (2012). Specifically, we expected participants exhibiting high levels of retrieval-induced forgetting to construct fewer negative simulations (and more positive simulations) of the future than participants exhibiting low levels of retrieval-induced forgetting.

EXPERIMENTAL STUDY

Method

Participants and design

A total of 132 individuals were recruited to participate in the experiment via the internet using Amazon's Mechanical Turk (Mturk), a website that allows a diverse population to sign up to complete small tasks for payment (for the validity of this methodology, see Buhrmester, Kwang, & Gosling, 2011; Gosling, Vazire, Srivastava, & John, 2004). The sample consisted of 86 women and 46 men (M years of age $= 35$, $SD = 10$). Only English-speaking individuals in the United States were allowed to participate. The valence (positive vs. negative) of the future events to be imagined was manipulated between subjects such that a randomly assigned half of the participants attempted to generate positive events, and the other randomly assigned half attempted to generate negative events. All participants completed the future thinking portion of the study first, followed by the retrieval-practice task to measure retrieval-induced forgetting. The entire study was completed via Mturk.

Measuring future event construction

All participants were presented with the same set of 20 neutral keywords (i.e., ball, bathroom, bite, blanket, book, bridge, candle, car, clock, envelope, ice, knife, leg, medicine, money, pool, ring, scarf, television, tree) to use as the basis for their generation of future events. Participants assigned to the positive condition were instructed to imagine specific personal future events that might make them feel "happy, proud, pleased, or gratified". In contrast, participants assigned to the negative condition were instructed to imagine specific personal future events that might make them feel "sad, embarrassed, or disappointed". Each keyword was presented and remained on the computer screen for 25 s while participants imagined and typed in a description of a corresponding future event. They were asked to imagine and write down only future events or scenarios that were novel, yet plausible, and not to write anything if nothing came to mind. Participants were also asked to use as many words as possible to describe each future event. The proportion of keywords (out of the set of 20) that elicited a future episodic event was calculated for each participant. Only events that were deemed by the rater, who was blind to each individual's retrieval-induced forgetting score, to be both episodic and appropriate to the target valence were coded as a successful episodic construction. One-word responses or instances in which the event was clearly nonepisodic (e.g., "I like cheese") or of the inappropriate valence ("I fell down and got bit by a dog" when asked to think of a positive future event) were not counted as successful constructions.

Measuring retrieval-induced forgetting

A measure of retrieval-induced forgetting was obtained for each participant using the same paradigm as that employed by Storm and Jobe (2012). Participants first studied a list of 48 category–exemplar pairs (e.g., *fruit:banana; metal:silver*), composed of six exemplars from each of eight categories, with pairs presented one at a time and

in a different random order for each subject. Then, during retrieval practice, participants attempted to retrieve new exemplars associated with half of the studied categories (semantic generation). The cues employed during retrieval practice consisted of a category name plus a two-letter-stem cue of an associated exemplar of relatively low taxonomic frequency (e.g., *fruit:gu___*, for *guava*). Three rounds of retrieval practice were conducted, with participants generating an appropriate response on 53% ($SD = 15\%$) of the trials. Then, after a 5-min delay (filled with playing Tetris), participants were tested on all 48 originally studied (but not practised) exemplars. In this final test, the cues employed were the category name plus the first letter of the exemplar (e.g., *fruit:b___*). Retrieval-induced forgetting was calculated by subtracting final recall performance for exemplars from practised categories from that for nonpractised categories. Thus, positive values indicate greater amounts of retrieval-induced forgetting, whereas negative values indicate lower amounts of retrieval-induced forgetting.

Results

Overall, a significant effect of retrieval-induced forgetting was observed: Unpractised exemplars from practised categories ($M = .62$, $SE = .02$) were recalled less well than unpractised exemplars from unpractised categories ($M = .69$, $SE = .01$), $t(131) = 7.32$, $p < .001$, $d = 0.64$. Additionally, a significant effect of valence was observed such that more future events ($M = .85$, $SE = .02$) were elicited in the negative condition than in the positive condition ($M = .79$, $SE = .02$), $t(130) = 2.14$, $p = .03$, $d = 0.37$.

A regression analysis examined the proportion of variance in future episodic thinking explained by valence, retrieval-induced forgetting, and the Valence × Retrieval-Induced Forgetting interaction. Valence was entered as a dummy variable indicating condition (positive vs. negative). The complete model was significant, $F(3, 128) = 4.33$, $p = .01$, $R^2 = .09$. More importantly, the interaction term explained significant variance above and beyond that explained by valence and

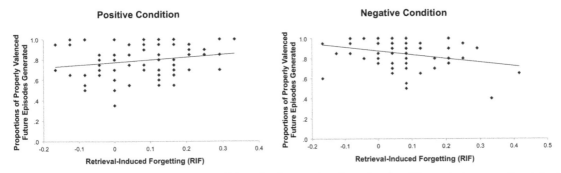

Figure 1. *Scatter plots (with best fitting regression lines) illustrating the proportions of properly valenced future episodes generated in relation to retrieval-induced forgetting. Generated future episodes for each valence were operationally defined as the proportion of keywords resulting in the generation of an episodic future event of that valence; values on the x-axis are raw retrieval-induced forgetting scores. The relation between retrieval-induced forgetting and the generation of negative future episodes is shown in the left panel, and the relation between retrieval-induced forgetting and the generation of positive future episodes is shown in the right panel.*

retrieval-induced forgetting alone. Specifically, the Valence × Retrieval-Induced Forgetting term accounted for significant additional variance, $F(1, 128) = 8.15$, $p = .005$, $\Delta R^2 = .06$, suggesting that the correlation between retrieval-induced forgetting and future episodic thinking was significantly different in the positive and negative valence conditions. As shown in Figure 1, a significant negative correlation was observed between retrieval-induced forgetting and episodic future thinking in the negative condition, $(r = -.31$, $p = .01$; Spearman's rho $= -.30$, $p = .02)$, whereas a nonsignificant positive correlation was observed between retrieval-induced forgetting and episodic future thinking in the positive condition $(r = .19$, $p = .11$; Spearman's rho $= .19$, $p = .13)$.[1] These results almost perfectly match the results reported by Storm and Jobe (2012; Experiment 1). Specifically, they observed Pearson's correlations of $-.31$ and $.17$ in the negative and positive autobiographical remembering conditions, respectively.

To explore the data further, we examined the number of words that participants used while reporting their episodic constructions. On average, participants in the negative condition used more words to describe each episodic future event $(M = 15.3$, $SE = 0.8)$ than did participants in the positive condition $(M = 13.5$, $SE = 0.5)$, $t(130) = 2.03$, $p < .05$, $d = 0.35$. Moreover, a significant negative correlation was observed in the negative condition such that participants exhibiting greater levels of retrieval-induced forgetting used fewer words than did participants exhibiting lower levels of retrieval-induced forgetting $(r = -.26$, $p = .04)$. No such correlation was observed in the positive condition $(r = -.02$, $p = .86)$.

GENERAL DISCUSSION

Recollection of the past and expectations of the future are often positively biased (Brown, MacLeod, Tata, & Goddard, 2002; Sharot et al., 2007; Walker et al., 2003; Weinstein, 1980), but not all individuals succeed in promoting and

[1]Spearman's rho is less susceptible to the influence of outliers than Pearson's r because it is calculated using rank order instead of actual scores. To further control for the influence of outliers we re-ran the regression analysis after removing subjects who exhibited retrieval-induced forgetting scores or episodic future thinking scores that deviated from the mean in a given condition by more than three standard deviations (only one subject in the negative condition needed to be removed). The results of this analysis mirrored those of the full analysis. Specifically, the complete model was significant, $F(3, 127) = 3.87$, $p = .01$, $R^2 = .08$, and the interaction term explained significant variance above and beyond that explained by valence and retrieval-induced forgetting alone, $F(1, 127) = 5.10$, $p = .03$, $\Delta R^2 = .04$.

maintaining this bias (e.g., Storm & Jobe, 2012; Taylor & Brown, 1988). The present study examined the extent to which individual differences in retrieval-induced forgetting are correlated with the capacity to imagine positive or negative episodic events taking place in the future. Specifically, participants were given neutral keywords and were asked to imagine either positive or negative future events associated with those keywords. Participants who exhibited greater levels of retrieval-induced forgetting were less prone to imagining negative events than were participants who exhibited reduced levels of retrieval-induced forgetting. This finding is consistent with earlier work by Storm and Jobe (2012), who found that individuals exhibiting greater levels of retrieval-induced forgetting were also less prone to remembering negative memories from the past than were individuals who exhibit reduced levels of retrieval-induced forgetting.

The present findings, along with those of Storm and Jobe (2012), are consistent with the idea that inhibition, or some other factor related to retrieval-induced forgetting, acts to prevent negative autobiographical experiences from coming to mind, with such an effect occurring regardless of whether people attempt to mentally time travel into the past or into the future. The fact that retrieval-induced forgetting correlated with episodic future thinking in the same way that it correlated with episodic remembering is also consistent with work showing that remembering the past and constructing the future rely on similar cognitive and neural processes (Schacter et al., 2007, 2008). By determining which memories of the past come to mind, the processes underlying or related to retrieval-induced forgetting may in turn determine the type of episodic simulations that can be constructed. According to the episodic simulation hypothesis, elements of past memories are flexibly extracted and recombined to construct simulations of future events. Thus, any mechanism that influences the accessibility of memories from the past should also influence the way in which one thinks about the future. On the other hand, it is also possible that the construction of negative episodic future events is limited in a more direct way that

is independent from what is, or is not, accessible in memory. There may be an implicit bias in the constructive process to avoid negative or self-threatening thoughts, for example, and inhibition or executive control may help to enforce this bias. A third possibility is that participants simply recasted past events as events occurring in the future, in which case it would not be surprising that we observed the same correlation as that in Storm and Jobe (2012). Unfortunately, because we did not collect a measure of novelty or of how distinctive a future event was from experienced past events, it is impossible to rule out this possibility.

Up until this point, our discussion has largely assumed that inhibition underlies retrieval-induced forgetting and that individual differences in retrieval-induced forgetting reflect individual differences in the capacity to inhibit nontarget items in memory. Although there is good evidence to support this assumption (e.g., M. C. Anderson, 2003; Murayama et al., 2014; Storm & Levy, 2012), some researchers believe that retrieval-induced forgetting can be sufficiently explained by noninhibitory mechanisms, such as associative interference or inappropriate contextual cueing (see, e.g., Jonker, Seli, & MacLeod, 2013; Raaijmakers & Jakab, 2013; Verde, 2012). According to most noninhibitory accounts, retrieval-induced forgetting occurs because retrieval practice strengthens a subset of items, and it is this strengthening that blocks the accessibility of other, nonstrengthened items at test. Interpreting the present results in this context, it is possible that individuals who exhibit greater levels of retrieval-induced forgetting recall and construct fewer negative episodic events than positive episodic events because such events are particularly prone to noninhibitory sources of forgetting. For example, although it is not immediately clear why this would be the case, negative events could be more susceptible to blocking than their positive counterparts.

Individual differences in retrieval-induced forgetting aside, it is somewhat surprising that participants imagined negative future events at a higher rate than they imagined positive future events—a finding that appears to be inconsistent with earlier work showing positivity biases in autobiographical memory and future thinking. Although we do not have a clear explanation for this finding, we can entertain several possibilities. First, studies have shown that negative information can be remembered more easily and with a greater sense of vividness and detail than positive information (e.g., Kensinger, 2009; Mickley & Kensinger, 2008), and our particular instructions may have emphasized the need to generate the type of vivid details during episodic simulation that are more easily associated to negative than to positive events. Moreover, limiting the time that participants had to respond on each trial might have favored the negative condition, as imagining a positive future event may have required placing an experience in a more elaborate context than imagining a negative future event. Finally, it may be important that our sample was recruited via MTurk. Not only might individuals who use MTurk be different from individuals who typically participate in laboratory studies, but the anonymity afforded by MTurk may provide participants greater freedom in the way in which they respond to autobiographical memory tests.

Concluding comment

It is important to emphasize that the correlation between retrieval-induced forgetting and the ability to generate episodic future events may be driven by any number of different variables, such as those related to attentional control or working memory capacity (e.g., Aslan & Bäuml, 2011; Schilling, Storm, & Anderson, 2014) or to psychiatric disorders, such as depression or anxiety (e.g., MacLeod & Cropley, 1995; MacLeod, Tata, Kentish, Carroll, & Hunter, 1997; Raune, MacLeod, & Holmes, 2005). Moreover, although our results suggest that retrieval-induced forgetting may play a role in promoting a positivity bias in autobiographical memory, because they are correlational, they do not on their own provide evidence for a causal relationship. Future research may seek to examine how manipulating a person's capacity to inhibit nontarget items (and thus retrieval-induced forgetting) influences the ways in which the past is

remembered, and the future is constructed (e.g., Sharot et al., 2012). Such necessary cautions notwithstanding, however, the present findings mesh nicely with theories that emphasize the importance of adaptive inhibitory processes that promote psychological well-being (Taylor, 1991), and the observed link between inhibitory control and imagining future events holds the promise of helping us to understand the mechanisms underlying imagining and planning for the future.

REFERENCES

Addis, D. R., Wong, A. T., & Schacter, D. L. (2007). Remembering the past and imagining the future: Common and distinct neural substrates during event construction and elaboration. *Neuropsychologia, 45,* 1363–1377. doi:10.1016/j.neuropsychologia.2006. 10.016

Addis, D. R., Wong, A. T., & Schacter, D. L. (2008). Age-related changes in the episodic simulation of future events. *Psychological Science, 19,* 33–41. doi:10.1111/j.1467-9280.2008.02043.x

Anderson, J. R., & Schooler, L. J. (1991). Reflections of the environment in memory. *Psychological Science, 2,* 396–408. doi:10.1111/j.1467-9280.1991.tb00174.x

Anderson, M. C. (2003). Rethinking interference theory: Executive control and the mechanisms of forgetting. *Journal of Memory and Language, 49,* 415–445. doi:10.1016/j.jml.2003.08.006

Anderson, M. C., Bjork, R. A., & Bjork, E. L. (1994). Remembering can cause forgetting: Retrieval dynamics in long-term memory. *Journal of Experimental Psychology: Learning, Memory, and Cognition, 20,* 1063–1087. http://dx.doi.org/10. 1037/0278-7393.20.5.1063

Aslan, A., & Bäuml, K.-H. T. (2011). Individual differences in working memory capacity predict retrieval-induced forgetting. *Journal of Experimental Psychology: Learning, Memory, and Cognition, 37,* 264–269.

Atance, C. M., & O'Neill, D. K. (2001). Episodic future thinking. Trends in Cognitive *Science, 5,* 533–539. doi:10.1016/S1364-6613(00)01804-0

Atance, C. M., & O'Neill, D. K. (2005). The emergence of episodic future thinking in humans. *Learning and Motivation, 36,* 126–144. doi:10.1016/j.lmot.2005. 02.003

Bjork, E. L., & Bjork, R. A. (1988). On the adaptive aspects of retrieval failure in autobiographical memory. In M. M. Gruneberg, P. E. Morris, & R. N. Sykes (Eds.), *Practical aspects of memory II* (pp. 283–288). London, UK: Wiley.

Bjork, E. L., Bjork, R. A., & Anderson, M. C. (1998). Varieties of goal-directed forgetting. In J. M. Golding & C. M. MacLeod (Eds.), *Intentional Forgetting: Interdisciplinary approaches* (pp. 103–137). Hillsdale, NJ: Lawrence Erlbaum, Associates.

Brown, G. P., MacLeod, A. K., Tata, P., & Goddard, L. (2002). Worry and the simulation of future outcomes. *Anxiety, Stress, & Coping: An International Journal, 15,* 1–17. doi:10.1080/10615800290007254

Buckner, R. L., & Carroll, D. C. (2007). Self-projection and the brain. *Trends in Cognitive Sciences, 11,* 49–57.

Buhrmester, M., Kwang, T., & Gosling, S. D. (2011). Amazon's mechanical Turk: A new source of inexpensive, yet high-quality, data? *Perspectives on Psychological Science, 6,* 3–5. doi:10.1177/174569161 0393980

Conway, M. A. (2005). Memory and the self. *Journal of Memory and Language, 53,* 594–628. doi:10.1016/j.jml.2005.08.005

Conway, M. A., & Pleydell-Pearce, C. W. (2000). The construction of autobiographical memories in the self-memory system. *Psychological Review, 107,* 261–288. http://dx.doi.org/10.1037/0033-295X.107.2.261

D'Argembeau, A., & Van der Linden, M. (2004). Phenomenal characteristics associated with projecting oneself back into the past and forward into the future: influence of valence and temporal distance. *Consciousness and Cognition, 13,* 844–858. doi:10.1016/j.concog.2004.07.007

D'Argembeau, A., & Van der Linden, M. (2006). Individual differences in the phenomenology of mental time travel: The effect of vivid visual imagery and emotion regulation strategies. *Consciousness and Cognition, 15,* 342–350. doi:10.1016/j.concog.2005.09.001

Davies, M., & Stone, T. (1995). *Mental simulation: Evaluations and applications.* Oxford, England: Blackwell.

Fiske, S. T., & Taylor, S. E. (1984). *Social cognition.* Reading, MA: Adison-Wesley.

Gosling, S. D., Vazire, S., Srivastava, S., & John, O. P. (2004). Should we trust Web-based studies? A comparative analysis of six preconceptions about Internet questionnaires. *American Psychologist, 59,* 93–104. http://dx.doi.org/10.1037/0003-066X.59.2.93

Holmes, D. S. (1970). Differential change in affective intensity and the forgetting of unpleasant personal experiences. *Journal of Personality and Social Psychology, 15,* 234–239. http://dx.doi.org/10.1037/h0029394

Jebb, R. C. (1909). *The rhetoric of Aristotle: A translation.* Cambridge: University Press.

Jonker, T. R., Seli, P., & MacLeod, C. M. (2013). Putting retrieval-induced forgetting in context: An inhibition-free, context-based account. *Psychological Review, 120,* 852–872. doi:10.1037/a0034246

Kahneman, D., & Miller, D. T. (1986). Norm theory: Comparing reality to its alternatives. *Psychological Review, 93,* 136–153. http://dx.doi.org/10.1037/0033-295X.93.2.136

Kensinger, E. A. (2009). Remembering the details: Effects of emotion. *Emotion Review, 1,* 99–113. doi:10.1177/1754073908100432

MacLeod, A. K., & Cropley, M. L. (1995). Depressive future-thinking: The role of valence and specificity. *Cognitive Therapy and Research, 19,* 35–50. doi:10.1007/bf02229675

MacLeod, A. K., Tata, P., Kentish, J., Carroll, F., & Hunter, E. (1997). Anxiety, depression, and explanation-based pessimism for future positive and negative events. *Clinical Psychology & Psychotherapy, 4,* 15–24. doi:10.1002/(SICI)1099-0879(199703)4:1<15::AID-CPP112>3.0.CO;2-#

Mickley, K. R., & Kensinger, E. A. (2008). Emotional valence influences the neural correlates associated with remembering and knowing. *Cognitive, Affective, and Behavioral Neuroscience, 8,* 143–152. doi:10.1016/j.plrev.2010.01.006

Murayama, K., Miyatsu, T., Buchli, D., & Storm, B. C. (2014). Forgetting as a consequence of retrieval: A meta-analytic review of retrieval-induced forgetting. *Psychological Bulletin, 140,* 1383–1409. http://dx.doi.org/10.1037/a0037505

Pham, L. B., & Taylor, S. E. (1999). From thought to action: Effects of process- versus outcome-based mental simulations on performance. *Personality and Social Psychology Bulletin, 25,* 250–260. doi:10.1177/0146167299025002010

Raaijmakers, J. G. W., & Jakab, E. (2013). Is forgetting caused by inhibition? *Current Directions in Psychological Science, 22,* 205–209. doi:10.1177/0963721412473472 cdps.sagepub.com

Ramadanovic, P. (2001). From Haunting to Trauma: Nietzsche's Active Forgetting and Blanchot's writing of the disaster. *Postmodern Culture, 11,* 30. doi:10.1353/pmc.2001.0005

Raune, D., MacLeod, A. K., & Holmes, E. A. (2005). The simulation heuristic and visual imagery in pessimism for negative events in anxiety. *Clinical Psychology & Psychotherapy*, *12*, 313–325. doi:10.1002/cpp.455

Schacter, D. L. (2001). *The seven sins of memory: How the mind forgets and remembers.* Boston and New York: Houghton Mifflin.

Schacter, D. L. (2012). Adaptive constructive processes and the future of memory. *American Psychologist*, *67*, 603–613. doi:10.1098/rstb.2007.2087

Schacter, D. L., & Addis, D. R. (2007a). Constructive memory: The ghost of past and future. *Nature*, *445*, 27. doi:10.1038/445027a

Schacter, D. L., & Addis, D. R. (2007b). The cognitive neuroscience of constructive memory: remembering the past and imagining the future. *Philosophical Transactions of the Royal Society B: Biological Sciences*, *362*, 773–786. doi:10.1098/rstb.2007.2087

Schacter, D. L., Addis, D. R., & Buckner, R. L. (2007). Remembering the past to imagine the future: The prospective brain. *Nature Reviews Neuroscience*, *8*, 657–661. doi:10.1038/nrn2213

Schacter, D. L., Addis, D. R., & Buckner, R. L. (2008). Episodic simulation of future events: concepts, data, and applications. *Annals of the New York Academy of Sciences*, *1124*, 39–60. doi:10.1196/annals.1440.001

Schacter, D. L., Guerin, S. A., & St. Jacques, P. L. (2011). Memory distortion: An adaptive perspective. *Trends in Cognitive Sciences*, *15*, 467–474. doi:10.1016/j.tics.2011.08.004

Schilling, C. J., Storm, B. C., & Anderson, M. C. (2014). Examining the costs and benefits of inhibition in memory retrieval. *Cognition*, *133*, 358–370.

Sharot, T. (2011). *The optimism bias: A tour of the irrationally positive brain.* New York: Vintage Books.

Sharot, T., Kanai, R., Marston, D., Korn, C. W., Rees, G., & Dolan, R. J. (2012). Selectively altering belief formation in the human brain. *Proceeding of the National Academy of Sciences*, *109*, 17058–17062.

Sharot, T., Riccardi, A. M., Raio, C. M., & Phelps, E. A. (2007). Neural mechanisms mediating optimism bias. *Nature*, *450*, 102–105. doi:10.1038/nature06280

Storm, B. C., Angello, G., Buchli, D. R., Koppel, R. H., Little, J. L., & Nestojko, J. F. (2015). A review of retrieval-induced forgetting in the contexts of learning, eye-witness memory, social cognition, autobiographical memory, and creative cognition. *Psychology of Learning and Motivation*, *62*, 141–194. doi:10.1016/bs.plm.2014.09.005.

Storm, B. C., & Jobe, T. A. (2012). Retrieval-induced forgetting predicts failure to recall negative autobiographical memories. *Psychological Science*, *23*, 1356–1363. doi:10.1177/0956797612443837

Storm, B. C., & Levy, B. J. (2012). A progress report on the inhibitory account of retrieval-induced forgetting. *Memory & Cognition*, *40*, 827–843. doi:10.3758/s13421-012-0211-7

Suddendorf, T., & Corballis, M. C. (2007). The evolution of foresight: what is mental time travel and is it unique to humans? *Behavioral and Brain Sciences*, *30*, 299–313. http://dx.doi.org/10.1017/S0140525X07001975

Szpunar, K. K. (2010). Episodic future thought: An emerging concept. *Perspectives on Psychological Science*, *5*, 142–162. doi:10.1177/1745691610362350

Szpunar, K. K., Addis, D. R., & Schacter, D. L. (2012). Memory for emotional simulations: Remembering a rosy future. *Psychological Science*, *23*, 24–29. doi:10.1177/0956797611422237

Szpunar, K. K., Watson, J. M., & McDermott, K. B. (2007). Neural substrates of envisioning the future. *Proceedings of the National Academy of Sciences of the United States of America*, *104*, 642–647. doi:10.1073/pnas.0610082104

Taylor, S. E. (1989). *Positive illusions: Creative self-deception and the healthy mind.* New York: Basic Books.

Taylor, S. E. (1991). Asymmetrical effects of positive and negative events: The mobilization-minimization hypothesis. *Psychological Bulletin*, *110*, 67–85. http://dx.doi.org/10.1037/0033-2909.110.1.67

Taylor, S. E., & Brown, J. D. (1988). Illusion and well-being: A social psychological perspective on mental health. *Psychological Bulletin*, *103*, 193–210. http://dx.doi.org/10.1037/0033-2909.103.2.193

Taylor, S. E., Pham, L. B., Rivkin, I. D., & Armor, D. A. (1998). Harnessing the imagination: Mental simulation, self-regulation, and coping. *American Psychologist*, *53*, 429–439. http://dx.doi.org/10.1037/0003-066X.53.4.429

Taylor, S. E., & Schneider, S. K. (1989). Coping and the simulation of events. *Social Cognition*, *7*, 174–194. http://dx.doi.org/10.1521/soco.1989.7.2.174

Thompson, R. H. (1930). An experimental study of memory as influenced by feeling tone. *Journal of Experimental Psychology*, *13*, 462–468.

Tulving, E. (1985). How many memory systems are there? *The American Psychologist*, *40*, 385–398. http://dx.doi.org/10.1037/0003-066X.40.4.385

Tulving, E. (2001). The origin of autonoesis in episodic memory. In H. L. Roediger, J. S. Nairne, I. Neath, A. M. Suprenant (Eds.), *The nature of remembering: Essays in honor of Robert G. Crowder* (pp. 17–34). Washington, DC: Am. Psychol. Assoc.

Tulving, E. (2002). Episodic memory: From mind to brain. *Annual Review of Psychology, 53,* 1–25. doi:10.1146/annurev.psych.53.100901.135114

Verde, M. F. (2012). Retrieval-induced forgetting and inhibition: A critical review. In B. H. Ross (Ed.), *The psychology of learning and motivation* (Vol. 56, pp. 47–80). San Diego, CA: Elsevier Academic.

Walker, W. R., & Skowronski, J. J. (2009). The fading affect bias: But what the hell is it for? *Applied Cognitive Psychology, 23,* 1122–1136. doi:10.1002/acp.1614

Walker, W. R., Skowronski, J. J., & Thompson, C. P. (2003). Life is pleasant and memory helps to keep it that way. *Review of General Psychology, 7,* 203–210. http://dx.doi.org/10.1037/1089-2680.7.2.203

Weinstein, N. D. (1980). Unrealistic optimism about future life events. *Journal of Personality and Social Psychology, 39,* 806–820. http://dx.doi.org/10.1037/0022-3514.39.5.806

Wheeler, M. A., Stuss, D. T., & Tulving, E. (1997). Toward a theory of episodic memory: The frontal lobes and autonoetic consciousness. *Psychological Bulletin, 121,* 331–354. doi:10.1037/0033-2909.121.3.331

Understanding deliberate practice in preschool-aged children

Jac T. M. Davis[1,2], Elizabeth Cullen[1], and Thomas Suddendorf[1]

[1]School of Psychology, University of Queensland, Brisbane, QLD, Australia
[2]Department of Psychology, University of Cambridge, Cambridge, UK

Deliberate practice is essential for skill acquisition and expertise and may be a direct consequence of episodic foresight. However, little is known about how deliberate practice develops in children. We present two experiments testing children's ability to selectively practise a behaviour that was going to be useful in future and to reason about the role of practice in skill formation. Five-year-olds demonstrated an explicit understanding of deliberate practice both in selectively choosing to practise a future-relevant skill and in predicting skill change in others based on their practice. Four-year-olds showed some capacities, but failed to demonstrate consistent understanding of the relationship between practice and skill improvement. Children's understanding of this relationship was significantly related to their understanding of how information leads to knowledge, suggesting that both may draw on similar cognitive developmental changes.

The development of episodic foresight—the ability to imagine future situations and to organize current action in light of anticipated events (Suddendorf & Moore, 2011)—gives children the chance to prepare for future opportunities and threats and, further, to change their future selves in deliberate ways. Humans can plan to acquire knowledge and skills that alter their future capacities. Indeed, the extraordinary diversity of human expertise is to a large extent due to the fact that people can decide what they want to get better at and practise accordingly (Suddendorf, 2013). The reliable acquisition of diverse skills is fundamental to humans' survival and success as a species, and understanding the mechanisms of learning and skill development has become increasingly important to theories of human evolution (Sterelny, 2012). Yet, very little is known about how children come to understand deliberate practice as a means of improving selected skills. Here we describe a first attempt to explore deliberate practice in preschoolers.

In a broad sense, deliberate practice may be defined as repeated actions that are conducted with the explicit goal to improve one's future capacities (e. g., Suddendorf, Brinums, & Imuta, in press). Though adults such as parents and teachers may encourage it, deliberate practice is a voluntary process that can be self-initiated (Donald, 1999).

We thank the parents and children who participated in this research, and members of the Early Cognitive Development Centre at the University of Queensland, in particular lab manager Sally Clark.

Thomas Suddendorf was supported by the Australian Research Council [grant number DP0770113].

Furthermore, engaging in deliberate practice is typically effortful and involves monitoring of progress (e.g., Ericsson, 2008). Through deliberate practice we can improve relatively simple motor skills, such as aimed throwing, and with much dedication, we can use it to master very complex skills such as the accomplished playing of a musical instrument.

Deliberate practice has been studied primarily in the context of the creation of elite adult experts (e.g., see Côté, Baker, & Abernethy, 2003; Ericsson, Charness, Feltovich, & Hoffman, 2006) and the conditions under which it is most effective (e.g., Dempster, 1996; Ericsson, Krampe, & Tesch-Römer, 1993). Deliberate practice in this field is often more narrowly defined as highly structured activities that were designed for the express purpose of improving performance through feedback. Ericsson et al. (1993), for instance, argue that practice must require effort, generate no immediate rewards, and be motivated by a future-oriented goal of improved performance independent of inherent enjoyment. This kind of practice is arguably more important to expertise and elite mastery of skills than any other factor, including innate talent (Ericsson et al., 1993; Helsen, Hodges, Van Winckel, & Starkes, 2000; Slobodoa, Davidson, Howe, & Moore, 1996; Ward, Hodges, Williams, & Starkes, 2004; but see Baker, Côté, & Deakin, 2005; Campitelli & Gobet, 2011; McPherson, 2005). In the current investigation, we focus not on the development of extraordinary skill, but rather on humans' more basic general capacity for voluntarily acquiring common skills.

Deliberate practice, in this broad sense of intentional repetition of a behaviour in the anticipation of improved future performance capacity, is required for everyday competence in skills such as riding a bicycle or learning to write. So conceived it is a much more fundamental construct than that conceptualized in the literature on expertise. It may not be particularly effortful and could well be immediately rewarding. Although more ordinary than the process of skill acquisition studied in elite experts, this basic capacity may have been essential to human evolutionary success (e.g.,

Donald, 1999; Suddendorf, 2013). Practice allows humans to rapidly adapt by acquiring skills that one foresees as useful at a future point in time. This flexible specialization may be a key component of the general human capacity for complex, goal-directed action, such as that implicated in Early Stone Age tool making (Stout, Toth, Schick, & Chaminade, 2008). Practice broadly defined thus deserves to be studied not only as a tool for elite skill formation, but as a fundamental human capacity.

Having defined deliberate practice, we may consider its developmental trajectory. To do so, we may look to research on children's development of similar cognitive functions. Understanding how skills can be changed may present children with an analogous cognitive problem to understanding how knowledge may be altered. Research on theory of mind development has shown that young children often struggle in explicitly understanding how knowledge changes as a result of experience. For instance, once they have knowledge of a fact, very young children tend to insist that they have always known what in fact they have just learned (Naito, 2003; Taylor, Esbensen, & Bennett, 1994). In a study by Gopnik and Graf (1988) children discovered objects hidden in a set of drawers by seeing the objects, being told what they were, or figuring out what they were from a clue. Children were then asked to identify both the object in the drawer and how they knew what was in there. While 3-year-old children responded randomly, older children (5 years) could reliably report how they acquired the new knowledge, indicating that they could separate their present state from a past one, and pinpoint the agent of change.

Given that thought about future skill improvement is essential to our definition of deliberate practice, the development of a capacity to travel mentally into the future (episodic foresight) is a critical prerequisite. Children learn to foresee different future scenarios and to choose to pursue one path, even if it is not immediately rewarding, over others that are. Verbal answers to questions about hypothetical future events and plans suggest that children's ability to imagine themselves in a future scenario emerges over the preschool years

(Busby Grant & Suddendorf, 2005; Hayne, Gross, McNamee, Fitzgibbon, & Tustin, 2011; Hudson, Shapiro, & Sosa, 1995; Quon & Atance, 2010) but significant developments continue to occur subsequently (e.g., Lagattuta, 2014; Suddendorf & Redshaw, 2013). Various lines of evidence suggest that episodic foresight is not an encapsulated ability, but the product of a range of sophisticated cognitive components (Suddendorf & Corballis, 2007), which mature along diverse developmental trajectories (for a review see Suddendorf & Redshaw, 2013). Several of these purported components, such as working memory capacity and executive functions, have also been implicated in the development of deliberate practice (Suddendorf et al., in press). Over the preschool years these components begin to come together to enable children to foresee what they will need and to select tools that will be useful at a specific future point in time (e.g., Atance, Louw, & Clayton, 2015; Atance & Meltzoff, 2005; Russell, Alexis, & Clayton, 2010; Scarf, Gross, Colombo, & Hayne, 2013; Suddendorf & Busby, 2005; Suddendorf, Nielsen, & van Gehlen, 2011).

To ascertain that children's selection in such tasks is driven by foresight and is not simply the result of other factors such as cueing or associative learning, Suddendorf et al. (2011) conducted a carefully controlled study in which children were presented with novel problems in one room (a box for which the correctly shaped key was broken, or a puppet wanting a specific food that was not available) before being taken to another room. After engaging in distracting activities for 15 minutes, children were given the opportunity to secure a solution to the problem from a set of options, to take back with them into the first room. While 3-year-olds selected at random, 4-year-olds picked the correct solution above chance levels. Both age groups performed well in instant versions of the tasks indicating that the problems themselves were conceptually easy. A control experiment confirmed that temporal rather than spatial displacement was critical to performance, and a follow-up study showed the same results when a delay was inserted between selection and implementation of the solution (Redshaw &

Suddendorf, 2013). Together, these results demonstrate that 4-year-olds can remember distinct problems sufficiently enough to secure their future solutions. Our first experiment adopted a similar experimental approach to examine whether preschool children not only begin to select tools with the future in mind, but also begin to selectively engage in acts that may improve their own capacity to obtain a reward at a future point in time.

Our second experiment subsequently examined young children's understanding of deliberate practice and how this related to performance on the practice task that we developed in Experiment 1. To assess understanding of the relation between practice and skill, children were told stories about characters that did or did not repeatedly engage in an activity and then had to make predictions about their future skills. We also asked children directly what they would do if they wanted to improve a skill, and we questioned parents about their children's grasp of the concept. We included Gopnik and Graf's (1988) measure of children's understanding of the relation between experience and knowledge, in order to examine potential parallels. Finally, we examined what factors might predict children's overall practice scores. Together, these are first attempts to chart young children's emerging understanding of deliberate practice.

EXPERIMENT 1

Method

Participants

Sixty children (24 females and 36 males) were recruited from the Early Cognitive Development Centre Database at the University of Queensland. There were 20 three-year-olds (mean age = 36 months, $SD = 0.72$ months), 20 four-year-olds (mean age = 48 months, $SD = 0.61$ months), and 20 five-year-olds (mean age = 60 months, $SD = 0.64$ months). All children in the study came from a medium socioeconomic background. Children were given a certificate and a small gift for participating.

Materials and procedure

In both Experiments 1 and 2, the order of presentation of all tasks and stimuli and the choice of target stimuli were randomized or counterbalanced to control for testing and order effects. The exception was the direct question in Experiment 2, which was administered at the end of the session to ensure that the content did not prime children to anticipate the correct answers to the other tasks.

Children were required to complete three different tasks to assess deliberate practice. These tasks each involved three phases: introduction to the problem in Room A, opportunity to practise in Room B, and return to the problem (Room A). The problems always included one target and three similar distractor items, such that playing with each item was intrinsically equally attractive, but only mastering one offered the promise of a sticker reward in future. Each of the items included slightly different challenges so that practising with the target item was the most effective way to prepare for the future opportunity to win the rewards. There were 5-minute delays between Phases 1 and 2, in which parts of the PPVT–III (Peabody Picture Vocabulary Test, Third Edition) were administered using the standardized procedure outlined in the associated manual (Dunn & Dunn, 1997). On the occasion that a child completed the PPVT–III before the allotted time, they were given a simple fishing puzzle to distract them until the 5 minutes was completed. All tasks were videotaped.

Circuit task. Room A was identified to the child as "Elmo's room" by virtue of a large plush Elmo doll seated on the table. The children were shown a large box of stickers, which were prizes that could be won. They were then introduced to four electronic circuits with different shaped wires (see Figure 1). The circuits were always lined up in the same order. Children were shown that when the wire loop touched the wire shape, the light would come on, and it was explained that the goal of the game was to get the loop from one side to the other without the light coming on. Children were then given the opportunity to try each circuit once so that they could understand the difficulty of the game and, potentially, the need for practice.

Children were then told that Elmo had a preference for one particular shape, and they were promised three stickers for completing that particular circuit. Children were informed that completion of the other circuits would not gain them any stickers. This was illustrated by the experimenter attempting each shape once and only winning stickers for the successful completion of Elmo's favourite shape. The children were then told that they would be going to another room but that they would be returning to Elmo's room later on. They were explicitly told that there would be another chance to try one of the electronic circuits upon return. Lastly, before leaving Room A, the child was reminded which shape was Elmo's favourite.

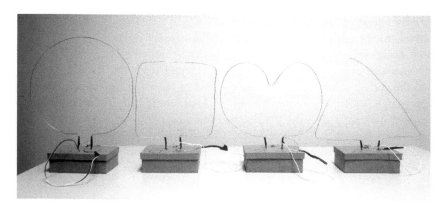

Figure 1. *Four electronic circuits with wires shaped in a circle, square, heart, or triangle.*

Children were then led to Room B (called "Winnie the Pooh room") and engaged in the PPVT–III for 5 min. After 5 min children were told that there was only a short amount of time left to play in this room as they would soon be returning to Elmo's room (Room A). Four electronic circuits, identical to the ones in Room A, were then revealed, and it was explained to the child that even though these circuits did not have any batteries, and no prizes could be won here, they could choose one to play with before returning to Elmo's room. Children were allowed to practise with their chosen circuit for one minute. The child was then asked, "Why did you choose that shape?" Upon return to Room A, children were again asked to pick one of the circuits and were allowed to choose three stickers from Elmo's box if they selected and successfully completed the target circuit. Children were again asked why they selected the shape they did.

Food task. The apparatus used was four different plastic foods attached to a string, which could be swung into differently shaped boxes on a wooden pole (see Figure 2). Children were informed what Elmo's favourite food was and that they would receive stickers if they could catch it. Other than these changes in the apparatus used and goal of the game, the procedure was identical to that described above for the circuit task.

Lizard task. The procedure for the lizard task was identical to that for the other two tasks. The apparatus used in this task was four battery-operated boxes with pictures of a lizard, which were covered with four different coloured transparent films (see Figure 3). Each lizard had a different-sized hole cut out of the film, and each hole was lined with a metal edge. The aim of this game was to use tweezers to remove a bug from the hole in the lizard without touching the sides.

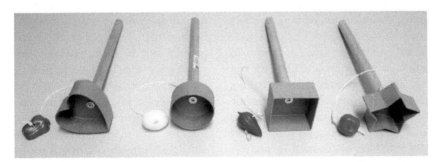

Figure 2. *Four items of wooden apparatus attached to a plastic food (capsicum, egg, strawberry, or apple) via a string.*

Figure 3. *Four battery-operated lizards, which were coloured in blue, pink, yellow, or green.*

Touching the sides led to a loud buzzing noise and would cause the box to vibrate. Children were told Elmo's favourite colour and that removing the bug from the lizard with that colour would win them a prize (stickers). The remainder of the procedure was identical to that used in the circuit and food tasks.

Analysis methods
To determine whether children passed a task at a rate significantly different to chance, we calculated binomial probabilities. These were constructed as the probability of at least k successes from N trials, where k was the number of children that passed the task and N the number of children that attempted the task, and the chance of selecting the correct shape in both rooms being $0.25 \times 0.25 = .0625$. We tested for variation among groups using chi-squared tests, as a preliminary test of variation other than expected by chance. To compare dichotomous outcomes (e.g., passing or failing a task) across two groups (e.g., 4-year-old and 5-year-old children), we calculated odds ratio effect sizes with corresponding p values. The odds ratio is the odds of an event (e.g., passing a task) in one group, compared to the odds of the event in the other group, and higher odds ratios indicate a larger difference between groups. We used multiple regression with ordinary least squares estimation to identify the variation in practice scores predicted by sets of predictor variables, as reported in the relevant sections of the results.

Results

Task order and reliability
The order in which children completed the tasks had no systematic effect on performance in the circuit task, $\chi^2(5, N = 60) = 9.75$, $p = .083$, the food task, $\chi^2(5, N = 60) = 6.99$, $p = .221$, or the lizard task, $\chi^2(5, N = 60) = 7.22$, $p = .205$. The PPVT score was used to calculate a mental age for each child, based on age-normed scores. Children's mental ages were on average slightly above their actual ages, indicating that the 3-year-olds (mean mental age = 40.65, $SD = 9.44$), 4-year-olds ($M = 60.65$, $SD = 13.68$), and 5-year-

olds ($M = 69.55$, $SD = 11.14$) tested in this experiment performed slightly better than average children on the PPVT.

Circuit task. To pass each task, children had to select the target item for practice in Room B and attempt to win awards upon return to Room A. Results of all three tasks are displayed in Figure 4. In the circuit task, 3-year-olds did not select the correct shape in both rooms significantly above chance (.0625; binomial $p = .126$), but 4-year-olds and 5-year-olds did ($p < .001$ for both groups). Pearson's chi-square analyses revealed a significant difference between the performances of 3-year-olds (15% correct), 4-year-olds (40% correct), and 5-year-olds (60% correct) on the circuit task, $\chi^2(2, N = 60) = 8.60$, $p = .014$. Of those who selected the correct circuit in both the practice and test rooms, no children explicitly identified practice as the reason for selecting the shape when asked why they had made that choice. Sixteen children who selected the correct option in both rooms said they had done so because that shape (or food, or colour) had been pointed out to them, saying, for example, "It's Elmo's favourite" (61-month-old male). In contrast, children who failed the task tended to justify their choices with statements unrelated to the future return to the problem saying, for instance, "It's my favourite shape" (61-month-old female). A similar pattern was seen across all tasks.

Food task. In the food task, 3-year-olds did not select the correct fruit significantly above chance (.0625; binomial $p = .126$), while 4-year-olds and 5-year-olds did ($p < .001$ for both groups). Pearson's chi-square analyses revealed a significant difference between the performances of 3-year-olds (15% correct), 4-year-olds (50% correct), and 5-year-olds (55% correct) on the food task, $\chi^2(2, N = 60) = 7.92$, $p = .019$.

Lizard task. In the lizard task, 3-year-olds, 4-year-olds, and 5-year-olds all selected the correct lizard significantly above chance (.0625; binomial $p = .001$, $p < .001$, and $p < .001$, respectively). Pearson's chi-square analyses did not reveal

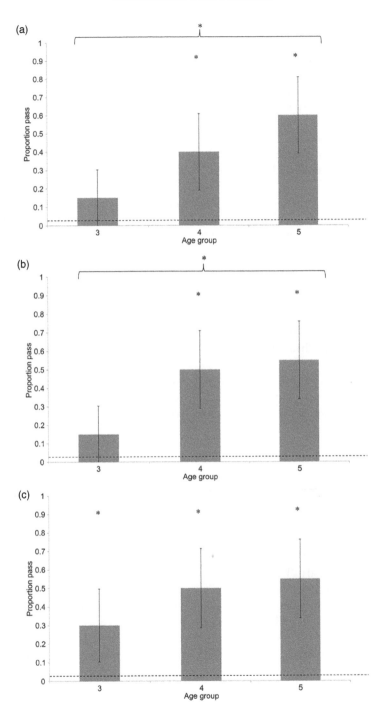

Figure 4. *Proportion of children who selected the correct object in both rooms for (a) the circuit task, (b) the food task, and (c) the lizard task. The dotted line represents the proportion of children who would be expected to pass the task if they were responding randomly. Error bars represent the upper and lower limits of the 95% confidence interval for the proportion. An asterisk is provided if children's performance was significantly different from chance (asterisk above bar) or significantly different between age groups (above bracket).*

significant differences between the performances of 3-year-olds (30% correct), 4-year-olds (50% correct), and 5-year-olds (65% correct) on the lizard task, $\chi^2(2, N = 60) = 4.94, p = .085$. However, 4-year-olds and 5-year-olds were far more likely to produce an explanation for their choice that was related to the task, such as "It's Elmo's favourite" than 3-year-olds were (50% of explanations in both 4-year-olds and 5-year-olds, compared to 5% of explanations in 3-year-olds).

Correlation of practice scores

Children's performances were highly correlated across all three tasks, such that children who passed one task were likely to pass the other tasks as well (two-way odds ratios from 1.04 to 18.25, *p* value from .002 to .042). Therefore, scores on the three tasks were summed to create an overall practice score for each child. Overall practice scores were significantly lower for 3-year-old children ($M = 0.60$, $SD = 0.82$) than for 4-year-olds ($M = 1.40$, $SD = 0.94$), $t(38) = 2.87$, $p = .007$, or 5-year-olds ($M = 1.80$, $SD = 1.20$), $t(38) = 3.70$, $p < .001$, although there was no significant difference between the performances of 4-year-olds and 5-year-olds, $t(38) = 1.17$, $p = .248$.

Multiple regression analysis

A multiple regression was conducted with the combined score as the dependent variable, and age and mental age as the predictor variables. The model explained between 18% and 21% of the variance in the combined practice scores, which was statistically significant, $R = .45$, $R^2 = .21$, $R^2_{adi} = .18$, $SEE = 1.00$, $F(2, 57) = 7.38$, $p = .001$. The only significant predictor of overall practice score was children's age group, such that older children had higher practice scores, $b = 0.50$ (95% confidence interval, CI[0.05, 0.96]), $p = .030$. Mental age did not predict overall practice scores, $b = 0.01$, 95% CI[−0.01, 0.03], $p = .553$.

Discussion

Children's performances on all three tasks were highly associated, meaning that the children who passed one task were likely to pass another and

providing evidence that all three tasks were measuring a similar underlying construct. The testing paradigm showed a clear age effect as expected, with children's performance on the practice tasks improving between the ages of 3 and 5 years. On all three tasks, four- and five-year-old children demonstrated that they could deliberately select the task relevant to an anticipated future situation. In contrast, three-year-old children performed at chance on two of the three tasks and, despite achieving above-chance scores on the third task, could not explain their choices.

This experiment contained several critical novel features and controls to ensure that what was being tested was in fact deliberate, future-oriented practice. The inclusion of multiple counterbalanced stimuli in each task (i.e., multiple shapes for the circuit task, multiple foods for the food task, and multiple colours for the lizard task) controlled for any intrinsic appeal of one shape, food, or colour. The use of different rooms and different (though identical) sets of stimuli for the practice and test trials were intended to prevent, as far as possible, children being prompted by any environmental cues.

However, it remains possible that children acted the way they did for reasons other than deliberate practice. The children that passed the task might have selected the object that had been made salient to them and associated with a reward, without explicitly understanding that practice improves their chances of future performance. In Experiment 2 we therefore focused on children's explicit conceptual understanding of the effect of practice on skill levels for themselves and others. Furthermore, a substantial amount of variation in children's performance on the tasks was not explained by age or verbal intelligence. We therefore decided to include additional measures of cognitive abilities that may account for variation in understanding of deliberate practice.

First, we included a measure of source memory. Understanding how practice leads to improved performance of a skill is potentially analogous to understanding how information leads to knowledge. Three-year-old children may not understand that people with skills have not always had these skills, just as they may consider that they have always known something they have just learned (Naito, 2003; Taylor et al., 1994) or be unable to report how they have obtained information (Gopnik & Graf, 1988). Second, since children's scores on the picture vocabulary test unexpectedly failed to predict their performance in Experiment 1, in Experiment 2 we examined whether a nonverbal measure of mental age would be associated with practice. Third, we measured a number of other control variables that could also potentially explain differences in children's practice behaviour, including enrolment in school or preschool, extracurricular activities requiring practice, older siblings, and parents' expectations of their child's practice.

In Experiment 2 we assessed children's explicit understanding of the effect of deliberate practice on future performance. We also tried to replicate one of the tasks of Experiment 1 and examine the links to various cognitive and control variables. We included only one of the three tasks from Experiment 1 to keep the overall testing time to under an hour, and because children's performance across the three tasks was highly correlated. In addition, because only the 4-year-old and 5-year-old children demonstrated consistent selection of the correct activity in Experiment 1, Experiment 2 was limited to children in these age groups.

EXPERIMENT 2

Method

Participants

Forty-three children (24 males and 19 females) between the ages of 4 years and 5 years were recruited through the same means as in Experiment 1. Data from one 5-year-old were excluded from the sample due to parental interaction during testing potentially biasing the results. The final sample comprised 21 four-year-olds (mean age = 48.70 months, SD = 1.90 months, range = 46–51 months) and 21 five-year-olds (mean age = 61.60 months, SD = 1.23 months, range = 59–63 months).

Materials and procedure

Participants completed five measures: the circuit task from Experiment 1, two novel story-based tasks with puppets, a source of knowledge task (Gopnik & Graf, 1988), a direct question about practice, and the Wechsler Preschool and Primary Scale of Intelligence III (WPPSI–III; Wechsler, 2002) Block Design task. The Block Design task was administered using the standard procedure during the delay period in the circuit task. If a child took longer than the allocated time to complete the Block Design task, extra time was allocated at the end of the testing session for him or her to complete the task.

Story tasks. The experimenter introduced the child to two puppets and told the child a story. Both puppets were not good at juggling, but both wanted to be good at it. One puppet juggled every day, and the other juggled once and did not juggle again after that. The experimenter explained that this happened until the puppets grew up ("*got big*") and introduced the child to two identical but larger puppets (see Figure 5). The experimenter checked that the children knew which puppet was which, and which had juggled every day and which had not. Children were then told that the grown-up puppets were having a juggling competition. The experimenter asked the child to identify which puppet would win the juggling competition and then to explain why.

To avoid effects of preexisting associations about practice and juggling, children were also told a second story with exactly the same structure, but featuring alien puppets and a made-up skill (*Woolikee*).

Source of knowledge. This task followed the procedure reported by Gopnik and Graf (1988). The experimenter showed children a set of six drawers (two across, three down) and told them that they would discover what was inside each drawer. The child opened and saw the contents of two drawers, was told the contents of two drawers,

Figure 5. *Puppets used in the story task (above: juggling; below: alien).*

and was given a clue to infer the contents of two drawers.

After the presentation of each object, the experimenter checked that the child remembered what was in the drawer and then asked the child how he or she knew what was in the drawer. The experimenter listed the three options for the source of knowledge and asked the child to choose one: "How do you know it's a [object]? Did you see it, did I tell you, or did you figure it out yourself from a clue?". The number of correct choices was summed for a total score ranging from zero to six.

Direct questions. At the end of the session, the experimenter asked the child directly, "If you want to get better at something, what do you think you should do?". If the child did not answer, the experimenter asked the question a second time. If the child still did not answer, he

or she was coded as having not answered that question, and the score was treated as missing. The parent filled out a short questionnaire asking about their perception of their child's understanding of deliberate practice, their child's enrolment in school and extracurricular activities, and demographic details.

Coding and reliability

Children were assigned a single score for each practice task, a score for the source of knowledge task, and a score for the Block Design task. A randomly selected subset (17%, $n = 7$) of the videotapes of the sessions was coded by an independent coder who was blind to the hypotheses of the study, in order to assess interrater reliability. Statistical analysis methods were similar to those in Experiment 1.

Results

Task order and reliability

The order in which children completed the tasks had no systematic effect on performance in the circuit task, $\chi^2(34, N = 42) = 36.06$, $p = .372$, the story task, $\chi^2(34, N = 42) = 35.88$, $p = .381$, or the direct question, $\chi^2(34, N = 42) = 39.88$, $p = .225$. Interrater reliability was excellent (Cronbach's kappa = .83; Fleiss, 1981; Landis & Koch, 1977). The dataset from the primary coder was used in all statistical analyses.

None of the demographic details collected (gender, older siblings, enrolment in extracurricular activities, school or preschool enrolment) predicted practice scores, so these were not included in the following analyses. Summary statistics for all tasks are shown in Table 1.

Story task. For the story task, the performance of 4-year-olds was not significantly different from that expected by chance (50%) on the alien version of the task ($p > .999$), but 4-year-olds performed above chance on the juggling version of the task ($p = .019$). Five-year-olds performed above chance both in the juggling version and in the alien version ($p = .001$ and $p = .001$, respectively).

As seen in Figure 6, 5-year-olds performed better than 4-year-olds across both versions of the story task. A binomial test revealed that across both story tasks, 4-year-olds as a group did not perform above chance (25%; $p = .256$), whereas 5-year-olds did ($p < .001$). That is, significantly more 5-year-olds (81%) than 4-year-olds (33%) selected the correct puppet in both the real and alien versions of the task, $\chi^2(1, N = 42) = 9.72$, $p = .002$, and the odds of a 5-year-old choosing the correct puppet in both versions were 8.5 times higher than the odds of a 4-year-old doing so (odds ratio = 8.50, 95% CI[2.06, 35.08], $p = .003$).

Of those who selected the correct puppet in both versions, 83% correctly identified practice or repetition as the reason for the winning puppet's success when asked why they had made that choice. For example, a 50-month-old male said, "Because he does it every day", and a 60-month-old male stated, "Because he practised more". In contrast, none of the children who failed to make the right choice made the correct reference and instead justified their choice with statements such as "Because I want him to win" (50-month-old male) and "He was being good" (61-month-old male).

Direct question. In the direct question, significantly more 5-year-olds (67%) than 4-year-olds (10%) correctly identified practice or repetition as a successful strategy for them personally to improve at a skill, $\chi^2(1, N = 42) = 14.54$, $p < .001$. Six of the 4-year-olds and one of the 5-year-olds did not

Table 1. *Descriptive statistics for task performance in Experiment 2, by age*

Task	4-year-olds			5-year-olds			Overall		
		95% CI			95% CI			95% CI	
	Proportion correct	UL	LL	Proportion correct	UL	LL	Proportion correct	UL	LL
Story task (overall)	.33	.13	.53	.81	.63	.99	.51	.35	.67
Direct question	.10	−.01	.30	.67	.49	.85	.38	.24	.52
Circuit task	.14	−.02	.30	.29	.09	.49	.21	.09	.33
	Mean	SD		Mean	SD		Mean	SD	
Block design	12.67	3.91		16.38	3.25		14.52	4.02	
Source of knowledge	3.33	1.65		5.76	0.63		4.55	1.74	

Note: Proportions and associated confidence intervals are presented for categorical variables, and means and standard deviations for continuous variables. 95% CI = the 95% confidence interval for the proportion; UL = upper limit of the 95% confidence interval; LL = lower limit of the 95% confidence interval. $N = 42$ (21 four-year-olds, 21 five-year-olds).

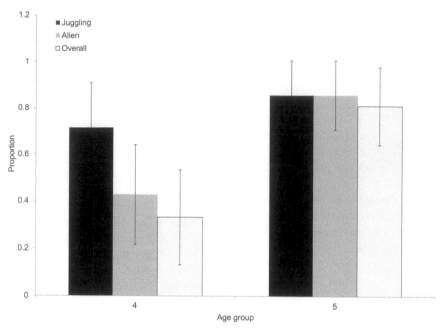

Figure 6. *Proportion of children who selected the correct puppet in juggling, alien, and both versions of the story task.*

give an answer. The odds of a 5-year-old answering correctly were 19 times higher than the odds of a 4-year-old doing so (odds ratio = 19.00, 95% CI [3.41, 105.73], p = .001). Correct explanations included statements such as "Practise a bit harder" (50-month-old male) and "Do it every day" (60-month-old female). Some incorrect explanations were nevertheless plausible answers to the question, such as "Ask someone" (50-month-old female) and "Concentrate" (61-month-old female).

Circuit task. In the circuit task, 4-year-olds did not select the correct shape in both rooms significantly above chance (.0625; binomial p = .140), while 5-year-olds did (p = .001). The 4-year-olds selected the correct shape above chance (.25) in the test room (Room A) but not in the practice room (Room B; binomial p = .029 and p = .332, respectively). In contrast, 5-year-olds selected the correct shape in both the test room (Room A) and the practice room (Room B; binomial p < .001 and p = .021, respectively). The difference between 4-year-olds and 5-year-olds was not statistically significant, $\chi^2(1, N = 42) = 1.27$, p = .454, and this

may have been due in part to low rates of passing the task overall (4-year-olds: 14% correct; 5-year-olds: 29% correct).

Only one 60-month-old male explicitly identified practice as the reason for selecting the circuit shape when asked why he had made that choice, saying, "Because it's the favourite one and to practise". Four children who selected the correct shape in both rooms did so because the shape had been pointed out to them. In contrast, none of the children who failed to make the right choice made verbal reference to the problem in Room A and instead justified their choice with statements such as, "I like yellow" (46-month-old male).

Correlation of practice scores
Children's performance on the direct question was significantly and positively related to performance on both the story task (overall odds ratio = 11.97, p = .011; 4-year-olds odds ratio = 1.27, p > .999; five-year-olds: odds ratio = 8.01, p = .088) and the circuit task (overall odds ratio = 25.00, p = .007; 4-year-olds: odds ratio = 10.95, p = .216; 5-year-olds: odds ratio = 4.51, p = .193). Performance on

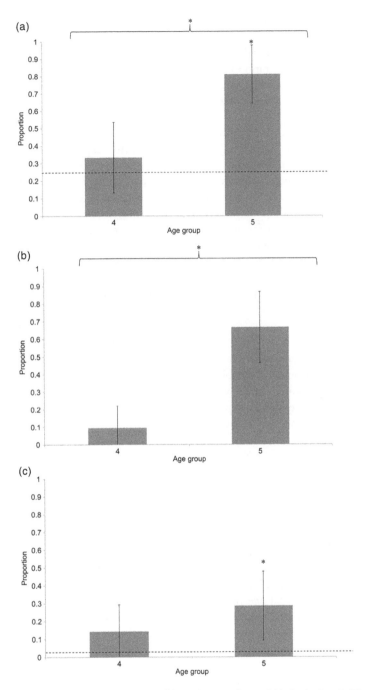

Figure 7. *Proportion of children who passed (a) the story task, (b) the direct question, and (c) the circuit task. The dotted line represents the proportion of children who would be expected to pass the task if they were responding randomly. Error bars represent the upper and lower limits of the 95% confidence interval for the proportion. An asterisk is provided if children's performance was significantly different from chance (asterisk above bar) or significantly different between age groups (above bracket).*

the story task was not significantly related to performance on the circuit task, but the direction of the association was positive (overall odds ratio = 1.41, $p = .717$; 4-year-olds: odds ratio = 0.48, $p = .619$; 5-year-olds: odds ratio = 1.96, $p > .999$). The results of the three tasks are summarized in Figure 7.

A composite measure was created by summing each child's scores on the direct question (1 for answers containing practice, 0 for any other answer), the story task (1 for correct selection on both juggling and alien versions, 0 for any other combination), and the circuit task (1 for selecting the correct shape in both rooms, 0 for any other selection), to give an overall practice score for each child (range = 0 to 3). The combined practice score was significantly higher for 5-year-olds ($M = 1.57$, $SD = 0.81$) than for 4-year-olds ($M = 0.86$, $SD = 0.48$), $t(40) = -3.48$, $p = .001$.

Multiple regression analysis

A multiple regression was conducted with the combined practice score as the dependent variable and age, block design score, and source of knowledge as the predictor variables. The model explained between 34% and 39% of the variance in the combined practice scores, which was statistically significant, $R = .62$, $R^2 = .39$, $R^2_{adj} = .34$, $SEE = 0.61$, $F(3, 38) = 8.04$, $p < .001$.

Source memory predicted a statistically significant percentage of the variance in the combined practice score, with larger scores on the source of knowledge task predicting larger practice scores such that for every extra correct answer on the source of knowledge task, a 0.17 unit increase in practice was predicted, $b = 0.17$, 95% CI[0.01, 0.32], $p = .036$. In addition, mental age predicted performance on the practice tasks, such that for every unit increase in block design score, a 0.06 unit increase in practice was predicted, $b = 0.06$, 95% CI[0.01, 0.12], $p = .024$. When source of knowledge and block design score were included in the model, there was no direct effect of age on combined practice score, $b = 0.10$, 95% CI[−0.47, 0.68], $p = .718$.

In the questionnaire, parents were asked to identify whether they thought their child understood the relation of practice to future performance. The majority of parents answered "yes" to this question: 95% of parents of 4-year-olds (18 of 19) and 85% of parents of 5-year-olds (17 of 20). A few parents elaborated in an open question, such as "yes, but is a perfectionist" (parent of a 51-month-old male), or "yes, but not to apply it" (parent of a 60-month-old male).

Parents' expectations about their children's understanding of practice were hence markedly different from most children's actual performance on the tasks. A series of chi-square tests for independence revealed no association between parents' expectations and their child's performance on the story task, $\chi^2(1, N = 42) = 2.42$, $p = .120$, the direct question, $\chi^2(1, N = 42) = 1.99$, $p = .348$, or the circuit task, $\chi^2(1, N = 42) = 0.88$, $p = .348$. A Welch's t test for equality of means revealed no significant difference between the combined practice scores of children whose parents thought they understood practice and children whose parents didn't think they understood practice, $t(38) = -1.02$, $p = .313$.

Discussion

The results of Experiment 2 revealed a substantial difference in explicit understanding of deliberate practice between 4-year-old and 5-year-old children. When directly asked what they should do to get better at a skill, only 10% of 4-year-olds identified practice or repetition as a strategy. In contrast, over two thirds of 5-year-olds answered that practice or rehearsal could improve their skills.

This gap was also evident when children were asked to judge who of two characters would be better at a skill, the one who practised or the one who did not. Over 80% of 5-year-olds picked the characters that had practised, whereas only a third of 4-year-olds did. The younger children demonstrated some understanding in the juggling condition, possibly because they had some learning history related to this skill, but they performed at chance in the novel "Woolikee" condition. In contrast, the older children could answer both versions of the question correctly. Overall, these results strongly suggest that by age 5 children tend to have some understanding of deliberate practice.

Experiment 2 also examined whether explicit understanding of deliberate practice was related to performance on one of the measures from Experiment 1: the circuit selection task. Results replicated the finding that 5-year-olds can perform above chance. However, the 4-year-olds did not (discussed further in the General Discussion, below). Importantly, children's performance was consistent across both verbal and behavioural tasks, suggesting that performance in the circuit task may be the result of real understanding about the need for practice in order to succeed at the task. Children who could spontaneously identify practice as a method for improving a skill could also identify the benefits of practice in the story task and could choose the correct activity to practise in the circuit task.

The better performance of 5-year-olds could potentially be explained by the fact that formal schooling starts at age 5, given that many school activities require structured practice. Interestingly, however, school enrolment did not predict practice scores independent of age. Furthermore, all 4-year-olds except for one were enrolled in some type of part-time formal education (preschool, kindergarten, or playgroup) in which they would have potentially received exposure to ideas about practice similar to those in formal schooling. These limitations in our sample prevented us from investigating further the potential relationship between formal education or training and understanding deliberate practice. Given that 5-year-olds would have had more exposure to formal education, this remains a potential driving factor that deserves future research attention.

Children's practice scores were predicted by their nonverbal intelligence, as expected. Furthermore, independent of intelligence, children who had high practice scores also performed well on the source memory task. That is, children who demonstrated understanding of how knowledge is gained also demonstrated understanding of how skills are gained. The potential implications of this finding are explored below in the General Discussion.

Surprisingly, parents as a group were inaccurate in their predictions of their children's understanding of practice. Parents overwhelmingly thought that their children could understand the relation of practice to performance, while particularly in the 4-year-old group, their children did poorly on multiple tests measuring such understanding. Although parental confidence is appealing in theory, problems may arise where this confidence is mismatched to children's actual ability. Children are increasingly expected to learn skills, such as reading, that require deliberate practice from an early age (Spencer, 2011). The results of Experiment 2 imply a potential scenario in which parents may expect children to understand and therefore spontaneously engage in deliberate practice of certain skills, yet these children may not yet grasp the very nature of the relation of practice to skills. This discrepancy could create frustration and conflict if parents were to misinterpret their child's lack of practice, for example as a lack of interest in the skill or laziness.

GENERAL DISCUSSION

Several novel tests for understanding deliberate practice were created for use in the current study. Consistent performance across tasks demonstrated that by age five, children have a basic understanding of the link between practice and skill formation, whereas younger children demonstrated little explicit comprehension.

Some capacity to think about future events is required before one can deliberately practise with the goal of changing one's future skills. The current results are therefore in line with recent research examining the emergence of episodic foresight over the preschool years (e.g., Suddendorf & Redshaw, 2013). In a previous study with a similar structure to our practice task, 4-year-olds, but not 3-year-olds, demonstrated a capacity to select an object in one situation with the goal of securing a solution to a future problem (Suddendorf et al., 2011). In the practice task of the current study, children could not take an object to the other room but a relevant experience:

they could select to practise the task that they anticipated attempting to solve in the other room. Over both experiments 5-year-olds made these prudent choices.

The 4-year-olds performed above chance in Experiment 1, and even 3-year-olds showed some competence on one of the tasks. However, in Experiment 2, 4-year-olds' performance was not replicated. This was surprising given that the circuit task was conducted essentially the same way in both experiments, and the order in which children encountered the tasks had no effect in either experiment. It is possible that this represents chance variation, especially given that Experiment 2 had less power, featuring only one of the three practice tasks of Experiment 1. It is also possible that 4-year-olds were using a different strategy to 5-year-olds, one that relied on situational cues rather than episodic foresight. The 4-year-olds selected the correct shape above chance in the test room, but not in the practice room. A potential explanation for this finding is that the 4-year-olds may indeed have been relying on cues in Room A to help them remember which was the preferred shape, rather than carrying the knowledge of what skill would be required from Room A to Room B. In contrast, 5-year-olds selected the correct shape in both the test room and the practice room, suggesting that the 5-year-olds may have been using a different strategy to the 4-year-olds, and potentially carrying their knowledge of the future requirements of the task between rooms. Overall, more research is required to examine 4-year-olds' capacity to select activities that improve relevant future skills.

Our Experiment 1 included three novel tasks, designed to test whether children would deliberately practise a particular version of a task in order to increase the likelihood of performing well at this task on a future occasion. However, the similarity of the four versions of each task may have been a problem, because learning skills on one version may have been transferable to the other. We selected to make each version very similar because we needed to avoid another problem: the possibility of one option being inherently more attractive to some children than others. This

similarity, however, may have meant that some children could have selected to play on one version of the task, knowing that this would improve their chances also for the future performance on the target task. Note, however, that none of the children indicated anything like such reasoning in their verbal explanations for their choices. Nonetheless, we appreciate that similarity may have been a problem, and future studies might hence want to give children more diverse options that are matched for attractiveness through pretests.

There are numerous other ways in which future studies may build on this basic paradigm to further examine the reasons for children's choices. For instance, a control condition could drop the third stage of the procedure, the return to the Room A, to investigate whether children in Room B would select the task that had been made salient in Room A, even when there was no point in practising this particular task for the future. One could also expand the duration of the option for practice and try to measure actual changes in skill levels as a function of more extensive practice. This would probably necessitate some careful tailoring of the practice tasks to children's initial skill level, because if a task is too easy, children have no need for practice, and if a task is too difficult, they may not be motivated to attempt it. In our tasks one could account for individual motor skill levels by, for instance, adjusting the size of the loop in the circuit task, or the size of the cup in the food task, so that the children reach a comparable level of competence in a pretest. Although there are many promising manipulations that one could pursue, we decided that the most important first follow-up would be to explore children's explicit understanding of deliberate practice as a means for improving skills—and to see whether this could explain their choices.

Our Experiment 2 therefore primarily addressed children's explicit understanding of deliberate practice and its role in skill formation. First, we asked children directly. In line with their poor performance on the circuit task, 4-year-old children primarily gave answers that failed to demonstrate a clear understanding of how skills improve. By contrast, a substantial proportion of the 5-year-old children identified practice as a method for improving skills. Children who gave such answers were more likely to select the correct circuit than those who did not, suggesting that children selected to practise the relevant task with the improvement of skill for the future problem in mind.

Five-year-olds' understanding of deliberate practice was confirmed by their performance on the story task. The children recognized that the puppet that had practised a skill was more likely to win a competition than the one that had not. Children's ability to identify patterns in others does not always translate into their understanding of what action to take themselves. For example, children's decisions in a delayed gratification paradigm differed according to whether they were deciding for themselves or for another person (Prencipe & Zelazo, 2005). By contrast, in the current study, children's understanding of the effect of practice on skills in others was associated with how they themselves would aim to change their skills. Evidently, by age 5 children can demonstrate some understanding that deliberate practice can lead to the improvement of skills.

We calculated overall practice scores across the tasks in Experiment 2 to investigate which factors may predict performance. In line with predictions, we found support for the influence of age and source of knowledge. Only by understanding the source of knowledge can one peruse sources with the intention of acquiring new knowledge. By the same token, only by understanding the link between practice and skill can one deliberately set out to practise with the goal of improving one's skill.

Deliberate practice is one important consequence of our capacity for episodic foresight (Suddendorf, 2013). Imaging future episodes is a critically important human skill (Schacter & Addis, 2007; Suddendorf & Corballis, 1997, 2007; Szpunar, 2010). It gives us power to shape our future selves—at least to a certain extent. Parents encourage children to act in ways that they think will be beneficial in the long run (e.g., to study and to practise), but children themselves seem to appreciate this power only from around the end of the preschool years. Various lines of

evidence suggest that the causal links between past, present, and future events become clearer to children then. In one study, for instance, children were told stories about one person who acquired objects or knowledge in the past and one who will acquire these in the future (Busby Grant & Suddendorf, 2010). By age 5 children could tell who has the object or knowledge in the present, whereas 4-year-olds responded inconsistently. Our findings are in line with this broader emerging picture suggesting that significant developments in children's capacity to travel mentally in time occur around age 4 and 5 (Atance & Jackson, 2009; Hayne et al., 2011; Russell et al., 2010; Scarf et al., 2013; Suddendorf & Busby, 2005; Suddendorf & Corballis, 1997; Suddendorf & Redshaw, 2013).

Neuroscientific experiments have shown that mental time travel, theory of mind, and related abilities such as wayfinding involve similar brain networks, chiefly associated with the default mode network (Buckner & Carroll, 2007; Schacter et al., 2012; Spreng & Grady, 2010; Spreng, Mar, & Kim, 2009). This evidence has led researchers to posit that these abilities form part of a general human ability to imagine the self in different scenarios and embed them into larger causal narratives (e.g., Suddendorf, 2013): in the past and future (mental time travel), in different locations (wayfinding), and in the situation of others (theory of mind). Deliberate practice may therefore depend on a form of self-projection (Buckner & Carroll, 2007; Ford, Driscoll, Shum, & Macaulay, 2012; Suddendorf & Corballis, 1997), because it involves anticipating a future need for a skill. Similarly, the source of knowledge task can be considered a test for self-projection into a personal past (e.g., Gopnik & Graf, 1988; Naito, 2003; Perner, Kloo, & Gornik, 2007). Future work could examine links between deliberate practice and these other self-projection abilities, especially other measures of episodic foresight.

Although a general capacity to deliberately practise may emerge at the end of the preschool years, whether individual children actually engage in deliberate practice is likely to be a function of diverse sociocultural factors. Children's parents, teachers, and coaches can and do ask children to practise, and they educate children on the importance of practice (e.g., "practice makes perfect"). Though we argue that episodic foresight is a prerequisite for deliberate practice, this is not to say that children may not engage in practice as a result of instruction. Such direct instruction may become increasingly important to children's engagement in practice over the preschool and early primary school years. Whether by following instruction or by simply repeating activities they enjoy, children gather experience in the relationship between repetitive actions and skill development, and they may subsequently choose to practise certain skills deliberately. Future work could therefore investigate the influence of education and experience on children's understanding of practice.

Many questions remain to be answered about the development of this fundamental human capacity. For instance, how and when do children appreciate that certain performances can be altered through practice, and others, such as throwing dice, cannot? How does children's development of basic cognitive functions such as executive control, semantic capacities, and working memory influence the development of practice? What are the long-term consequences of individual differences in deliberate practice and of sociocultural factors that encourage or discourage its use?

Humans show an extraordinary diversity in skills and expertise, and much of that variety is the result of people deciding what knowledge and skills to acquire and what capacities to perfect through practice. The current study is only a first step towards examining the development of deliberate practice. We hope that future studies may extend this line of research, especially with diverse, ecologically valid, behavioural tests for practice. The preliminary results of our experiments suggest that over the preschool years children may acquire a basic capacity for, and understanding of, deliberate practice: They can begin to think about what skills they want to have and to take steps to shape their future self accordingly.

REFERENCES

Atance, C. M., & Jackson, L. K. (2009). The development and coherence of future-oriented behaviors during the preschool years. *Journal of Experimental Child Psychology*, *102*(4), 379–391. doi:10.1016/j.jecp.2009.01.001

Atance, C. M., Louw, A., & Clayton, N. S. (2015). Thinking ahead about where something is needed: New insights about episodic foresight in preschoolers. *Journal of Experimental Child Psychology*, *129*, 98–109. doi:10.1016/j.jecp.2014.09.001

Atance, C. M., & Meltzoff, A. N. (2005). My future self: Young children's ability to anticipate and explain future states. *Cognitive Development*, *20*(3), 341–361. doi:10.1016/j.cogdev.2005.05.001

Baker, J., Côté, J., & Deakin, J. (2005). Expertise in ultra-endurance triathletes early sport involvement, training structure, and the theory of deliberate practice. *Journal of Applied Sport Psychology*, *17*(1), 64–78. doi:10.1080/10413200590907577

Buckner, R. L., & Carroll, D. C. (2007). Self-projection and the brain. *Trends in Cognitive Sciences*, *11*(2), 49–57. doi:10.1016/j.tics.2006.11.004

Busby Grant, J., & Suddendorf, T. (2005). Recalling yesterday and predicting tomorrow. *Cognitive Development*, *20*(3), 362–372. doi:10.1016/j.cogdev.2005.05.002

Busby Grant, J., & Suddendorf, T. (2010). Young children's ability to distinguish past and future changes in physical and mental states. *British Journal of Developmental Psychology*, *28*(4), 853–870. doi:10.1348/026151009X482930

Campitelli, G., & Gobet, F. (2011). Deliberate practice: Necessary but not sufficient. *Current Directions in Psychological Science*, *20*(5), 280–285. doi:10.1177/0963721411421922

Côté, J., Baker, J., & Abernethy, B. (2003). Chapter 4: From play to practice: A developmental framework for the acquisition of expertise in team sports. In J. L. Starkes & K. A. Ericsson (Eds.), *Expert performance in sports: Advances in research on sport expertise* (pp. 89–113). Champaign, IL: Human Kinetics.

Dempster, F. N. (1996). Chapter 9: Distributing and managing the conditions of encoding and practice. In E. L. Bjork & R. A. Bjork (Eds.), *Memory* (pp. 317–344). San Diego: Academic Press.

Donald, M. (1999). Preconditions for the evolution of protolanguage. In M. C. Corballis & S. E. G. Lea (Eds.), *The descent of mind: Psychological perspectives on hominid evolution* (pp. 138–154). Oxford: Oxford University Press.

Dunn, L. M., & Dunn, L. M. (1997). *Peabody picture vocabulary test* (3rd ed). Circle Pines, MN: American Guidance Service.

Ericsson, K. A. (2008). Deliberate practice and acquisition of expert performance: A general overview. *Academic Emergency Medicine, 15*(11), 988–994. doi:10.1111/j.1553-2712.2008.00227.x

Ericsson, K. A., Charness, N., Feltovich, P. J., & Hoffman, R. R. (eds.) (2006). *The Cambridge handbook of expertise and expert performance*. Cambridge: Cambridge University Press.

Ericsson, K. A., Krampe, R. T., & Tesch-Römer, C. (1993). The role of deliberate practice in the acquisition of expert performance. *Psychological Review, 100*(3), 363–406. doi:10.1037/0033–295X.100.3.363

Fleiss, J. L. (1981). *Statistical methods for rates and proportions* (2nd ed.). New York: Wiley.

Ford, R. M., Driscoll, T., Shum, D., & Macaulay, C. E. (2012). Executive and theory-of-mind contributions to event-based prospective memory in children: Exploring the self-projection hypothesis. *Journal of Experimental Child Psychology, 111*(3), 468–489. doi:10.1016/j.jecp.2011.10.006

Gopnik, A., & Graf, P. (1988). Knowing how you know: Young children's ability to identify and remember the source of their beliefs. *Child Development, 59*(5), 1366–1371. doi:10.2307/1130499

Hayne, H., Gross, J., McNamee, S., Fitzgibbon, O., & Tustin, K. (2011). Episodic memory and episodic foresight in 3- and 5-year-old children. *Cognitive Development, 26*(4), 343–355. doi:10.1016/j.cogdev.2011.09.006

Helsen, W. F., Hodges, N. J., Van Winckel, J., & Starkes, J. L. (2000). The roles of talent, physical precocity and practice in the development of soccer expertise. *Journal of Sports Sciences, 18*(9), 727–736. doi:10.1080/02640410050120104

Hudson, J. A., Shapiro, L. R., & Sosa, B. B. (1995). Planning in the real world: Preschool children's scripts and plans for familiar events. *Child Development, 66*(4), 984–998. doi:10.1111/j.1467-8624.1995.tb00917.x

Lagattuta, K. H. (2014). Linking past, present, and future: Children's ability to connect mental states and emotions across time. *Child Development Perspectives, 8*(2), 90–95. doi:10.1111/cdep.12065

Landis, J. R., & Koch, G. G. (1977). The measurement of observer agreement for categorical data. *Biometrics, 33*(1), 159–174. doi:10.2307/2529310

McPherson, G. E. (2005). From child to musician: Skill development during the beginning stages of learning an instrument. *Psychology of Music, 33*(1), 5–35. doi:10.1177/0305735605048012

Naito, M. (2003). The relationship between theory of mind and episodic memory: Evidence for the development of autonoetic consciousness. *Journal of Experimental Child Psychology, 85*(4), 312–336. doi:10.1016/S0022-0965(03)00075-4

Perner, J., Kloo, D., & Gornik, E. (2007). Episodic memory development: Theory of mind is part of re-experiencing experienced events. *Infant and Child Development, 16*(5), 471–490. doi:10.1002/icd.517

Prencipe, A., & Zelazo, P. D. (2005). Development of affective decision making for self and other: Evidence for the integration of first- and third-person perspectives. *Psychological Science, 16*(7), 501–505. doi:10.1111/j.0956-7976.2005.01564.x

Quon, E., & Atance, C. (2010). A comparison of preschoolers' memory, knowledge, and anticipation of events. *Journal of Cognition and Development, 11*(1), 37–60. doi:10.1080/15248370903453576

Redshaw, J., & Suddendorf, T. (2013). Foresight beyond the very next event: Four-year-olds can link past and deferred future episodes. *Frontiers in Psychology, 4*, 404–0. doi:10.3389/fpsyg.2013.00404

Russell, J., Alexis, D., & Clayton, N. (2010). Episodic future thinking in 3- to 5-year-old children: The ability to think of what will be needed from a different point of view. *Cognition, 114*(1), 56–71. doi:10.1016/j.cognition.2009.08.013

Scarf, D., Gross, J., Colombo, M., & Hayne, H. (2013). To have and to hold: Episodic memory in 3- and 4-year-old children. *Developmental Psychobiology, 55*(2), 125–132. doi:10.1002/dev.21004

Schacter, D. L., & Addis, D. R. (2007). The cognitive neuroscience of constructive memory: Remembering the past and imagining the future. *Philosophical Transactions of the Royal Society B: 2007 2087*. doi:10.1098/rstb.2007.2087

Schacter, D. L., Addis, D. R., Hassabis, D., Martin, V. C., Spreng, N., & Szpunar, K. K. (2012). The future of memory: Remembering, imagining, and the brain. *Neuron, 76*(4), 677–694. doi:10.1016/j.neuron.2012.11.001

Slobodoa, J. A., Davidson, J. W., Howe, M. J. A., & Moore, D. G. (1996). The role of practice in the development of performing musicians. *British Journal of Psychology, 87*(2), 287–309. doi:10.1111/j.2044-8295.1996.tb02591.x

Spencer, T. (2011). Learning to read in the wake of reform: Young children's experiences with scientifically based reading curriculum. *Perspectives on Urban Education, Spring, 8*(2), 41–50.

Spreng, R. N., & Grady, C. L. (2010). Patterns of brain activity supporting autobiographical memory, prospection, and theory of mind, and their relationship to the default mode network. *Journal of Cognitive Neuroscience, 22*(6), 1112–1123. doi:10.1162/jocn.2009.21282

Spreng, R. N., Mar, R. A., & Kim, A. S. N. (2009). The common neural basis of autobiographical memory, prospection, navigation, theory of mind, and the default mode: A quantitative meta-analysis. *Journal of Cognitive Neuroscience, 21*(3), 489–510. doi:10.1162/jocn.2008.21029

Sterelny, K. (2012). *The evolved apprentice: How evolution made humans unique.* Cambridge, MA: MIT Press.

Stout, D., Toth, N., Schick, K., & Chaminade, T. (2008). Neural correlates of Early Stone Age toolmaking: Technology, language and cognition in human evolution. *Philosophical Transactions of the Royal Society B: Biological Sciences, 363,* 1939–1949. doi:10.1098/rstb.2008.0001

Suddendorf, T. (2013). *The gap – The science of what separates us from other animals.* New York: Basic Books.

Suddendorf, T., Brinums, M., & Imuta, K. (in press). Shaping one's future self – The development of deliberate practice. In S. B. Klein, K. Michaelian, & K. K. Szpunar (Eds.), *Seeing the future: Theoretical perspectives on future-oriented mental time travel.* London: Oxford University Press.

Suddendorf, T., & Busby, J. (2005). Making decisions with the future in mind: Developmental and comparative identification of mental time travel. *Learning and Motivation, 36*(2), 110–125. doi:10.1016/j.lmot.2005.02.010

Suddendorf, T., & Corballis, M. C. (1997). Mental time travel and the evolution of the human mind. *Genetic, Social, and General Psychology Monographs, 123,* 133–167.

Suddendorf, T., & Corballis, M. C. (2007). The evolution of foresight: What is mental time travel, and is it unique to humans? *Behavioural and Brain Sciences, 30*(3), 299–351. doi:10.1017/S0140525X07001975

Suddendorf, T., & Moore, C. (2011). Introduction to special issue: the development of episodic foresight. *Cognitive Development, 26*(4), 295–298. doi:10.1016/j.cogdev.2011.09.001

Suddendorf, T., Nielsen, M., & von Gehlen, R. (2011). Children's capacity to remember a novel problem and to secure its future solution. *Developmental Science, 14*(1), 26–33. doi:10.1111/j.1467-7687.2010.00950.x

Suddendorf, T., & Redshaw, J. (2013). The development of mental scenario building and episodic foresight. *Annals of the New York Academy of Science: The Year in Cognitive Neuroscience, 1296,* 135–153.

Szpunar, K. K. (2010). Episodic future thought: An emerging concept. *Perspectives on Psychological Science, 5*(2), 142–162. doi:10.1177/1745691610362350

Taylor, M., Esbensen, B. M., & Bennett, R. T. (1994). Children's understanding of knowledge acquisition: The tendency for children to report that they have always known what they have just learned. *Child Development, 65*(5), 1581–1604.

Ward, P., Hodges, N. J., Williams, A. M., & Starkes, J. L. (2004). Deliberate practice and expert performance: Defining the path to excellence. In A. M. Williams & N. J. Hodges (Eds.), *Skill acquisition in sport: Research, theory and practice* (pp. 231–258). London: Routledge.

Wechsler, D. (2002). *Wechsler preschool & primary scale of intelligence* (3rd ed.). San Antonio, TX: The Psychological Corporation.

Autonoetic consciousness: Reconsidering the role of episodic memory in future-oriented self-projection

Stanley B. Klein

Department of Psychological and Brain Sciences, University of California, Santa Barbara, CA, USA

Following the seminal work of Ingvar (1985. "Memory for the future": An essay on the temporal organization of conscious awareness. Human Neurobiology, 4, 127–136), Suddendorf (1994. The discovery of the fourth dimension: Mental time travel and human evolution. Master's thesis. University of Waikato, Hamilton, New Zealand), and Tulving (1985. Memory and consciousness. Canadian Psychology/PsychologieCanadienne, 26, 1–12), exploration of the ability to anticipate and prepare for future contingencies that cannot be known with certainty has grown into a thriving research enterprise. A fundamental tenet of this line of inquiry is that future-oriented mental time travel, in most of its presentations, is underwritten by a property or an extension of episodic recollection. However, a careful conceptual analysis of exactly how episodic memory functions in this capacity has yet to be undertaken. In this paper I conduct such an analysis. Based on conceptual, phenomenological, and empirical considerations, I conclude that the autonoetic component of episodic memory, not episodic memory per se, is the causally determinative factor enabling an individual to project him or herself into a personal future.

The central nervous system enables its owner to prepare for contingencies that experience suggests will probably be encountered (e.g., Klein, Cosmides, Tooby, & Chance, 2002; Pezzulo, 2008; Suddendorf & Corballis, 1997). Such anticipatory orientation clearly is an adaptive priority: Confronted with the uncertainties that inevitably attend one's environment—even those possessing considerable structure and order—flexible, adaptive strategies greatly benefit an organism's survivability and hence its reproductive success (e.g., Klein, 2013a; Klein, Robertson, & Delton, 2010; Suddendorf, 1994; Tulving, 2005). The temporal scope and imaginative complexity of one's ability to anticipate and plan for contingencies that cannot be known with certainty is an obvious target for natural selection (e.g., Bischof-Koehler, 1985; Klein, Cosmides, et al., 2002; Suddendorf, 2013; Suddendorf & Corballis, 1997).

MEMORY AND FUTURE-ORIENTED MENTAL TIME TRAVEL: A VERY BRIEF HISTORY

Not surprisingly, psychology has taken a strong interest in future-oriented abilities. Attention to the effects (both adaptive and maladaptive) of temporal orientation on behaviour was in full display from the 1940s through the late 1960s. Much of this consisted in evaluating the effects of subjective temporality on variables of concern primarily to clinical, developmental, personality/

social psychologists (e.g., goals, motivation, personality, psychopathology; for a review see Cottle & Klineberg, 1974). However, by the 1970s interest had waned—work that remained largely focused on questions pertaining to estimations of the duration of objective temporal intervals (a topic whose origins trace to the birth of psychophysics; for review see Fraisse, 1963).

Following a period of relative neglect, the psychology of subjective temporality was reinvigorated by three largely conceptual meditations—Tulving (1985), Ingvar (1985), and Suddendorf (1994). However, the questions addressed had changed: Influenced by the theoretical commitments of the "cognitive revolution", inquiry now was trained on the relation between memory and future-oriented thought. In short order several empirical papers followed (e.g., Dalla Barba, Cappelletti, Signorini, & Denes, 1997; Klein, Loftus, & Kihlstrom, 2002; Williams, Ellis, Tyers, Healy, Rose, & MacLeod, 1996). It was not long before research on memory and what had come to be called future-oriented mental time travel (FMTT) had grown into a thriving enterprise: By the close of the first decade of the new millennium well over one hundred articles had appeared in scholarly venues. Interest shows no sign of abating—as attested to by the increasing pace and broadening scope of the questions being asked, as well as special issues (e.g., the present collection), symposia (e.g., Society for Personality and Social Psychology [SPSP], 2015), and edited volumes (e.g., Michaelian, Klein, & Szpunar, forthcoming).

The role of memory in imagining the future

Given the hindsight provided by contemporary perspectives on neurobehavioural functionality, it seems obvious that our capacity to anticipate and plan is underwritten by access to memories that are relevant to situational demands. However, the link between these two opposite-facing temporal faculties has a very long, fluctuating history. Memory, initially accorded a position of prominence in FMTT, subsequently was deposed. It would be 2000 years before it reclaimed that place of distinction.

The earliest known writing on the relation between memory and subjective temporality dates from the eighth century BC. In his Theogony (West, trans. 1988), Hesiod mentions that one's ability to subjectively transcend objective time is made possible by the faculty of human memory. Whatever currency this idea had on Greek thought in antiquity largely is unknown (what remains of Ionian and Greek philosophical thought from this period largely consists in fragmentary records). What is known is that Hesiod's views subsequently were silenced by the imposing voice of Dante's "master of them who know"—Aristotle (384–322 BC). In his monumental treatment of memory—De Memoria—Aristotle is adamant that the future is known by acts of anticipation, not by acts of memory. "The object of memory is the past" (cited in Sorabji, 1972, p. 13).

Aristotle's pronouncement dominated the intellectual landscape for approximately two millennia (e.g., Coleman, 1992; Klein, 2013a). We thus find Augustine of Hippo (354–430 AD) declaring: "The time of present things past is memory, the time of present things present is direct experience and the time of present things future is expectation" (The Confessions, 1997, Book 11, chapter 20, heading 26). Although scholastic authorities of the Middle Ages proposed emendations to the concept, memory's relation to the past was secure, serving as a stable resting place for scholarly discourse (for reviews see Coleman, 1992; Klein, in press).

It was not until the nineteenth century that memory attained the status of an object of scientific inquiry (e.g., Ebbinghaus, 1885), providing a new perspective from which to evaluate the conceptual warrant of the "received doctrine". Bradley (1887), influenced by Darwinian principles of natural selection, adopted a stance diametrically opposed to the one that had dominated discourse for nearly 2000 years: Rather than saddle memory to the past, he proposed (echoing the long-forgotten insights of Hesiod, though probably for different reasons)—that memory must, of adaptive necessity, be oriented toward the future.

Bradley's (1887) observations proved prophetic—though a rapprochement between memory and

FMTT had to wait another 100 years. However, by the mid-1980s psychologists had begun to consider a radical possibility: that the evolved function of memory was to focus thought and behaviour on the future rather than on the past (for reviews see Boyer, 2009; Klein, 2013b; Tulving, 2005). Within this framework, one particular type of memory—episodic—was taken to play a foundational role in most forms of FMTT (for recent reviews see Addis & Schacter, 2012; Klein, 2013a, 2013b; Schacter et al., 2012; Suddendorf & Corballis, 2007; Szpunar, 2010).[1]

THE ROLE OF EPISODIC MEMORY IN CONTEMPORARY TREATMENTS OF FMTT

As the connection between episodic memory and FMTT became widely accepted, scholarly treatments—impressive both in quantity and in diversity—began to populate the academic landscape. They included, but were not limited to, examination of neural correlates, developmental trajectory, evolutionary considerations, the specificity and detail of imagined scenarios, and psychopathological (e.g., amnesia, schizophrenia, and depression) implications (for a recent review see Klein, 2013a). Such terms as "episodic future thought" (Atance & O'Neill, 2005; Race, Keane, & Verfaellie, 2011; Schacter & Addis, 2007; Szpunar & McDermott, 2008) "episodic simulation/construction" (Addis, Cheng, Roberts, & Schacter, 2011; Hassabis & Maguire, 2007; Schacter, Addis, & Buckner, 2008), "episodic

self-projection" (e.g., Buckner & Carroll, 2007), and "episodic foresight" (e.g., Attance & Sommerville, 2014; Suddendorf, 2010) became the lingua franca of the field.[2]

One possible explanation for the tight focus on episodic memory is that—with two notable exceptions (Atance & O'Neill, 2001; Klein, Loftus, et al., 2002)—initial research and theory examined only the effects of episodic memory on FMTT (Dalla Barba et al., 1997; Suddendorf & Corballis, 1997; Tulving, 1985; Wheeler, Stuss, & Tulving, 1997; Williams et al., 1996). Another possibility (discussed in the section "The problem of conceptual underspecification and theory construction") is that a relation between episodic memory and FMTT simply makes sense (i.e., it has assumed the role of a scientific precommitment—that is, an unstated, but intuitively plausible, presumption that plays a formative role in the questions we pose to nature).

Given this state of affairs, a careful conceptual analysis of exactly what the term "episodic memory" picks out would seem basic to any treatment bestowing on it a position of causal prominence. Unfortunately, most papers on FMTT appear to take for granted that its definitional status is sufficiently well established that explicit explication is unnecessary (most work relies—either explicitly or implicitly—on Tulving's pre-1985 treatment of the construct). Accordingly, before I tackle the relation between episodic memory and FMTT, it would seem prudent to provide an up-to-date treatment of what the term "episodic memory" references.

---

[1]However, review papers published toward the end the first decade of the 21st century voiced concern over the possibility that episodic exclusivity might be an unnecessary constraint on the memorial underpinnings of FMTT (e.g., Addis & Schacter, 2012; Klein, 2013a; Irish et al., 2012; Kwan et al., 2012). In fact, a considerable number of recent publications (reviewed in Klein, 2013a) provide clear support for Klein, Loftus, and Kihlstrom's (2002) demonstration that semantic memory also underwrites certain forms (mostly nonpersonal) of FMTT.

[2]Theoretical and investigative attention remains largely trained on the contributions of episodic memory to future-oriented thought and behavior—despite increasing evidence for the role played by a range of recently evolved and late-developing cognitive capacities (e.g., systems of knowledge, executive function, scene construction, temporal self-projection, imagination; e.g., Arzy, Collette, Ionata, Fornari, & Blanke, 2009; Craver, Kwan, Steindam, & Rosenbaum, 2014; Irish, Addis, Hodges, & Piguet, 2012; Irish & Piguet, 2013; Kwan et al., 2012; Maguire & Mullally, 2013; Manning, Denkova, & Unterberger, 2013; Mullaley, Vargha-Khadem, & Maguire, 2014; Schacter et al., 2012, Suddendorf, 2010; Zeithamova, Schlichting, & Preston, 2012).

So, what is episodic memory?

As initially conceptualized, episodic memory was held to provide its owner with a record of the temporal, spatial, and self-referential features of the context in which learning originally transpired. Semantic memory (the other component of the declarative system of long-term memory), by contrast, lacked these features: Its offerings were experienced as knowledge devoid of the contextual elements surrounding its acquisition (e.g., Tulving, 1972, 1983).

An obvious implication of this distinction was that these two types of memory are associated with different temporal phenomenology. Episodic memory, in virtue of its contextual properties, makes it possible for an occurrent mental state to be directly experienced as a re-presentation of events that occurred in one's past. By contrast, the temporal experience associated with semantic memory is restricted to the "here and now": Memorial content is given to awareness as present. Though one can know this content was acquired in the past via an act of inference, a prereflective feeling of reexperiencing the act of acquisition is not part of its given presentation.

These temporal distinctions were fully appreciated by Tulving, and in 1985 he made them the basis for distinguishing between episodic and semantic memory (Tulving, 1985; see also Tulving, 2002, 2005; Wheeler et al., 1997). Focusing attention on the type temporal subjectivity present at retrieval, Tulving proposed that episodic memory is characterized by autonoetic consciousness, while semantic memory entails a form of consciousness he labelled noetic (e.g., Szpunar & Tulving, 2011; Tulving, 1985, 2002, 2005; Wheeler et al., 1997).

Autonoesis, noesis, and mental time travel

Mental time travel refers to the possibility that a first-person perspective can be located at subjective times other than the present. It is manifest in memory when (a) one remembers a past happening as if one were experiencing it again, and (b) in anticipation when one projects oneself into a future experience (for example by imagining what X will be like).

Borrowing terminology from the writings of early twentieth century phenomenologists, Tulving (1985) drew a distinction between two modes of consciousness, which he called autonoetic and noetic (Tulving also identified a mode of consciousness—which he called anoetic—but since it is held to play no role in subjective temporality, it is not discussed herein). A person who possesses autonoetic consciousness "is capable of becoming aware of her own past as well as her own future; she is capable of mental time travel, roaming at will over what has happened as readily as over what might happen, independently of physical laws that govern the universe" (Tulving, 1985, p. 5).

In its role in episodic memory, Tulving described autonoetic consciousness as enabling one to revisit earlier experience, " ... a unique awareness of re-experiencing here and now something that happened before, at another time and in another place" (Tulving, 1993, p. 68; for related views see Suddendorf & Corballis, 1997, 2007; Szpunar, 2010; Wheeler et al., 1997). For Tulving, autonoesis is a source of a proprietary phenomenology: "It is autonoetic consciousness that confers the special phenomenal flavor to the remembering of past events, the flavor that distinguishes remembering from other kinds of awareness, such as those characterizing perceiving, thinking, imagining, or dreaming" (Tulving, 1985, p. 3). Importantly for our purposes (see the section, "Autonoetic consciousness is not intrinsic to episodic memory"), autonoesis "does *not* reside in memory traces as such; it emerges as the phenomenally *apprehended* product of the episodic memory system . . . in ways that are as mysterious as the emergence of other kinds of consciousness from brain activity" (Tulving, 2005, p. 17; emphasis added).

Tulving distinguished autonetic from the noetic form of consciousness. Noetic consciousness "allows an organism to be aware of, and to cognitively operate on, objects and events, and relations among objects and events, in the absence of these objects and events" (Tulving, 1985, p. 3). An individual whose memorial experience is noetic retrieves information " . . . in the absence of a feeling of re-experiencing the past" (Szpunar, 2010, p. 144).

Thus, noetic consciousness does not provide its owner with a subjective feeling that she or he is mentally traveling back in time to the events and experiences that gave birth to the content in awareness. She or he may infer from subsequent analysis that content given to awareness (e.g., "I know that I saw Jimi Hendrix in concert when I was in High School") refers to the past (e.g., "Although I cannot recollect being at the concert, I know that I was in High School between 1966 and 1970— so I must have seen him in the late 1960s"), or, by logical implication, that the content in awareness must have "come from somewhere" (e.g., "I know that the sun is approximately 93 million miles from earth. I must have learned this fact at some point in my past, though I no longer remember where or when. Most likely in Junior High School"). But these are acts of inference and interpretation contingently joined to the noetic state (provided the individual chooses, or is motivated, to construct such linkage). By contrast, the feeling of subjective time travel is intrinsic to autonoesis—that is, it is prereflectively given, requiring no additional conceptual gymnastics for its realization in awareness.

This is not to say that noesis provides no basis for mental time travel. On the contrary, as Klein, Loftus, et al. (2002) have shown, noetic consciousness enables a form of mental time travel (which the authors called "known time"—i.e., an appreciation of time as chronology) in which temporal knowledge is the product of inferential or interpretive acts, rather than presented directly to awareness (as is the case with autonoesis; for extensive discussion, see Klein, 2014b, in press).

Until recently, noetic forms of mental time travel had received relatively little empirical and theoretical attention. However, in the past 5 years it has become clear that certain forms of mental time travel—particularly ones that enable a person to consider the future chronologically rather than in terms of personal preliving—are enabled noetic consciousness accompanied by interpretive

temporal analyses (for recent reviews see Addis & Schacter, 2012; Klein, 2013a).

Thus, of the modes of consciousness identified by Tulving, only autonoesis provides a prereflective feeling of personal temporal experience. It is directly given to awareness and does not require any further considerations or deliberations to justify one's feeling that the content in awareness is connected to the past or future (e.g., Klein, 2013a; Markowitsch, 2003; Tulving, 2005).

With regard to types of declarative long-term memory, autonoetic and noetic consciousness appear isomorphic with the temporal commitments assumed to characterize episodic and semantic memory, respectively (e.g., Klein, 2013a; Markowitsch, 2003; Tulving, 1985, 2002; Wheeler et al., 1997). For present purposes, the crucial point is that FMTT researchers, influenced by the autonoetic properties of episodic recollection, typically have assumed that episodic, not semantic, memory underpins our ability to imagine the future (see the section "The role of episodic memory in contemporary treatments of FMTT"). While recent work suggests that this exclusivity of focus is overly restrictive (e.g., Footnote 1; for reviews see Addis & Schacter, 2012; Klein, 2013a, 2014b), the more fundamental issue is whether episodic memory plays any role in FMTT. In this paper I argue that evidence for such a determinative role largely is lacking. Instead, as I hope to show, a strong case can be made for the proposition that the autonoetic component of episodic memory (rather than episodic memory per se) is the causally relevant player in projection into personal future scenarios.

Autonoesis and episodic memory

The reformulation of episodic and semantic memory in terms of temporal subjectivity avoids a number of messy findings that, over the years, have chipped away at the methodological warrant of relying on the temporal, spatial, and self-referential features of retrieved content to distinguish systems of memory (a practice that continues to characterize memory research). For instance, the contention that episodic, but not semantic, memory entails a self-referential component has given way to the well-documented finding that semantic memory also can be self-referential (for reviews see Grilli & Verfaellie, 2014; Klein, 2004; Klein & Lax, 2010; Klein & Loftus, 1993; Renoult, Davidson, Palombo, Moscovich, & Levine, 2012).

In addition, semantic memory is fully capable of providing spatial and temporal information (e.g., "I know that John Lennon was born on 9th October 1940 in Liverpool, UK, although I no longer remember the occasion in which I acquired that knowledge"; for recent reviews see Grilli & Verfaellie, 2014; Klein & Gangi, 2010; Klein & Lax, 2010; Martinelli, Sperduti, & Piolino, 2013). Thus, the core constituents of episodic memory as initially proposed (time, space, and self) also can be found in semantic memory. Accordingly, there are neither logical nor evidential bases for asserting that these systems can be distinguished by analysis of memory content.[3]

This further is demonstrated by cases in which patients congenitally deprived of (or having lost access to) episodic memory, can be taught (or retaught) the temporal, spatial, and self-referential details of their life-narratives—albeit details lacking a feeling of temporally reexperiencing the events and circumstances that they reference (this phenomenological lacuna, of course, assumes that their pathology targets autonoetic consciousness rather than processes mediating the acquisition, storage, or retrieval of acquired content).

For example, patient J.V. suffered neural pathology resulting in the loss of premorbid personal content as well as autonoetic accompaniment, rendering him incapable of engaging in acts of episodic recollection (Stuss & Guzman, 1988). Nonetheless, he successfully relearned many temporal and spatial details of his personal past—although this content was experienced as factual knowledge (which it was!), rather than as a personal reliving. Thus, despite profound episodic impairment, J.V. was able to reacquire his personal narrative via intact semantic memory function.

A similar pattern of lost and relearned personal knowledge is seen in the case of patient M.L. (Levine et al., 1998). A brain trauma left M.L. densely amnesic for episodic memories predating his injury. Despite the severity of impairment, he was able to "relearn significant facts his own past" (Levine et al., 1998, p. 1956). However, that knowledge was not coupled with a feeling of reacquaintance with the act of acquisition. As expected, subsequent testing revealed that M.L.'s autonoetic consciousness was seriously compromised—thus explaining his inability to experientially refer relearned content to its point of origin.

Demonstrations of an intact ability to (re)acquire premorbid personal content in juxtaposition with autonoetic impairment is found scattered throughout the literature (e.g., Bindschaedler, Perer-Faver, Maeder, Hirsbrunner, & Clarke, 2011; Broman, Rose, Hotson, & Casey, 1997; Gadian et al., 2000; Guillery-Girard, Martins, Parisot-Carbuccia, & Eustache, 2004; Markowitsch & Staniloiu, 2013; for a review see Klein, in press). Although in many—though not all—cases, relearned material shows less detail than the original, this difference can be accommodated by consideration of the known effects of time in storage on content detail and specificity (for example, multiple trace theory; e.g., Nadel, Hupbach, Gomez, & Newman-Smith, 2012; Nadel & Moscovitch, 1997; for a related view see Dalla Barba, 2002). Although discussion would take us far afield, the relevance of multiple trace theory (and kindred treatments) to the issues at hand can be found in Klein (2013c).

Given these considerations, a conceptually, empirically, and phenomenologically nuanced

---

[3]It might be objected that other hallmarks presumed to characterize episodic memory (e.g., complexity and coherence) regularly are found when individuals describe memory experience. However, as is discussed in the sections "The system-neutrality of stored content" and "The causal foundation of FMTT: An argument for autonoesis", there is no rational or evidential basis for the expectation that episodic memory necessarily provides more complex and coherent content than semantic memory. As such, studies attempting to identify the contributions of episodic memory exclusively from an analysis of reported content suffer from the logical error of assuming in advance (e.g., contextual detail = episodic recollection) what they are attempting to demonstrate (e.g., episodic recollection = contextual detail).

definition of episodic memory can be stated. In its most simple form, episodic memory is a type of mental experience. More precisely, it is not the content of an experience, but the manner in which that content is experienced. That manner entails a special mode of temporal subjectivity—one that provides the experiencer with a phenomenological relation to his or her past not conferred by other forms of memory (these ideas receive fuller treatment in Klein, 2013c, in press).

On this view, episodic memory, though dependent on the integrity of a set of subexperiential processes (encoding, storage, and retrieval), is not their inevitable product. These same processes can give rise to a variety of mental experiences (e.g., thought, imagination, belief, desire, inference, plans, attitudes, hope, fear; for discussion, see Klein, in press). To qualify as an episodic memory, content must be subjected to autonoetic consciousness at retrieval (while it might appear that I am equating retrieval with episodic memory, this would be an incorrect reading. The act of retrieval is largely subexperiential. It is a process whose workings can result in memory, but also can result in nonmemorial states such as belief, thought, desire, and so on. It is the act of conjoining of retrieved content with a particular mode of temporal subjectivity that makes experienced content "memory content").

Episodic memory (i.e., recollection) thus consists in two separate, but mutually dependent, parts. First, to qualify as an act of episodic memory a mental state must be causally linked to experiences that the person formerly enjoyed. Second, episodic memory is not simply from the past; it is a special way of being about the past (e.g., Klein, 2013b). To qualify as an act of memory, the content in awareness must present itself as a reexperience of a previously entertained experience (e.g., Klein, 2013c; Markowitsch, 2003; Tulving, 1985, 2005; Wheeler, 2005; Wheeler et al., 1997). This feeling of reexperiencing is prereflectively given to awareness by a concomitant act of autonoesis at retrieval (for evidence, see the section "Autonoetic consciousness is not intrinsic to episodic memory"), rather than as the product of inference or interpretation (of course,

if inference subsequently were to evoke—in some unknown manner—autonoetic accompaniment, the content then would be taken as an act of recollection. By contrast if—as often is the case—inference resulted only in an analytic determination that occurrent content issued from past experience, the experience associated with that content would be noetic—that is, one of knowledge or belief, but not recollection).

In summary, the retrieval of content is not sufficient to make its experience an episodic recollection. To so qualify, content must be joined with a prereflective mode of temporal subjectivity (i.e., autonoesis). A practical extension of this position is that content analysis, absent consideration of the mode of temporality in which content is presented to awareness, does not provide a reliable basis for diagnosing the system of memory from which it issues (see Klein, 2013c, in press, and Footnote 3).

The system-neutrality of stored content

Based on these considerations, the presumption that content can be apportioned into episodic or semantic memory based on analysis of its referential properties and contextual detail is called to question. Specifically, on the view presented, there is no episodic system of memory as traditionally construed (e.g., Foster & Jelicic, 1999; Schacter & Tulving, 1994). Rather, there is learned content that is stored in a system-neutral format and is available at retrieval to a variety of experiential outcomes, only one of which is recollection (e.g., judging, categorizing, deciding, believing, imagining, desiring, intending, planning, thinking, recognizing, searching, navigating, hope, fear). The designation "episodic" meaningfully applies only after content has been conjoined with autonoetic consciousness during an act of retrieval (e.g., Klein, 2013c, in press).

In short, the position taken in this paper (see also Klein, 2013c, in press) is that there is no "episodic content" per se (see the sections under "The role of episodic memory in contemporary treatments of FMTT"). Rather, there is "content" that can be experienced in a mental state referred to as "episodic" provided that content is juxtaposed (at retrieval; see the section "Autonoetic consciousness is not intrinsic to episodic memory") with autonoetic consciousness. But the same (or very similar) content need not be indicative of episodic memory. For example, experienced content lacking autonoetic accompaniment can be taken as semantic knowledge despite having contextual features and details typically (but mistakenly) assumed diagnostic of episodic recollection (e.g., Klein, 2013c, in press; Klein & Nichols, 2013).

On this view, the predicate "episodic" used in conjunction with a variety of future-oriented thought (e.g., episodic foresight, episodic scene construction, episodic self-projection; see the section "The role of episodic memory in contemporary treatments of FMTT") is of questionable utility. What makes an occurrent mental state episodic (rather than, say, an act of thought or imagination) is that it enables a direct, noninferential feeling of reacquaintance with one's past (e.g., Klein, 2013c, in press; Tulving, 1985, 2002; Wheeler et al., 1997). It does this by linking system-agnostic content with past-oriented autonoetic consciousness during the act of retrieval (e.g., Klein, 2013c, in press).

This decidedly is not the phenomenology naturally associated with mental states involving planning and anticipating (although such phenomenology can be elicited when participants are given instructions to report memorial experience during lab-based investigations of FMTT; e.g., Anderson, 2012; Arnold, McDermott, & Szpunar, 2011). Under nonlaboratory conditions, the aspect of autonoetic consciousness elicited by acts of anticipation and planning often is 180 degrees displaced from that found when one undergoes recollective experience (positioning the person toward what will happen, not what previously transpired; e.g., Boyer, 2009; Klein, 2013b, 2014b; Tulving, 2005). Indeed, as is discussed in the section "Autonoetic consciousness is not intrinsic to episodic memory", it is not clear what adaptive function a temporal orientation toward the past (i.e., that associated with episodic recollection) serves with respect to future-oriented mentation.

THE PROBLEM OF CONCEPTUAL UNDERSPECIFICATION AND THEORY CONSTRUCTION

If we try to solve a problem by means of a notion that does not apply, we cannot help going wrong. (Descartes, 1970, p. 138)

As Heisenberg (1958/1999, p. 58) sagely observes "What we observe is not nature itself but nature exposed to our method of questioning". From this it follows that, "Asking the right question is frequently more than halfway to the solution of the problem" (Heisenberg, 1958/1999, p. 35).

The scientific method thus construed is more than simply posing questions to nature and waiting for her to "push back". To receive answers possessing the resolution necessary to fine-tune our understanding of the object of inquiry, the questions we ask must be the "right" ones. This requires careful analytic treatment of the issues of interest as well as nuanced consideration of the epistemic warrant of concepts receiving methodological consideration (e.g., Klee, 1997).

As noted in the section "The role of episodic memory in contemporary treatments of FMTT", investigation of the part played by episodic memory in FMTT has exploded over the past 15 years. Most of this work (though there are exceptions; e.g., Klein, 2013a; Maguire & Mullalley, 2013; Schacter et al., 2012; Suddendorf, 2010; Szpunar, 2010) has been characterized by a relatively tight experimental focus—emphasizing such questions as similarities and differences (both cognitive and anatomical) between episodic memory and episodic future thought, the developmental trajectory of our capacity to imagine the future, clinical impairments of this ability, the ability of animals to perform tasks requiring FMTT, and what we can learn about the neural substrates of FMTT from radiological analysis. In contrast, sustained treatment of key theoretical presuppositions (e.g., that reexperiencing of the past is the basis by which we imagine the future) has been left primarily to philosophers (e.g., Byrne, 2010; Cornish, 2011; Hoerl, 2008; Mathen, 2010).

One consequence of this apportioning of labour is that the epistemological warrant of our models of

FMTT remains underspecified. By offloading the task of detailed conceptual analysis to philosophers—who often lack a full appreciation of the complexity of the empiricism on which they train their analytic skills—theoretical structures are erected largely on the basis of empirical outcomes. In consequence, they often provide inadequate conceptual grounding for the principles they embody (for discussion see Klein, 2013a). Newell summarized the problem more than 40 years ago:

> As I examine the fate of our [empirical efforts], looking at those already in existence as a guide to how they fare and shape the course of science, it seems to me that clarity is never achieved. Matters simply become muddier and muddier as we go down through time. Thus far from providing the rungs of a ladder by which psychology gradually climbs to clarity, this form of conceptual structure leads rather to an ever increasing pile of issues, which we weary of or become diverted from, but never settle. (Newell, 1973, pp. 288–289; brackets added for expositional clarity)

It is my contention that the task of scrutinizing the theoretical precommitments that (often implicitly) guide our investigations of FMTT is (a) an essential, but relatively underappreciated, aspect of the process of formulating the "right" questions to address to nature, and (b) a task that needs to be undertaken by those occupying the experimental trenches—psychological FMTT investigators (clearly, a fully collaborative effort between psychology and philosophy would be ideal).

Rethinking the role of episodic memory in FMTT

In what follows, I focus on an assumption treated as virtually axiomatic in contemporary FMTT research—that is that episodic memory has a special causal potency in regard to future-oriented personal thought. As I hope to show, this assumption is more the product of reasonable stipulation than conceptual analysis.

For example, what adaptive advantage does the experiencing of reliving one's past (i.e., episodic memory) have for constructing future-oriented plans and scenarios? Wouldn't retrieved content known to be from one's past, but lacking a noninferential feeling of having previously been experienced (i.e., semantic memory), be as useful? If not, why not?

As Szpunar and Tulving (2011) argue, it is the autonoetic component of episodic memory that enables a person to travel backward and forward in time (e.g., Szpunar & Tulving, 2011). If that is the case, why not posit autonoesis, rather than episodic memory (which provides the experience of reliving, not preliving), as the causally determinative factor in FMTT?

Drawing on the theoretical considerations presented in the sections under "The role of episodic memory in contemporary treatments of FMTT", I next attempt to show that the connection between episodic memory and FMTT is based more on theoretical precommitments than conceptually and evidentially grounded argument. When such analysis is undertaken, I believe it becomes clear that it is autonoetic consciousness (at least that facet of autonoesis that enables one to imagine a personal future), not episodic memory, that is the causally relevant factor in most forms of FMTT.

THE CAUSAL FOUNDATION OF FMTT: AN ARGUMENT FOR AUTONOESIS

Detecting the footprints of episodic memory from the reported properties of memorial content—a standard tactic of FMTT research—is, as we have seen, fraught with interpretive difficulties (e.g., the section "So, what is episodic memory?"). First, content containing self-referential, temporal, and spatial properties can be associated with episodic and semantic memory (albeit the former is more likely to represent these contextual elements as they were experienced during the act of acquisition).

Second, the coherence and complexity of reported memory content—often employed as an index of episodicity (e.g., Anderson, 2012; Hurley, Maguire, & Vargha-Khadem, 2011; Race et al., 2011; Squire et al., 2010; for a discussion and critique see Arnold et al., 2011)—has neither rational nor empirical justification. While relearned personal histories (e.g., "So, what is episodic memory?") often—though not invariably—possess fewer details than episodic recollections, content complexity is an unreliable mark of memory status. The content of semantic memory can show considerable intricacy and narrative coherence (for example, knowing the rules for how to behave and what to expect in a restaurant). Conversely, episodic memory can yield content of extreme simplicity (for instance, recollecting a single word from a list; e.g., Gardiner, 2001).

Third, there is no obvious adaptive advantage to retrieving content conjoined with a directly given, prereflective feeling that it references a personal experience from one's past (i.e., episodic memory). Such knowledge can easily be gleaned from temporal markers embodied in the content (e.g., since I know I attended Stanford in the early 1970s, this is must be part of my past; e.g., the case of R.B. see next section) without a need for the additional experience of reliving. In short, it is unclear what part episodic memory (as opposed to content retrieved) plays in the formation of future-oriented plans and scenarios.

Autonoetic consciousness, by contrast, captures a fundamental aspect of the phenomenology associated with mental time travel (e.g., Arnold et al., 2011; Markowitsch & Staniloiu, 2011; Suddendorf, 1994; Szpunar & Tulving, 2011; Tulving, 1985, 2005; Wheeler et al., 1997; for reviews see Markowitsch, 2003; Tulving, 2005; Wheeler, 2005). In this paper I take the position that the enabling factor in FMTT is not the content provided by an act of retrieval, but rather the autonoetic consciousness that accompanies that content (in particular, the facet of autonoesis aimed at the future). While evidentiary grounding provided by stored content may be needed for certain forms of FMTT (reviewed in Klein, 2013a), it never is sufficient. In fact, for some forms of future-oriented mentation it is not required at all (e.g., Klein, 2013d).

Autonoesis, by contrast, is always necessary (at least for projection into a possible personal future; see the section "Autonoesis, noesis, and mental time travel"). Seen in this light, the case can (and will) be made that it is the autonoetic component of episodic memory, not episodic memory per se, that enables one to navigate a personal future. Accordingly, placing episodic memory in determinative juxtaposition with FMTT is an instance of trying "to solve a problem by means of a notion that does not apply". To make this case, however, I first need to show that autonoetic consciousness is a contingent rather than necessary feature of retrieved content.

Autonoetic consciousness is not intrinsic to episodic memory

So, in what does relation between autonoetic consciousness and episodic memory consist? One possibility is that autonoetic consciousness is intrinsic to "episodic content". On this view, episodic memory is the outcome of retrieving autonoetically endowed content. In contrast, a relational interpretation holds that the association between autonoesis and episodic "content" is a matter of contingency (i.e., circumstance) rather than (bio)logical necessity. On this view, the bond between content and autonoesis (resulting in an episodic memory experience) is forged at retrieval.

The available evidence, though not plentiful, favours the relational view—that is, what makes an experience a memory experience is not the nature of the content given to awareness, but the mode of consciousness associated with that content during its retrieval. Consider, for example, the case of patient R.B. (e.g., Klein, 2013b, 2013c; Klein & Nichols, 2013). Following an automotive accident, R.B. exhibited a very rare —though not unique (e.g., Lane, 2012; for review see Klein, 2014a)—memory problem: While fully capable of describing events from his life with the rich contextual detail traditionally associated with episodic recollection, he did not experience this content as episodic memory. Rather, lacking the warmth, intimacy (e.g., James, 1890), and feeling of reliving associated with recollection, it was felt to be known from a third-person perspective. As is seen below, R.B.'s impairment appears to have compromised neither his autonoetic ability (he was not stuck in time and was able to formulate detailed plans) nor stored content (he could produce richly detailed representations of his past). Instead, what appear to have come undone were the mechanisms that enable autonoetic consciousness to bond with retrieved content, making possible recollective experience (for discussion see Klein, 2014a).

For example, in response to a request to remember a specific time involving his experiences while a student at Massachusetts Institute of Technology (MIT), R.B. replied:

When I remember the scene with my friends, studying, I remember myself walking into the room . . . and . . . other things I did and felt [details are recounted]. But it feels like something I didn't experience . . . [like something I] was told about by someone else. It's all quite puzzling.

He continues:

I can picture the scene perfectly clearly . . . studying with my friends in our study lounge . . . but it has the feeling of imagining . . . like something my parents [might have] described from their college days. It [the memory] not feel like it was something that really had been a part of my life. Intellectually I suppose I never doubted that it was . . . perhaps because there was such continuity of memories that fit a pattern that lead up to the present time. But that in itself did not help change the feeling of [lost] ownership.

Asked to describe childhood memories, R.B. responded:

I . . . [am] remembering scenes, not facts . . . I am recalling scenes . . . that is . . . I can clearly recall a scene of me at the beach in New London with my family as a child [he then describes the scene in rich contextual detail]. But the feeling is that the scene is not my memory . . . as if I was looking at a photo of someone else's vacation.

Memory of recent events showed a similar dissociation between content and feelings of reliving:

I remember eating pizza at XXX in Isla Vista about a month before [his accident], but the memory belongs to someone else. But knowing I like pizza in the present . . . now . . . is owned by me . . . when I recall memories from my past I intellectually know they are about me. It just does not feel like it . . . when I remember scenes from before [the accident] they do not feel as if they happened to me—though intellectually I know they did.

R.B.'s memory reports (treated more fully in Klein, 2013c, and Klein & Nichols, 2013) show all the presumed characteristics of episodic recollection, save one important thing—they lack autonoetic accompaniment. They contain detailed temporal, spatial, and self-referential elements that correctly track

(all of R.B.'s memories were substantiated by third parties) the manner in which the original learning transpired. What is missing are (a) the feeling that the content present in awareness is a reliving of what previously took place (i.e., recollections), and (b) a nonanalytic confidence that the events remembered actually did take place (for discussion of confidence/certainty and episodic recollection, see Klein, 2014b). Rather, R.B. treated retrieved content as things he simply knew or believed he should know (R.B. reports he relies on inferential processes to decide whether content in awareness could be something he personally experienced). It is important to note that R.B. eventually recovered his ability to conjoin content with autonoetic consciousness, at which time the "same" content now was experienced as a recollection.

While this case stimulates a host of fascinating questions about self and memory (some of which are addressed in Klein, 2012, 2013d, 2014a), the important points for present purposes are (a) autonoetic consciousness is not an intrinsic property of retrieved content (see also Tulving, 2005), (b) content that contains all of the criterial features and richness of detail associated with episodic recollection can be present in awareness, yet not experienced as a personal reliving, and (c) the same (or largely indistinguishable) content can be taken as "inferentially from my past" or "directly from my past" depending on the functional integrity of the mechanisms (presently unknown) that conjoin experienced content and autonoetic consciousness at retrieval.

TAKING STOCK: A BRIEF SYNOPSIS OF EPISODIC MEMORY, AUTONOESIS, AND FMTT

As the evidence presented hopefully makes clear, there is no logical argument or empirical support for the idea that the "who, where, and when" of past experience is unique to episodic memory. While the fact that episodic and semantic memory share properties is not a "death sentence" for partitioning them into distinct categories, it highlights the difficulties faced by investigators who rely on "time, place, and self" as the basis for classification. Simply put, these criteria are insufficient to the task for which they (too) frequently have been enlisted.

By contrast, the autonoetic/noetic criterion (and its assessment by "remember/know" tasks; e.g., Gardiner, 2001; Tulving, 1985) captures a fundamental feature of memory phenomenology, providing a rationally sound and empirically grounded means for identifying types of memorial experience. A strong implication of the relational view (see the section "Autonoetic consciousness is not intrinsic to episodic memory") is that prior to (or in the absence of) a concurrent act of autonoesis, retrieved content is system-neutral. Depending on the type of subjective temporality associated with content during retrieval, the same content can be experienced as episodic or semantic memory (for extended discussions see Klein, 2013c, in press). An implication of this view is that, prior to retrieval, content is subjectively atemporal: While it may contain chronological referents—for example, I saw Hendrix when I was in High School—this information is derivative, acquired in virtue of subsequent analysis rather than prereflectively given.

Since some forms of FMTT travel do not require access to previously acquired content (e.g., personal diachronicity; Klein, 2013d), atemporal, system-agnostic content cannot be necessary for FMTT in all its manifestations.[4] However, even when content is required, in the absence of a sense of subjective temporality it is left without temporal compass. In short, it is the future-oriented aspect of autonoetic consciousness, not the act of recollection (i.e., content + past-oriented autonoesis), that serves as the platform from which we project ourselves into a personal future.

AUTONOESIS, NOT EPISODIC MEMORY, ENABLES OUR ABILITY TO TRAVEL INTO THE PERSONAL FUTURE: EMPIRICAL FINDINGS

As I hope to have shown, a variety of theoretical considerations provide traction for the position that autonoetic consciousness—not the episodic memories in which it normally manifests—provides the neurocognitive scaffolding necessary to navigate one's future. More, it is not autonoesis in toto, but rather that aspect that takes the future as its temporal pole.

This is not to imply that episodic memory cannot play any role in FMTT. Rather, it means that episodic memory is not a necessary constituent of most forms of temporal self-projection (e.g., Klein, 2013d) and that the term "episodic", used in reference to future-oriented mentation, does more to obfuscate than illuminate the neurocognitive operations mediating future-oriented imaginings involving the self (Mathen, 2010).

Some support for these ideas comes from a recent study by Klein, Robertson, and Delton (2010). Participants were shown a list of objects (e.g., matches, television) and were asked to decide whether these were objects (a) they remembered taking on a previous camping trip (the "episodic memory" condition; note that participants all were pretested to ensure they had clear recollections of camping), (b) likely to be found on a generic camping trip (the "semantic memory" condition), or (c) they might plan to take in preparation for a future camping excursion (the "FMTT" condition).

While participants in all three conditions identified the same subset of objects as camping-relevant, there were important mnemonic differences—for example, a subsequent test of memory

---

[4]To fully appreciate the temporal commitments of FMTT and the diversity of its manifestations, one must recognize the difference between temporal experience conceived as a constant flow from future to present to past, with temporal designators continually changing ontological status (e.g., what once was future now is present, what once was present now is past, etc.), and temporal experience as a fixed, earlier–later (or before–after) chronology in which temporal placement of an event is invariant (e.g., 4th April 1982 is, and always will be, prior to 4th April 1983). These two modes of temporal conceptualization are not logically reducible, one to the other (e.g., Loizou, 1986; McTaggart, 1908; for an opposing view, see Cornish, 2011). Moreover, they map reasonably well onto the types of FMTT assumed to depend on autonoetic and noetic consciousness, respectively. Fuller discussion can be found in Klein (2013a) and Klein, Loftus, and Kihlstrom (2002) as well as Dalla Barba (2002).

for the objects they made decisions about revealed that participants in both the episodic and semantic conditions differed from those in the FMTT condition both in amount and type of items recalled (e.g., the episodic and semantic memory conditions produced statistically equivalent recall, but both recalled significantly fewer items than did participants in the FMTT condition). Additional differences between past (e.g., episodic memory) and future (e.g., episodic foresight) oriented mentation also are discussed at length by Suddendorf (2010).

The finding that a number of observable differences emerge when participants are required either to access episodic memory or to plan for the future shows that episodic recollection and FMTT are not coextensive. Of course, FMTT still may require recollective acts, but its realization may be layered with additional processes that account for the obtained dissimilarities. Accordingly, such results do not constitute a knock-down against an episodic basis for FMTT. However, neither do they offer any support. At best, the findings presented by Klein et al. (2010) sanction the conclusion that, under the experimental conditions utilized, episodic recollection shows greater affinity to semantic processing than to future-oriented processing with regard to measures of retention. Fortunately, findings from individuals suffering impairments of recollective ability provide additional reasons to question the role of episodic memory in FMTT.

Data from patients suffering episodic memory impairment

One useful way to place the constituents of a system (e.g., FMTT) on view is to examine them when the system to which they belong has broken down. A system's constituents—normally masked by the fluid manner in which they work together to affect a common end—are laid bare as the whole of which they are part unravels (e.g., Klein, Rozendal, & Cosmides, 2002; Rosenbaum, Gilboa, & Moscovitch, 2014). Accordingly, in what follows I draw on evidence from individuals suffering impairments of episodic memory and autonoetic consciousness.

A unique perspective on the relation between autonoesis, episodic memory, and FMTT is provided by (rare) occasions in which autonoetic consciousness and mental content remain intact but their ability to bond at retrieval is compromised (for extended discussion, see Chapter 5 of Klein, 2014a). Under these circumstances, the autonoetic model of FMTT predicts that the patient's ability to construct plans for his or her future should remain intact despite the loss of episodic recollective ability (presumably resulting from an inability to forge a connection between autonoetic consciousness and retrieved content).

This is exactly what is found. For example, patient R.B. (see the section "Autonoetic consciousness is not intrinsic to episodic memory") had no difficulty forming highly detailed, often personal, plans (Klein & Nichols, 2013). For instance, R.B. reports:

During the "un-owned" period I was able to plan for the future. Although my working memory impairment). . . made it challenging. When I slowly returned to work, it was hard to plan a complex strategy. I had to think of useful things to do and then do them. The best compensation I found was to separate the planning of the strategy from the execution. It worked best if I made a list of "Things To Do".

R.B. thus maintained access to content (often self-referential), but this content, broken free of its autonoetic moorings, was unable to be realized as episodic recollection. Nonetheless, due to his intact autonoetic ability, he was capable of constructing personally relevant scenarios to guide future thought and behaviour (although issues with working memory made it a challenge). In R. B. we clearly see the enabling effects of autonoetic consciousness on FMTT in the face of a virtually complete breakdown in recollective experience.

Consider next the case of Zasetsky, a Russian soldier in World War II. As a result of battle, he suffered massive neural damage to areas controlling higher cortical functions such as the analysis, synthesis, and organization of complex associations: He was rendered aphasic, perceptually and proprioceptively disoriented, and hemianopic. Most relevant for present purposes, he also experienced total impairment of both anterograde and retrograde episodic memory function (though he maintained some semantic function). Eventually, under the tutelage of Luria and others, Zasetsky slowly and painfully regained a rudimentary ability to read, write, and perform basic bodily functions. Consequently, he was able to provide a record of his thoughts and feelings, eventuating in a book documenting his experiences (Luria, 1972).

Although there are many remarkable aspects of this case study, I focus on one with direct relevance to the topic at hand. Despite monumental episodic memory dysfunction, Zasetsky maintained the ability (and desire) to plan for his personal future. He was aware of his deficits and was greatly troubled by their effects. To address his misfortune, he formulated clear goals to improve his situation and expressed unmistakable motivation to carry them forward. Indeed, it was his intact ability to imagine himself in a better life that gave him the strength to undertake the arduous rehabilitative programme that eventually made it possible for him to regain partial contact with the external world.

In short, though lacking episodic memory, Zasetsky clearly was not stuck in the present. The future was real for him, and he went to considerable efforts to ensure that it would be more congenial than the situation in which he found himself after battle. Here we have another case of an individual who, lacking episodic memory (and, based on Luria's observations, likely to have serious issues with content availability and/or accessibility) nonetheless could orient toward and plan for a personal future by drawing on the meagre cognitive resources he still possessed conjoined with intact future-oriented autonoetic consciousness.

Accessibility to stored content absent autonoetic accompaniment compromises one's ability to imagine and plan for the personal future

Amnesic patient H.M. provides a similar take-away message, albeit for rather different reasons. As a result of a surgical resection of his medial temporal lobes performed in his mid-20s, H.M. was left profoundly amnesic for events experienced following his procedure (e.g., Scoville & Milner, 1957). In contrast to his severe anterograde impairment, memory for events preceding surgery was partially spared. For example, he remembered his father's

gun collection in rich detail: "There were pistols—.32, .38, and .44 caliber—and there were at least two rifles, both .22 caliber, one of them fitted with a scope. Behind the house, the field rose into woods and there was no road" (Hilts, 1995, p. 84). H.M. remembers seeing birds—even a pheasant—but not being allowed to use his father's weapons to hunt, except for an occasional squirrel. He fondly remembers keeping his "eyes open" and listening for "chirping noises" that might signal the presence of parentally sanctioned prey. In addition, he could provide memories of hunting and fishing trips made with his father and his own friends.

Although the status of his autonoetic consciousness was not tested, aspects of H.M.'s case suggest that his autonoetic abilities were severely compromised by surgery (although there still is discussion about whether his preserved memories were episodic, the consensus now is that they were the result of his largely intact semantic memory function; e.g., Corkin, 2013; Squire, 2009). For example, in an interview with Hilts (1995), H.M. presents the picture of an individual who is severely disoriented in time:

Hilts: "How old are you?"
H.M.: "Well, I don't know."
Hilts: "How old do you feel?"
H.M.: "I don't remember the year now . . . I think I am about thirty-three" (he was 50).
Hilts: "So you feel like you are thirty-three?"
H.M.: "I feel like I am, but it's sort of a natural deduction." (i.e., an act of inference ratherthan a prereflective sense of subjective temporality).
Hilts: "How old were you when you had your operation?"
H.M.: "Gee, I don't know."
Hilts: "If you were to tell me the date that sounds most likely to you that you had youroperation, what date would you choose?"
H.M.: "I think of '78, right off" (it was 1953).

Additional conversation leaves little doubt that H. M. lives a subjective existence in which his sense of being a temporal continuant is severely compromised in both chronological directions (e.g., it is not uncommon for him to comment that he feels confined to life in the present; Corkin, 2013).

Thus, despite H.M.'s ability to retrieve some contextually rich content from his past, his lack of autonoetic accompaniment has devastating consequences for his capacity to project himself into the future. H.M. can do nothing without specific directions from others. When asked about his ability to anticipate a personal future, he offers that he cannot even imagine "what I should be doing next" (Hilts, 1995, p. 119). As Corkin (2013) observes, H.M. does not make predictions about his future; if pushed to do so he typically fails to respond.

A similar conclusion is reached from examination of the case of patient D.B. (e.g., Klein, Loftus, et al., 2002: Klein, Rozendal, et al., 2002). Following an anoxic episode, D.B. suffered a complete loss of both anterograde and retrograde episodic function. Testing, however, revealed that at least some of this was due to autonoetic impairment (e.g., he was able to accurately recount a few specific incidents from his past, but unable to situate them in the proper temporal context). Testing revealed severe temporal disorientation (e.g., Klein, 2013d; Klein, Rozendal, et al., 2002).

As anticipated, despite preservation of some learned content, D.B. was unable to make any plans for his future. Though he successfully answered questions about impersonal future happenings, this presumably was due to his partially intact knowledge of public events and his preserved understanding of the language of time—for example, the meaning of the words such as "past", "present", and "future" (e.g., Klein, 2013d).

Summary of empirical findings

The take-away message is that patients showing partial or complete loss of episodic function can still navigate a personal future—provided their autonoetic abilities remain intact. In addition, patients showing access to stored content, but unable to associate that content with autonoetic consciousness (as the result of either a loss of autonoetic ability or failure to connect content with autonoesis at retrieval), are unable to imagine themselves in future-oriented scenarios.

In short, while content may be a necessary component for some forms of FMTT (e.g., Klein, 2013a), it is not sufficient to enable a person to escape the confines of the subjective present.

Future-oriented personal imaginings require future-oriented autonoetic consciousness. The content to which that mode of consciousness is conjoined may be a necessary condition for some forms of FMTT, but it is never sufficient.

Based on these considerations, it would be interesting to test amnesic patients suffering from impaired access to content (due to pathologies targeting acquisition, storage or retrieval), but possessing intact autonoetic abilities. A direct prediction is that such individuals would show little or no impairment of FMTT (depending, of course, on the extent to which content plays a role in the particular manifestation of FMTT under scrutiny), despite presenting varying degrees of episodic dysfunction.

CONCLUSIONS: THE RELATION BETWEEN EPISODIC MEMORY AND FMTT REVISITED

We now know that mental time travel into the past can be thought to consist in two conceptually distinct and separately measurable processes: "knowing" and "remembering" (e.g., Tulving, 1985). A large number of experiments have shown that these two behavioural measures reflect partially overlapping, but different sets of processes (for a review see Dunn, 2004). These processes typically are associated with noetic and autonoetic consciousness, respectively.

Before these two kinds of ecphory (e.g., Schacter, 1982; Tulving, 1983) were discovered, they almost always were taken to issue from single faculty (often labelled "recognition", though recent evidence suggests "remembering" and "knowing" apply to recall as well; e.g., Rybash, 1999), and numerous theories were woven around that faculty (e.g., Brown, 1976). The analytical and experimental separation of "remembering" (autonoesis) and "knowing" (noesis) has proven quite fruitful (both conceptually and experimentally) and thus can be seen as a step forward in addressing the "right sort" of questions to nature.

In research on FMTT, there has, to date, been no comparable division—although it is easy to imagine the conceptual and empirical utility of postulating one. While isolated hints in that direction can be found in the literature (e.g., Arnold et al., 2011; D'Argembeau & Van der Linden, 2004; Klein, 2013a), a sustained treatment of the issue is (to the best of my knowledge) still waiting to make an appearance.

The goal of the present paper was to provide such treatment. To gain traction, I chose to focus on the type of memory most commonly taken to be causally relevant to future-oriented self-projection—that is, episodic. Based on the analyses presented, I conclude that a strong case can be made that it is the autonoetic constituent of episodic memory (in particular, its future-oriented presentation), rather than episodic memory per se (which draws on past-oriented aspects of autonoesis), that constitutes the necessary condition for enabling temporal self-projection into the future.

Clearly, I have not addressed the phenomenon of FMTT in its fullness. For example, as originally theorized by Atance and O'Neill (2001) and empirically captured by Klein, Loftus, et al. (2002), some forms of FMTT (primarily nonpersonal) are based semantic memory—a proposition that, following nearly a decade of empirical neglect, has again assumed scientific respectability. In addition, as discussed by Klein (2013a), Schacter et al., (2012), and Suddendorf, Addis, and Corballis, (2009), among others, different forms of self-projection rely on factors other than (or in addition to) retrieved content. The nature and degree of involvement of FMTT with these factors remain to be more fully addressed.

A question that plagues (or at least should) most empirical treatments of the role of episodic memory in FMTT is "what exactly sanctions this presumed relation?". One commonly held (though seldom voiced) answer is that "episodic content" serves as the foundation on which we construct future-oriented self-projections (e.g., Wheeler et al., 1997). But, as the evidence and arguments offered in this paper suggest, when carefully scrutinized this idea is less than compelling.

Another line of evidence marshalled in support of an episodic/FMTT connection is that these two phenomena share many neural substrates (primarily in the medial temporal lobes: e.g., Addis, Wong, & Schacter, 2007; Arzy, Collette, Ionata, Fornari, & Blanke, 2009; Race et al., 2011; Schacter & Addis, 2007; Verfaellie, Race, & Keane, 2013). This apparently confers a degree of respectability on the hypothesis that episodic memory (somehow) is involved in FMTT.

However, as argument and evidence presented in this paper show, this inference requires more support than a demonstration of neuroanatomical overlap. For example, in what way or ways do these shared structures contribute to episodic memory (which consists in a number of causally determinative constituents; e.g., Klein, German, Cosmides, & Gabriel, 2004)? Are they involved in memory experience? Or do they store the content that subsequently can be recruited by memories (or by other processes) that play a role in FMTT?

Moreover, since memory is an experience, it would seem that for episodic memory to underwrite FMTT, individuals should have recollective experiences while formulating future-oriented scenarios. But, to the best of my knowledge, naturalistic studies of the relation between recollective experience and future-oriented imaginings have yet to be conducted (also see Footnote 1).

Interestingly, radiological analyses and neuroanatomical data suggest that autonoetic consciousness is associated with structures in the frontal, not temporal, lobes (e.g., Abraham, Schubotz, & von Cramon, 2008; Piolino et al., 2007; Tulving & Szpunar, 2012; Wheeler et al., 1997). Finding that separate neural networks are associated with content storage and autonoetic consciousness provides provisional (though far from conclusive) support for the idea that memory is the experienced outcome of temporal processes acting on content (which is assumed to be stored in structures in the temporal lobes: for review see Gabreili, 1998) during retrieval. Although content associated with the medial temporal lobes may eventuate, on retrieval, in a memory experience, the processes that affect this transformation—as might be expected on the basis of the relational model (see the section "Autonoetic consciousness is not intrinsic to episodic memory")—appear to be elsewhere in the brain (recent work suggests the parietal cortex also may be involved in autonoesis; e.g., Nyberg, Kim, Habib, Levine, & Tulving, 2010).

Clearly much remains in question about the role of memory in FMTT. But this much is clear: It is not sufficient to place the predicate "episodic" in a two-part relation with achievement words (e.g., foresight, simulation, projection) in the absence of careful reflection on exactly what the term "episodic memory" picks out. If our rapidly growing body of empiricism about future-oriented temporality—an ability whose complexity and reach separate us from the remainder of the animal kingdom (e.g., Klein, 2013a; Suddendorf, 2013)—is to attain meaningful direction, a number of challenging issues will require the sustained, critical analysis of front-line researchers.

REFERENCES

Abraham, A., Schubotz, R. I., & von Cramon, Y. (2008). Thinking about the future versus the past in personal and non-personal contexts. *Brain Research, 1233*, 106–119.

Addis, D. R., Cheng, T., Roberts, R. P., & Schacter, D. L. (2011). Hippocampal contributions to the episodic simulation of specific and general future events. *Hippocampus, 21*, 1045–1052.

Addis, D. R., & Schacter, D. L. (2012). The hippocampus and imaging the future: Where do we stand? Frontiers in Human Neuroscience, 5, Article 173.

Addis, D. R., Wong, A. T., & Schacter, D. L. (2007). Remembering the past and imagining the future: Common and distinct neural substrates during event construction and elaboration. *Neuropsychologia, 45*, 1363–1377.

Anderson, R. J. (2012). Imagining novel futures: The roles of event plausibility andfamiliarity. *Memory, 20*, 443–451.

Arnold, K. M., McDermott, K. B., & Szpunar, K. K. (2011). Individual differences in timeperspective predict autonoetic experience. *Consciousness and Cognition, 20*, 712–719.

Arzy, S., Collette, S., Ionata, S., Fornari, E., & Blanke, O. (2009). Subjective mental time travel: The functional architecture of projecting the self to the past

and future. *European Journal of Neuroscience, 30*, 2009–2017.

Atance, C. M., & O'Neill, D. K. (2001). Episodic future thinking. *Trends in Cognitive Sciences, 5*, 533–539.

Attance, C. M., & Sommerville, J. A. (2014). Assessing the role of memory in preschooler'sepisodic performance on episodic foresight tasks. *Memory, 22*, 118–128.

Bindschaedler, C., Perer-Faver, C., Maeder, P., Hirsbrunner, T., & Clarke, S. (2011). Growing up with bilateral hippocampal atrophy: From childhood to teenage. *Cortex, 47*, 931–944.

Bischof-Koehler, D. (1985). On the phylogeny of human motivation. In L. H. Eckensberger & E. D. Lnatermann (Eds.), *Emotion and reflexivitaet* (pp. 3–47). Vienna: Urban & Schwarzenberg.

Boyer, P. (2009). What are memories for? Functions of recall in cognition and culture. In P. Boyer & J. V. Wertsch (Eds.), *Memory in mind and culture* (pp. 3–28). Cambridge, UK: Cambridge University Press.

Broman, M., Rose, A. L., Hotson, G., & Casey, C. M. (1997). Severe anterograde amnesia with onset in childhood as a result of anoxic encephalopathy. *Brain, 120*, 417–433.

Brown, J. (1976). *Recall and recognition*. New York, NY: John Wiley & Sons Ltd.

Bradley, F. H. (1887). Why do we remember forwards and not backwards? *Mind, 12*, 579–582.

Buckner, R. L., & Carroll, D. C. (2007). Self-projection and the brain. *Trends in Cognitive Sciences, 11*, 49–57.

Byrne, A. (2010). Recollection, prediction , imagination. *Philosophical Studies, 148*, 15–26.

Coleman, J. (1992). *Ancient & medieval memories*. New York, NY: CambridgeUniversity Press.

Corkin, S. (2013). *Permanent present tense: The unforgettable life of amnesic patient H. M.* New York, NY: Basic Books.

Cornish, D. (2011). Earlier and later if and only if past, present and future. *Philosophy, 86*, 41–58.

Cottle, T. J., & Klineberg, S. L. (1974). *The present of things future: Explorations in the human experience of time*. London, UK: Collier Macmillan Publishers.

Craver, C. F., Kwan, D., Steindam, C., & Rosenbaum, R. S. (2014). Individuals with episodic amnesia are not stuck in time. *Neuropsychologia, 57*, 191–195.

Dalla Barba, G. (2002). *Memory, consciousness and temporality*. Norwell, MA: Kluwer Academic Publishers.

Dalla Barba, G., Cappelletti, J. Y., Signorini, M., & Denes, G. (1997). Confabulation: Remembering another' past, planning ?another' future. *Neurocase, 3*, 425–436.

D'Argembeau, A., & van der Linden, M. (2004). Phenomenal characteristics associated withprojecting back into the past and forward into the future: Influence of valence andtemporal distance. *Consciousness and Cognition, 13*, 844–858.

Descartes, R. (1970). *Philosophical letters*. Oxford, UK: Clarendon Press. (A. Kenny, Ed. and Trans.).

Dunn, J. C. (2004). Remember-know: A matter of confidence. *Psychological Review 111*, 524–542.

Ebbinghaus, H. (1885/1913). *Memory: A contribution to experimental psychology*. New York, NY: Teacher's College, Columbia University. (Translated by H. A. Ruger & C. Bussenius).

Foster, J. K., & Jelicic, M. (1999). *Memory: Systems, process, or function?* New York, NY: Oxford University Press.

Fraisse, P. (1963). *The psychology of time*. London, UK: Harper & Row Publishers.

Gabrieli, J. D. E. (1998). *Cognitive neuroscience of human memory. Annual Review of Psychology*, 49, 87–115.

Gadian, D. G., Aicardi, J., Watkins, K. E., Porter, D. A., Mishkin, M., & Vargha-Khadem, F. (2000). Developmental amnesia associated with early hypoxic-ischaemic injury. *Brain, 123*, 499–507.

Gardiner, J. M. (2001). Episodic memory and autonoetic consciousness: A first-person approach. *Philosophical Transactions of the Royal Society B, 356*, 1351–1361.

Grilli, M. D., & Verfaellie, M. (2014). Personal semantic memory: Insights fromneuropsychological research. *Neuropsychologia, 61*, 56–64.

Guillery-Girard, B., Martins, S., Parisot-Carbuccia, D., & Eustache, F. (2004). Semantic acquisition in childhood amnesic syndrome: A prospective study. *NeuroReport, 15*, 377–381.

Hassabis, D., & Maguire, E. A. (2007). Deconstructing episodic memory with construction. *Trends in Cognitive Sciences, 11*, 299–306.

Heisenberg, W. (1958/1999). *Physics and philosophy*. Amherst, NY: Prometheus Books.

Hilts, P. J. (1995). *Memory's ghost: The strange tale of Mr. M*. New York, NY: Simon &Schuster.

Hoerl, C. (2008). On being stuck in time. *Phenomenology and the Cognitive Sciences, 7*, 485–500.

Hurley, N. C., Maguire, E. A., & Vargha-Khadem, F. (2011). Patient HC with developmental amnesia can construct future scenarios. *Neuropsychologia, 49*, 3620–3628.

Ingvar, D. H. (1985). "Memory for the future": An essay on the temporal organization ofconscious awareness. *Human Neurobiology, 4*, 127–136.

Irish, M., Addis, D. R., Hodges, J. R., & Piguet, O. (2012). Considering the role ofsemantic memory in episodic future thinking: Evidence from semantic dementia. *Brain, 135*, 2178–2191.

Irish, M., & Piguet, O. (2013). The pivotal role of semantic memory in remembering the past and imagining the future. *Frontiers in Behavioral Neuroscience.* doi:10.3389/fnbeh.2013.00027.

James, W. (1890). *Principles of psychology (Vol.1)*. New York, NY: Henry Holt and Company.

Klee, R. (1997). *Introduction to the philosophy of science: Cutting nature at its seams*. New York, NY: Oxford University Press.

Klein, S. B. (2004). The cognitive neuroscience of knowing one's self. In M. A. Gazzaniga (Ed.), *The Cognitive Neurosciences III* (pp. 1007–1089). Cambridge, MA: MIT Press.

Klein, S. B. (2012). The self and its brain. *Social Cognition, 30*, 474–518.

Klein, S. B. (2013a). The complex act of projecting oneself into the future. *WIREs CognitiveSciences, 4*, 63–79.

Klein. S. B. (2013b). The temporal orientation of memory: It's time for a change ofdirection. *Journal of Applied Research in Memory and Cognition, 4*, 222–234.

Klein, S. B. (2013c). Making the case that episodic recollection is attributable to operationsoccurring at retrieval rather than to content stored in a dedicated subsystem of long-term memory. *Frontiers in Behavioral Neuroscience 7*, 3. doi:103389/fnbeh.2013.00003

Klein, S. B. (2013d). Sameness and the self: Philosophical and psychological considerations. *Frontiers in Psychology: Perception Science.* doi:10.3389/fpsyg.2014.00029.

Klein, S. B. (2014a). *The two selves: Metaphysical commitments and functionalindependence*. New York, NY: Oxford University Press.

Klein, S. B. (2014b). Autonoesis and belief in a personal past: An evolutionary theory ofepisodic memory indices. *Review of Philosophy and Psychology, 5*, 427–447.

Klein, S. B. (in press). What memory is. *WIREs Cognitive Science.*

Klein, S. B., Cosmides, L., Tooby, J., & Chance, S. (2002). Decisions and the evolution of memory: Multiple systems, multiple functions. *Psychological Review, 109*, 306–329.

Klein, S. B., & Gangi, C. E. (2010). The multiplicity of self: Neuropsychological evidence and its implications for the self as a construct in psychological research.

The Year in Cognitive Neuroscience 2010: Annals of the New York Academy of Sciences, 1191, 1–15.

Klein, S. B., German, T. P., Cosmides, L., & Gabriel, R. (2004). A theory of autobiographicalmemory: Necessary components and disorders resulting from their loss. *Social Cognition, 22*, 460–490.

Klein, S. B., & Lax, M. L. (2010). The unanticipated resilience of trait self-knowledge in the face of neural damage. *Memory, 18*, 918–948.

Klein, S. B., & Loftus, J. (1993). The mental representation of trait and autobiographicalKnowledge about the self. In T. K. Srull & R. S. Wyer (Eds.), *Advances in social cognition* (Vol. 5, pp. 1–49). Hillsdale, NJ: Erlbaum.

Klein, S. B., Loftus, J., & Kihlstrom, J. F. (2002). Memory and temporal experience: The effects of episodic memory loss on an amnesic patient's ability to remember the past and imagine the future. *Social Cognition, 20*, 353–379.

Klein, S. B., & Nichols, S. (2013). Memory and the sense of personal identity. *Mind, 121*, 677–702.

Klein, S. B., Robertson, T. E., & Delton, A. W. (2010). Facing the future: Memory as anevolved system for planning future acts. *Memory & Cognition, 38*, 13–22.

Klein, S. B., Rozendal, K., & Cosmides, L. (2002). A social-cognitive neuroscience analysis of the self. *Social Cognition, 20*, 105–135.

Kwan, D., Craver, C. F., Green, L., Mycrson, J., Boyer, P., & Rosenbaum, R. S. (2012). Future decision-making without episodic mental time travel. *Hippocampus, 22*, 1215–1219.

Lane, T. (2012). Toward an explanatory framework for mental ownership. *Phenomenologyand the Cognitive Sciences, 11*, 251–286.

Levine, B., Black, E., Cabeza, R., Sinden, M., McIntosh, A. R., Toth, J. P., Tulgin, E., & Stuss, D. T. (1998). Episodic memory and the self in a case of isolated retrograde amnesia. *Brain, 121*, 1951–1973.

Loizou, A. (1986). *The reality of time*. Aldershot, UK: Gower Publishing Co.

Luria, A. R. (1972). *The man with a shattered world*. Cambridge, MA: Harvard University Press.

Maguire, E. A., & Mullally, S. L. (2013). The hippocampus: A manifesto for change. *Journal of Experimental Psychology: General, 142*, 1180–1189.

Manning, L., Denkova, E., & Unterberger, L. (2013). Autobiographical significance in past and future public semantic memory: A case-study. *Cortex, 49*, 2007–2020.

Markowitsch, H. J. (2003). Autonoetic consciousness. In T. Kircher & A. David (Eds.), *Theself in neuroscience and psychiatry* (pp. 180–196). Cambridge, UK: CambridgeUniversity Press.

Markowitsch, H. J., & Staniliou, A. (2011). Memory, autonoetic consciousness, and the self. *Consciousness and Cognition, 20*, 16–39.

Markowitsch, H. J., & Staniloiu, A. (2013). The impairment of recollection in functional amnesic states. *Cortex, 49*, 1494–1510.

Martinelli, P., Sperduti, M., & Piolino, P. (2013). Neural substrates of the self- memory system: New insights from a meta-analysis. *Human Brain Mapping, 34*, 1515–1529.

Mathen, M. (2010). Is memory preservative? *Philosophical Studies, 148*, 3–14.

McTaggart, J. M. E. (1908). The unreality of time. *Mind, 68*, 457–484.

Michaelian, K., Klein, S. B., & Szpunar, K. K. (forthcoming) *Seeing the future: Theoretical perspectives on future-oriented mental time travel*. Oxford, UK: Oxford University Press.

Mullaley, S. L., Varga-Khadem, F., & Maguire, E. A. (2014). Scene construction indevelopmental amnesia: An fMRI study. *Neuropsychologia, 52*, 1–10.

Nadel, L., Hupbach, A., Gomez, R., & Newman-Smith, K. (2012). Memory formation, consolidation and transformation. *Neuroscience and Biobehavioral Reviews, 36*, 1640–1645.

Nadel, L., & Moscovitch, M. (1997). Memory consolidation, retrograde amnesia and the hippocampal complex. *Current Opinion in Neurobiology, 7*, 217–227.

Newell, A. (1973). You can't play 20 questions with nature and win: Projective comments on the papers of this symposium. In W. G. Chase (Ed.), *Visual information processing* (pp. 283–308). San Francisco, CA: Academic Press.

Nyberg, L. Kim, A. S. N., Habib, R., Levine, B., & Tulving, E. (2010). Consciousness ofsubjective time in the brain. *Proceedings of the National Academy of Sciences, 107*, 22356–22359.

Pezzulo, G. (2008). Coordinating with the future: The anticipatory nature of representation. *Minds & Machines, 18*, 179–225.

Piolino, P., Desgranges, B., Manning, L., North, P., Jokic, C., & Eustache, F. (2007). Autobiographical memory: The sense of recollection and executive functions after severe traumatic brain injury. *Cortex, 43*, 176–195.

Race, E., Keane, M. N., Verfaellie, M. (2011). Medial temporal lobe damage causes deficits in episodic memory and episodic future thinking not attributable to deficits in narrative construction. *Journal of Neuroscience, 31,* 10262–10269.

Renoult, L., Davidson, P. S. R., Palombo, D. J., Moscovitch, M., & Levine, B. (2012). Personal semantics: At the crossroads of semantic and episodic memory. *Trends in Cognitive Sciences, 16,* 550–558.

Rosenbaum, R. S., Gilboa, A., & Moscovitch, M. (2014). Case studies continue to illuminate the cognitive neuroscience of memory. *Annals of the New York Academy of Sciences.* doi:10.1111/nyas.12467

Rybash, J. (1999). Aging and the autobiographical memory: The long and bumpy road. *Journal of Adult Development, 6,* 1–10.

Saint Augustine. (1997). *The confessions.* Hyde park, NY: New City Press. (Translated by M. Boulding).

Schacter, D. L. (1982). *The stranger behind the engram: Theories of memory and the science of psychology.* Hillsdale, NJ: Lawrence Erlbaum Associates.

Schacter, D. L., & Addis, D. R. (2007). The cognitive neuroscience of constructive memory: Remembering the past and imagining the future. *Philosophical Transactions of the Royal Society B, 362,* 773–786.

Schacter, D. L., Addis, D. R., Hassabis, D., Martin, V. C., Spreng, R. N., & Szpunar, K. K. (2012). The future of memory: Remembering, imagining, and the brain. *Neuron, 76,* 677–694.

Schacter, D. L., Addis, D. R., & Buckner, R. L. (2008). Episodic simulation of future events: Concepts , data, and applications. *Annals of the New York Academy of Science, 1124,* 39–60.

Scoville, W. B., & Milner, B. (1957). Loss of recent memory after bilateral hippocampallesions. *The Journal of Neurology, Neurosurgery and Psychiatry, 20,* 11–21.

Sorabji, R. (1972). *Aristotle on memory.* Providence, RI: Brown University Press.

SPSP Annual Convention. (2015, 26–28 February). *Subjective Time & Mental Time Travel.* Long Beach Convention Center, Long Beach, California.

Squire, L. R. (2009). The legacy of patient H. M. for neuroscience. *Neuron, 61,* 6–9.

Squire, L. R., van der Horst, A. S., McDuff, G. R., Frascino, J. C., Hopkins, R. O., & Maudlin, K. N. (2010). Role of the hippocampus in remembering the past and imagining the future. *Proceedings of the National Academy of Sciences, 107,* 19044–19048.

Stuss, D. T., & Guzman, D. A. (1988). Severe remote mmeory loss with minimal anterograde amnesia: A clincial note. *Brain and Cognition, 8,* 21–30.

Suddendorf, T. (1994). The discovery of the fourth dimension: Mental time travel and human evolution. (Master's Thesis). University of Waikato, Hamilton, New Zealand.

Suddendorf, T. (2010). Episodic memory versus episodic foresight: Similarities and differences. *WIREs Cognitive Science, 1,* 99–107.

Suddendorf, T. (2013). *The gap: The science of what separates us from other animals.* New York, NY: Basic Books.

Suddendorf, T., Addis, D. R., & Corbaillis, M. C. (2009). Mental time travel and theshaping of the mind. *Philosophical Transactions of the Royal Society B, 364,* 1317–1324.

Suddendorf, T., & Corballis, M. C. (1997). Mental time travel and the evolution of the human mind. *Genetic, Social, and General Psychology Monographs, 123,* 133–167.

Suddendorf, T., & Corballis, M. C. (2007) The evolution of foresight: What is mental time travel, and is it unique to humans? *Behavioral and Brain Sciences, 30,* 299–313.

Szpunar, K. K. (2010). Episodic future thought: An emerging concept. *Perspectives on Psychological Science, 5,* 142–162.

Szpunar, K. K., & McDermott, K. B. (2008). Episodic future thought and its relation to remembering: Evidence from ratings of subjective experience. *Consciousness and Cognition, 17,* 330–334.

Szpunar, K. K., & Tulving, E. (2011). Varieties of future experience. In M. Bar (Ed.), *Predictions and the brain: Using our past to generate a future* (pp. 3–12). New York, NY: Oxford University Press.

Tulving, E. (1972). Episodic and semantic memory. In E. Tulving & W. Donaldson (Eds.), *Organization of memory* (pp. 381–403). New York: Academic Press.

Tulving, E. (1983). *Elements of episodic memory.* New York: Oxford University Press.

Tulving, E. (1985). Memory and consciousness. *Canadian Psychology/ PsychologieCanadienne, 26,* 1–12.

Tulving, E. (1993). What is episodic memory? *Current Directions in Psychological Science, 2,* 67–70.

Tulving, E. (2002). Episodic memory: From mind to brain. *Annual Review of Psychology, 53,* 1–25.

Tulving, E. (2005). Episodic memory and autonoesis: Uniquely human? In H. S. Terrace & J. Metcalfe (Eds.). *The missing link in cognition: Origins of self-reflective consciousness* (pp. 3–56). Oxford, UK: Oxford University Press.

Tulving, E., & Szpunar, K. K. (2012). Does the future exist? In B. Levine & F. I. M. Craik (Eds.), *Mind and the frontal lobes: Cognition, behavior, and brain imaging* (pp. 248–263). New York, NY: Oxford University Press.

Verfaellie, M., Race, E., & Keane, M. M. (2013). Medial temporal lobe contributions to future thinking: Evidence from neuroimaging and amnesia. *Psychologica Belgica, 52*, 77–94.

Wheeler, M. A. (2005). Theories of memory and consciousness. In E. Tulving & F. I. M. Craik (Eds.). *The Oxford handbook of memory* (pp. 597–608). Oxford, UK: Oxford University Press.

Wheeler, M. A., Stuss, D. T., & Tulving, E. (1997). Toward a theory of episodic memory: Thefrontal lobes and autonoetic consciousness. *Psychological Bulletin, 121*, 331–354.

Williams, J. M. C., Ellis, N. C., Tyers, C. Healy, H., Rose, G., & MacLeod, A. K. (1996). The specificity of autobiographical memory and imageability of the future. *Memory & Cognition, 24*, 116–125.

Zeithamova, D., Schlichting, M. L., & Preston, A. R. (2012). The hippocampus and inferential reasoning: Building memories to navigate the future. *Frontiers in Human Neuroscience, 6*, 1–14.

Index